Ansible for Real-Life Automation

A complete Ansible handbook filled with practical
IT automation use cases

Gineesh Madapparambath

BIRMINGHAM—MUMBAI

Ansible for Real-Life Automation

Copyright © 2022 Packt Publishing

Group Product Manager: Rahul Nair

Publishing Product Manager: Meeta Rajani

Senior Content Development Editor: Sayali Pingale

Technical Editor: Shruthi Shetty

Copy Editor: Safis Editing

Project Manager: Neil Dmello

Proofreader: Safis Editing

Indexer: Hemangini Bari

Production Designer: Shyam Sundar Korumilli

Marketing Coordinator: Nimisha Dua

Senior Marketing Coordinator: Sanjana Gupta

First published: September 2022

Production reference: 1020922

Published by Packt Publishing Ltd.
Livery Place
35 Livery Street
Birmingham
B3 2PB, UK.

978-1-80323-541-7

www.packt.com

To my wife, Deepthy, for supporting and motivating me as always.
To my son, Abhay, for allowing me to take time away from playing with him
to write the book. To my parents and my ever-supportive friends, for their
motivation and help.

- Gineesh Madapparambath

Contributors

About the author

Gineesh Madapparambath has over 15 years of experience in IT service management and consultancy with experience in planning, deploying, and supporting Linux-based projects.

He has designed, developed, and deployed automation solutions based on Ansible and Ansible Automation Platform (formerly Ansible Tower) for bare metal and virtual server building, patching, container management, network operations, and custom monitoring. Gineesh has coordinated, designed, and deployed servers in data centers globally and has cross-cultural experience in classic, private cloud (OpenStack and VMware), and public cloud environments (AWS, Azure, and Google Cloud Platform).

Gineesh has handled multiple roles such as systems engineer, automation specialist, infrastructure designer, and content author. His primary focus is on IT and application automation using Ansible, containerization using OpenShift (and Kubernetes), and infrastructure automation using Terraform.

About the reviewers

Vijay Jadhav is a solutions architect based out of Mumbai, India, with over two decades of experience in different roles in the IT industry. For the last 10 years, he has been working as an SME in cloud computing designing cloud-native applications using microservices-based architecture, and automated provisioning and de-provisioning of cloud resources using IaaC tools such as Ansible and Terraform. He currently works as a cloud architect at Cisco driving the innovation and adoption of next-gen technologies.

Vijay is a husband and a dad to an 8-year-old son. He spends his spare time (if such a thing does truly exist) watching popular movies and web series on OTT platforms.

Sean Cavanaugh is a senior principal technical marketing manager for Red Hat Ansible Automation Platform, where he brings over 12 years of experience in building and automating computer networks. Sean previously worked for both Cumulus Networks (acquired by Nvidia) and Cisco Systems where he helped customers deploy, manage, and automate their network infrastructures. He resides in Chapel Hill, NC, with his wife and children, and tweets from @IpvSean.

I'd like to thank my wife, who is truly my partner. She has her own career and workload outside the IT industry but is always committed to helping me succeed in all my endeavors. I have had a really exciting career getting to work with very intelligent people that constantly challenge me and make me a better person. Thank you to all my co-workers past and present that have helped me along my path.

Sreejith Anujan is a cloud technology professional with more than 15 years of experience in on-premises data center solutions and 10 years of experience in working with public cloud providers. He enjoys working with customers on their enablement plans to upskill the technical team on container and automation tooling. In his current role as a principal instructor within Red Hat, Sreejith is responsible for designing and delivering custom and tailored technology training and workshops to strategic customers across the Asia-Pacific region.

I would like to show my gratitude to the free software and open source community volunteers who have helped me in my journey so far. The sense of collaboration and free contribution are two values that make me cherish being part of those communities. A huge shout-out to my family, especially my wife for supporting our kids while I was busy traveling across the APAC region for my professional commitments!

Table of Contents

3

Automating Your Daily Jobs   47

4

Exploring Collaboration in Automation Development   81

Part 2: Finding Use Cases and Integrations

5

6

7

Managing Your Virtualization and Cloud Platforms   155

8

Helping the Database Team with Automation   193

9

Implementing Automation in a DevOps Workflow   217

10

Managing Containers Using Ansible   249

11

Managing Kubernetes Using Ansible   275

Part 3: Managing Your Automation Development Flow with Best Practices

14

15

16

Ansible Automation Best Practices for Production   417

Preface

Automation is the key to IT modernization, and using the right automation tool is a crucial step in the automation journey for organizations. Ansible is an open source automation software that you can use for automating most of your operations with IT and application infrastructure components including servers, storage, network, and application platforms. Ansible is one of the most well-known open source automation tools in the IT world right now and has strong community support, with more than 5,000 active contributors around the world.

Ansible is not a *learn-by-reading* technology. This book will help you to understand and practice the automation capabilities of Ansible with actual playbooks, configurations, and practical examples. It will help you to understand the basics of Ansible automation and slowly, you will learn how to use Ansible for automating your day-to-day tasks.

You will learn real-life IT automation use cases with practical examples, such as simple system reports, security scanning, and weekly rebooting Linux machines. After that, the book will teach you how to implement collaboration in Ansible automation and how to automate other devices and platforms, such as Microsoft Windows, network devices, VMware, AWS, and GCP. You will also learn how to use Ansible in a DevOps workflow with Jenkins integration and container and application management on Kubernetes. To expand your knowledge further on enterprise automation, the book will also teach you about Red Hat Ansible Automation Platform, secret management, and Ansible integration with other tools, such as Jira and ServiceNow. There are chapters in this book that cover how to automate non-supported devices and platforms using raw commands and API calls using Ansible. Before the book concludes, you will also explore the Ansible best practices for storing managed node information, variables, credentials, and playbooks for production environments.

Upon finishing *Ansible for Real-Life Automation*, you will have the skills to find automation use cases in your work environment and design and deploy automation solutions using Ansible.

Who this book is for

This book is intended for systems engineers and DevOps engineers who want to use Ansible as their automation tool. The book provides references and practical examples to start IT automation within your work environment.

What this book covers

Chapter 1, Ansible Automation – Introduction, gives you an introduction to Ansible and teaches you how to install and configure Ansible and configure managed nodes.

Chapter 2, Starting with Simple Automation, teaches you how to identify manual tasks to automate, find suitable Ansible modules, and use text editors for Ansible.

Chapter 3, Automating Your Daily Jobs, is the chapter where you started developing Ansible playbooks for real-life use cases, such as system information gathering, system reboot, and security scanning. You will also learn about Ansible collections, secrets in Ansible, and automating notifications.

Chapter 4, Exploring Collaboration in Automation Development, teaches you about the importance of version control in IT automation and the best practices to use source control management for storing Ansible artifacts.

Chapter 5, Expanding Your Automation Landscape, covers the methods to find automation use cases from your workplace and check the feasibility of automation. This chapter also teaches you how to use a dynamic inventory in Ansible.

Chapter 6, Automating Microsoft Windows and Network Devices, is where you discover the possibilities to automate Microsoft Windows and network devices using Ansible. This chapter will cover practical examples for Windows automation, VyOS fact gathering, and access control list creation on a Cisco ASA device.

Chapter 7, Managing Your Virtualization and Cloud Platforms, teaches you about **Infrastructure as Code (IaC)** and the methods to use Ansible as an IaC tool with practical examples for managing the VMware, AWS, and GCP platforms.

Chapter 8, Helping the Database Team with Automation, covers an introduction to database operations, such as installing the database server and creating databases, tables, and database users.

Chapter 9, Implementing Automation in a DevOps Workflow, is where you will be introduced to DevOps and the usage of Ansible in a DevOps workflow, with practical examples for reducing deployment time and managing application load balancers and rolling updates. The chapter also covers how to integrate Ansible with Terraform.

Chapter 10, Managing Containers Using Ansible, continues the look at Ansible and DevOps by covering the methods to manage application containers using Ansible. This chapter covers practical use cases such as deploying container hosts, using Ansible in CI/CD pipelines, building containers, and managing multi-container applications.

Chapter 11, Managing Kubernetes Using Ansible, goes into more container-based use cases with an introduction to Kubernetes and the method to manage Kubernetes clusters and applications using Ansible. The chapter teaches you how to deploy, manage, and scale containerized applications on the Kubernetes platform.

Chapter 12, *Integrating Ansible with Your Tools*, covers the introduction to the enterprise automation tool Red Hat **Ansible Automation Platform** (**AAP**). This chapter teaches you methods for using AAP for automation with practical examples, such as database management, Jenkins integration, and Slack notification.

Chapter 13, *Using Ansible for Secret Management*, explains the methods to handle sensitive data in automation using Ansible Vault and how to use encrypted data in Ansible playbooks.

Chapter 14, *Keeping Automation Simple and Efficient*, teaches you about the survey forms in Ansible automation controller and workflow job templates. This chapter also covers security automation and integrating Ansible with monitoring tools.

Chapter 15, *Automating Non-Standard Platforms and Operations*, teaches you the automate to handle non-supported platforms using raw commands and API commands. This chapter also teaches you how to develop custom Ansible modules for when there are no modules available.

Chapter 16, *Ansible Automation Best Practices for Production*, is where you learn the production best practices for Ansible, such as organizing Ansible artifacts, inventories, and variables. This chapter also covers the best practices for credential management and playbook development.

To get the most out of this book

In this book, I will first guide you through the installation and deployment of the Ansible automation tool, and later, I will explain some real IT use cases and methods to use Ansible for automating such operations. Since the focus of the book is on different automation use cases, some of the chapters might have additional technical requirements, such as basic knowledge of a specific technology or access to a lab environment (such as a Kubernetes cluster). For this, I have also covered the methods to arrange the development environment if you want to practice. Always refer to the Ansible documentation at `https://docs.ansible.com` and other provided links in the chapters for further learning.

Software/hardware covered in the book	OS requirements
`Ansible 2.9`	Red Hat Enterprise Linux 8 (RHEL8)
`ansible-core 2.11 later`	Microsoft Windows 2019 (for Windows use case)
Red Hat Ansible Automation Platform 2.1	VyOS (for the network automation use case)

For testing and development, you can get no-cost RHEL (`https://developers.redhat.com/articles/faqs-no-cost-red-hat-enterprise-linux`) subscriptions. It is also possible to replace RHEL8 with other operating systems, such as Fedora, CentOS, or Ubuntu, but you might need to adjust some of the commands and modules in the playbook.

If you are reading a soft copy or digital version of this book, it is advised to type the commands and develop the playbooks by yourself rather than copy-pasting from the book. However, you can access the code, snippets, and playbooks from the book's GitHub repository (a link is available in the next section) for reference.

Download the example code files

You can download the example code files for this book from GitHub at `https://github.com/PacktPublishing/Ansible-for-Real-life-Automation`. If there's an update to the code, it will be updated on the existing GitHub repository. You can point out any problems or issues in the code samples and submit any questions related to the book by raising issue tickets in this GitHub repository.

We also have other code bundles from our rich catalog of books and videos available at `https://github.com/PacktPublishing/`. Check them out!

Download the color images

We also provide a PDF file that has color images of the screenshots & diagrams used in this book. We recommend you download this file for your reference, to experience the color specifics in the syntax highlighting of the code snippets, as well as the exact appearances of the user interfaces. also provide a PDF file that has color images of the screenshots/diagrams used in this book. You can download it here: `https://packt.link/TVh0m`.

Conventions used

There are a number of text conventions used throughout this book.

`Code in text`: Indicates code words in text, database table names, folder names, filenames, file extensions, pathnames, dummy URLs, user input, and Twitter handles. Here is an example: "Configure the `KUBECONFIG` environment variable as our `kubeconfig` filename is different (`/home/ansible/.kube/minikube-config`) from the default filename (`/home/ansible/.kube/config`)."

A block of code is set as follows:

```
[ansible@ansible Chapter-11]$ export KUBECONFIG=$KUBECONFIG:/
home/ansible/.kube/minikube-config
```

Some of the code snippets and outputs are displayed as images for better readability and brevity.

Bold: Indicates a new term, an important word, or words that you see onscreen. For example, words in menus or dialog boxes appear in the text like this. Here is an example: "Update your job template and add new vault credentials by going to **Job Template** | **Edit**, then clicking on the *Search* button near **Credential**."

> Tips or Important Notes
> Appear like this.

Get in touch

Feedback from our readers is always welcome.

General feedback: If you have questions about any aspect of this book, mention the book title in the subject of your message and email us at customercare@packtpub.com.

Errata: Although we have taken every care to ensure the accuracy of our content, mistakes do happen. If you have found a mistake in this book, we would be grateful if you would report this to us. Please visit www.packtpub.com/support/errata, selecting your book, clicking on the Errata Submission Form link, and entering the details.

Piracy: If you come across any illegal copies of our works in any form on the Internet, we would be grateful if you would provide us with the location address or website name. Please contact us at copyright@packt.com with a link to the material.

If you are interested in becoming an author: If there is a topic that you have expertise in and you are interested in either writing or contributing to a book, please visit authors.packtpub.com.

Share Your Thoughts

Once you've read *Ansible for Real Life Automation*, we'd love to hear your thoughts! Scan the QR code below to go straight to the Amazon review page for this book and share your feedback.

https://packt.link/r/1803235411

Your review is important to us and the tech community and will help us make sure we're delivering excellent quality content.

Part 1

Using Ansible as Your Automation Tool

In this part, you will get a clear idea of how to get started with Ansible automation and automate your basic daily jobs.

This part of the book comprises the following chapters:

- *Chapter 1, Ansible Automation – Introduction*
- *Chapter 2, Starting with Simple Automation*
- *Chapter 3, Automating Your Daily Jobs*
- *Chapter 4, Exploring Collaboration in Automation Development*

1

Ansible Automation – Introduction

Ansible is open source automation and orchestration software that can be used for automating most of your operations with IT infrastructure components including servers, storage, networks, and application platforms. Ansible is one of the most popular automation tools in the IT world now and has strong community support with more than 5,000 contributors around the world.

In this chapter we are going to cover the following topics:

- What is Ansible? Where should I use this automation tool?
- Deploying Ansible
- Configuring your managed nodes

As of today, Ansible is only available on Linux/Unix platforms, but that doesn't mean you cannot use Ansible to automate other **operating systems** (**OSs**) or devices. It is possible to use Ansible to automate almost all components involved in the IT infrastructure, as there are thousands of supported modules available to support Ansible automation.

Technical requirements

The following are the technical requirements to proceed with this chapter:

- A basic understanding of the Linux OS and how to handle basic operations in Linux
- One or more Linux machines

The codes and snippets used in the chapter are tested in **Red Hat Enterprise Linux 8** (**RHEL8**). All the Ansible code, Ansible playbooks, commands, and snippets for this chapter can be found in the GitHub repository at `https://github.com/PacktPublishing/Ansible-for-Real-life-Automation/tree/main/Chapter-01`.

Hello engineers!

The primary role of a systems engineer is building and managing IT infrastructure for hosting applications and their data. In the olden days, the number of applications used was a lot less, hence the infrastructure size. As the applications and components grew, the IT infrastructure also grew, and systems engineers and system administrators started experiencing resource conjunction. In other ways, systems engineers are spending more time on building, maintaining, and supporting the infrastructure rather than spending time on improving the infrastructure designs and optimizing them.

For the support team, 90% of the event tickets are simple fixes including disk space full, user account locked, volumes not mounted, and so on. But the support engineer still needs to manually log in to each and every server and fix the issues one by one.

The task can be fixing a low disk space issue on servers, installing some packages, patching OSs, creating virtual machines, or resetting a user password; engineers are doing the same job repeatedly for multiple systems, and this led to the invention of automated operations. Initially, the solution for automation was custom scripts developed and maintained by individual engineers, but it was never a real solution for the enterprises as there was no collaboration, maintenance, or accountability for such custom automation scripts. If the developer leaves the organization, the script will become an orphan and the next engineer will create their own custom scripts.

With the introduction of DevOps methodologies and practices, developers, systems engineers, operations teams, and other platform teams started working together, the boundaries between them became thinner, and a better accountable ecosystem evolved. Everyone started building and maintaining the applications and the underlying IT infrastructure, which, in turn, made the automation use case list bigger and more complex.

What is Ansible? Where should I use this tool?

Ansible is an open source automation tool that was written and released by Michael DeHaan on February 20, 2012. In 2013, **Ansible, Inc.** (originally **AnsibleWorks, Inc.**) was founded by Michael DeHaan, Timothy Gerla, and Saïd Ziouani, and their intention was to commercially support and sponsor Ansible. In 2015, Ansible was acquired by Red Hat, and Red Hat supports and promotes Ansible as per the expectations of the open source community.

As of today, the Ansible control node is only available for Linux/Unix based platforms (most of the general-purpose OSs, such as Red Hat Enterprise Linux, CentOS, Fedora, Debian, or Ubuntu) and you cannot install it on Windows natively (it is possible to use Windows Subsystem for Linux or virtual machines for the same). This does not mean that you cannot use Ansible to automate your Windows operations. It is possible to use the Ansible control node on Linux and manage your Windows machines together, with other devices and platforms such as network devices, firewall devices, cloud platforms, and container platforms. There are more than 3,200 Ansible modules (as of today) available to use and, for Windows alone, there are more than 100 Ansible modules to automate Windows OS-based operations.

> **Ansible-Supported Windows OSs**
>
> Ansible can manage desktop OSs including Windows 7, 8.1, and 10, and server OSs including Windows Server 2008, 2008 R2, 2012, 2012 R2, 2016, and 2019. Refer to `https://docs.ansible.com/ansible/latest/user_guide/windows_setup.html#host-requirements` for more details.

The community version of Ansible is free to use like other open source software, but there is also a product offering from Red Hat based on Ansible called Red Hat **Ansible Automation Platform**, which is available with a paid subscription. Use either the community version of Ansible or the Red Hat-supported version with a subscription. Ansible Automation Platform is for enterprise use with functionalities such as **role-based access control (RBAC)**, **graphical user interface (GUI)**, **Application Programming Interface (API)**, redundancy, and scalability. Consider these options when you expand your automation use cases with a bigger team with many engineers working on automation and when you need auditing, tracing, and other integrations. Read more about Ansible Automation Platform at `https://www.ansible.com/products/automation-platform`.

> **Red Hat Ansible Automation Platform**
>
> The enterprise automation product from Red Hat was known as **Ansible Tower** until the announcement of the Red Hat Ansible Automation Platform in September 2019 (`https://www.ansible.com/blog/introducing-red-hat-ansible-automation-platform`). The components inside Ansible Automation Platform were renamed with more meaningful names, such as automation controller and execution environment. Read more about Ansible Automation Platform at `https://www.redhat.com/en/technologies/management/ansible`.
>
> Ansible documentation is available at `https://docs.ansible.com`.

Prerequisites

We will write automation steps in a YAML file called an **Ansible playbook**. Ansible will parse the playbook file and execute the tasks on target machines:

- You should know the basics of Linux; as I mentioned earlier, it is possible to install Ansible on a Linux or Unix machine only at the current time. That does not mean that you should be a Linux subject matter expert, but you need to be able to handle basic operations in Linux, such as file management and file editing.

- You need to understand the YAML syntax, which is easy, as YAML is one of the easiest human-readable file formats.

Ansible is based on the Python programming language, but you don't need to learn Python or any kind of programming language to develop Ansible automation playbooks.

> What is YAML?
>
> **YAML Ain't Markup Language** (**YAML**) is a human-readable language format used for most modern tools and software, including Ansible, Kubernetes, and OpenShift. YAML is often used as a file format for application data and configuration, for example.

Ansible control node requirements

There are no specific hardware requirements for the Ansible control node machine. It is possible to use a machine with 512 MB memory and one **virtual central processing unit** (**vCPU**). Follow some standard **virtual machine** (**VM**) specifications, such as 4 GB or higher, as you may need more memory when you have more managed nodes and more tasks to run in parallel. For the disk, you may follow standard VM specifications, as Ansible is a small program and does not require much disk space. Use any Linux/Unix machine with Python 3.8 or newer installed.

> **Ansible Control Node Requirements**
>
> Find the Ansible control node requirements at `https://docs.ansible.com/ansible/latest/installation_guide/intro_installation.html#control-node-requirements`.

Ansible managed node requirements

The target nodes (managed nodes) should be installed with Python 2.6 or later (3.x is recommended) to execute the task. For Windows machines, you need to install PowerShell and .NET.

Ansible is also supported for the following:

- Network devices
- VMware
- Public clouds such as AWS, Azure, and GCP
- Security devices

> **Python 2.x EOL**
>
> Install Ansible on a machine with Python 2.x, but Python 2.x is already **end-of-life** (**EOL**) by January 1, 2020, and it is best practice to use the latest Python version. For more detailsabout Python 2 end of life, visit `https://www.python.org/doc/sunset-python-2/`.

Ansible is agentless

There are two types of machines involved in Ansible automation. The machine in which you install the Ansible program is known as the **Ansible control node**. The control node can be any machine, a dedicated server, or even your workstation, and it will have your Ansible playbook and other configurations. Then, the machines or devices that you want to automate are known as **managed hosts**. You will run the Ansible jobs and playbooks from the control node and the jobs will be executed on the target nodes or managed nodes.

The following diagram shows the basic components of Ansible:

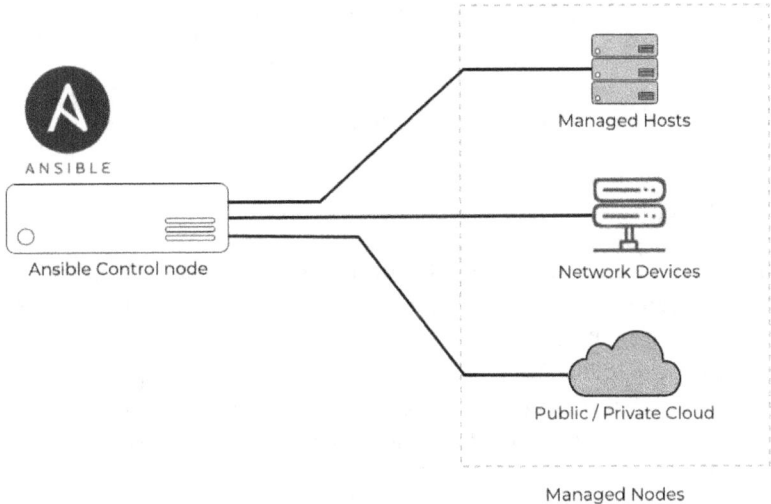

Figure 1.1 – Ansible and components

Ansible is agentless and you do not need to install any kind of agents on the managed nodes. Ansible uses default connection methods to communicate with managed nodes, such as ssh, WinRM, http, or other appropriate protocols. During the onboarding, you need to configure the credentials for the managed nodes, such as an SSH credential, with SSH keys, or an SSL certificate for WinRM connection. This is a one-time setup, and it is possible to configure or change this anytime. It is possible to use the same or different credentials for different managed nodes, and configure this for individual nodes or a group of managed nodes. You will learn about managed nodes and inventory in the next sections of this chapter.

Ansible architecture

The following diagram shows the Ansible internals and its components' structure:

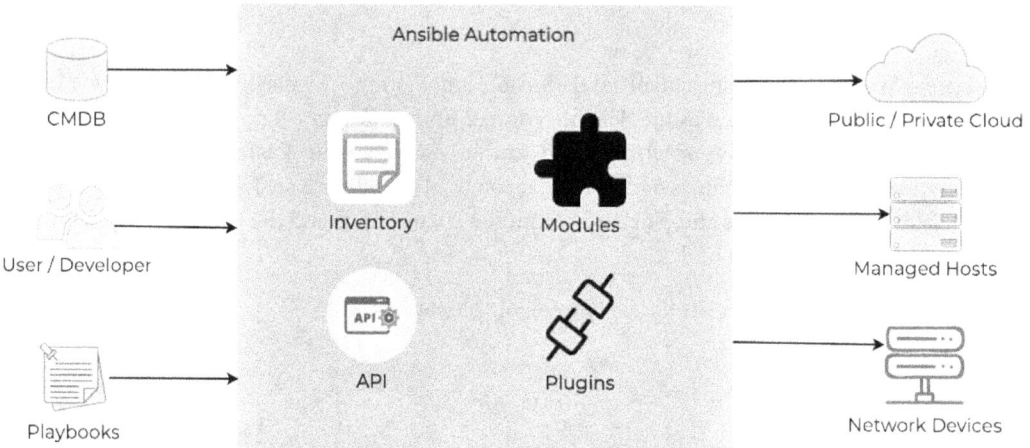

Figure 1.2 – Ansible and components

Ansible inventory

The Ansible inventory is a file or script that will provide the details about the managed nodes, including the hostname, connection methods, credential to use, and many other details. It is possible to pass the inventory to Ansible using static inventory files, dynamic inventory scripts, or using the **configuration management database** (**CMDB**). The CMDB is the same CMDB that can provide the managed nodes information. It is best practice to integrate CMDB with Ansible in an environment to avoid frequent updates on the static inventory files, but this is not a mandatory component.

It is possible to add any number of managed nodes inside the inventory file, as follows:

```
[nodes]
node1 ansible_host=192.168.56.25

[web]
node1 ansible_host=192.168.56.25
node2 ansible_host=192.168.56.24

[loadbalancer]
node3 ansible_host=192.168.56.45

[windows]
win2019 ansible_host=192.168.56.22

[nodes:vars]
ansible_ssh_private_key_file=/home/ansible/.ssh/id_rsa
```

Figure 1.3 – Ansible inventory with managed nodes

It is best practice to separate the managed nodes information in multiple inventory files based on the criticality, server types, and environment. You will learn more about inventory best practices in *Chapter 16, Storing Remote Host Information – Inventory Best Practices*.

Dynamic inventory plugins will collect the details of managed nodes from your virtualization platforms such as VMware, OpenStack, AWS, Azure, and GCP, or from other container platforms such as Kubernetes. There are more than 40 dynamic inventory plugins available to use in the Ansible GitHub repository. Use them if needed or create your own dynamic inventory scripts if those are not suitable for your requirements.

> **Ansible Dynamic Inventory**
>
> For more details about the Ansible dynamic inventory, look it up at `https://docs.ansible.com/ansible/latest/user_guide/intro_dynamic_inventory.html` and `https://github.com/ansible/ansible/tree/stable-2.9/contrib/inventory`.

Ansible plugins

Ansible plugins are small pieces of code that help to enable a flexible and expandable architecture. You have Ansible executable, and add plugins as needed for other features and capabilities like any other software. There are different types of plugins in Ansible such as **connection plugins**, **action plugins**, **become plugins**, and **inventory plugins**. For example, the default connection plugin you will be using is called `ssh`, and it is possible to use connection plugins called `docker` or `buildah` for connecting to containers. If you need to, install and use these plugins.

> **Ansible Plugins**
>
> Read more about Ansible plugins at `https://docs.ansible.com/ansible/latest/plugins/plugins.html`.

Ansible modules

An Ansible module is a piece of reusable and standalone script that can be used to achieve some specific tasks. Modules provide a defined interface with options to accept arguments and return information to Ansible in JSON format. When you execute a task using a module, the module script will be executed on the target machine using Python or using PowerShell for Windows machines.

For example, the following Ansible content is using a `ping` module to verify the connectivity to the target machine and another task with the `yum` module for installing `httpd package` on a managed node using the `yum` package manager:

● ● ●

```
- name: Ping to managed node
  ping:
- name: Install httpd package
  yum:
    name: httpd
    state: latest
```

Figure 1.4 – Ansible ping module and yum module

I said earlier that we can automate network devices and firewall devices using Ansible, but, we all know that we cannot install Python or PowerShell on those devices. Unlike most of the Ansible modules, network modules do not run on network devices. Instead, these modules will be executed from the Ansible control node itself and run the appropriate commands on target network devices to achieve the task. From the user's point of view, there is no difference in the execution of network modules as you still use them like any other modules. It is possible to manage the network devices like other Linux machines and Windows machines but with different connection methods, such as `network_cli`, `netconf`, and `httpapi`.

> **Read: How Network Automation Is Different**
>
> Go to `https://docs.ansible.com/ansible/latest/network/getting_started/network_differences.html` to learn more about network device automation.

Ansible content collections

Before version 2.10, Ansible was a big package with all modules and libraries inside, but the community grew very fast and thousands of new modules were contributed to Ansible. Whenever there is a new module or new version of a module available from the community or vendors, then users need to wait for the next release of Ansible to get the updated module. To resolve this dependency, a new way of distribution has started in which Ansible modules are separated from the Ansible base and distributed as **Ansible content collections**, or simply **Ansible collections**. You have the choice of installing Ansible alone or installing the Ansible package including Ansible collections. If you need to manage some different set of nodes or systems (for example, managing the VMware private cloud or automating Fortigate firewall devices), then install the required Ansible collection and use it. This modularity allows you to install only the required modules and plugins rather than all available modules.

The following diagram shows the transition of the Ansible collection from old Ansible:

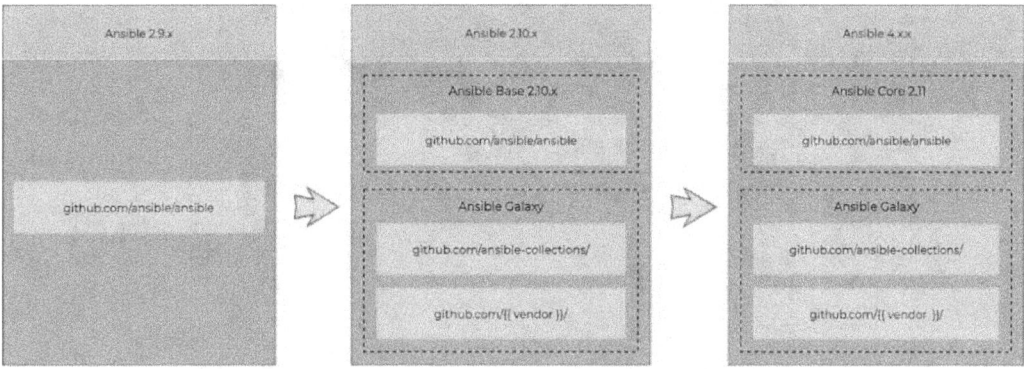

Figure 1.5 – Ansible to Ansible core and Ansible collection transformation

Also, note some of the changes in the Ansible base and Ansible collection restructuring:

- Before version 2.9.x: The package included `ansible` and all the Ansible modules and other plugins.

- From version 2.10++: Ansible was renamed to `ansible-base` and modules were moved to Ansible collections (vendors and communities).

- From version 2.11++: `ansible-base` was renamed to `ansible-core`.

- When you install Ansible 3.x or Ansible 4.x, then you are installing the Ansible community package, which includes `ansible-core` and all the Ansible collections (vendor and community).

> **Restructuring the Ansible Project**
>
> Read the blog post `ansible.com/blog/thoughts-on-restructuring-the-ansible-project` and `ansible.com/blog/the-future-of-ansible-content-delivery` to know more about Ansible collection transition.

Ansible playbook

The Ansible playbook is a simple file written in YAML format with the instruction list of automation tasks. The following diagram explains the components, syntax, and structure of an Ansible playbook:

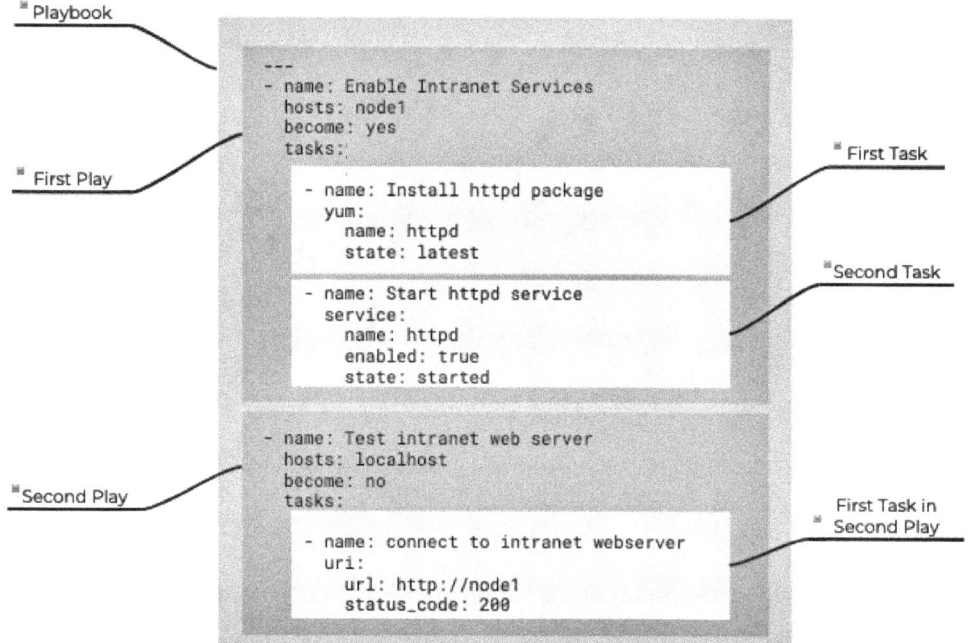

Figure 1.6 – Ansible playbook basic components

Each task in the **play** runs a module to do some specific job on the target node. You can have multiple tasks in a play and multiple plays inside a single playbook, as shown in the diagram. The plays, tasks, and module arguments are identified by the indentation in YAML format. In this example, we have a playbook with two plays.

In the first play, `Enable Intranet Services`, we are running the tasks against `node1` – see the `hosts: node1` line.

Under the tasks, see two tasks:

- The first task is using the `yum` module and installing the `httpd` package.

- The second task is using the `service` module to start the `httpd` service.

Then, we have a second play, `Test intranet web server`, in the playbook, in which you have only one task:

- Using the `uri` module to test the intranet web server

Ansible will parse this playbook and execute the tasks one by one on the target nodes.

Ansible use cases

It is possible to build, manage, or maintain almost all components in IT infrastructure using Ansible:

- Servers and storages
- Network devices, firewalls, **software-defined networks (SDNs)**, and load balancers
- Application components
- Containers and container platforms
- Database applications
- Public cloud and private cloud

You need to find the use cases from your day-to-day work and find the highest number of tasks you are repeating every day and consider those tasks as automation candidates. The following is a sample list of such use cases:

- OS patching
- Application and package deployment
- Orchestration of application
- Configuration management
- Infrastructure provisioning
- Continuous delivery
- Security and compliance auditing and remediation
- Database provisioning and management

In the following section, we will learn how to install and configure Ansible.

Installing Ansible

There are multiple ways to install Ansible on your system, such as using the default package manager (`yum`, `apt`, `dnf`), using Python's `pip`, or installing the source code itself. Depending on your environment and restrictions, follow any of the methods provided in the official documentation (`https://docs.ansible.com/ansible/latest/installation_guide/intro_installation.html`). If you are using the OS package manager, such as `dnf`, `yum`, or `apt`, then you need `sudo` privileges (or `root` access) to install Ansible.

Please note, you do not require `root` or `sudo` access on the Ansible control node for using Ansible. In the following snippet, `sudo` access is used for installing and updating packages.

Verify dependencies

As you learned that Ansible needs Python to work, you need to check the Python installation first:

```
● ● ●

[ansible@ansible ~]$ sudo dnf list installed python3*
Updating Subscription Management repositories.
Installed Packages
python3-bind.noarch                      32:9.11.26-3.el8              @rhel8-appstream-media
python3-chardet.noarch                   3.0.4-7.el8                   @anaconda
.
.
..<output omitted for brevity>..
python36.x86_64                          3.6.8-2.module+el8.1.0+3334+5cb623d7    @rhel8-appstream-media

## also verify the version of Python
[ansible@ansible ~]$ python3 -V
Python 3.6.8
```

Figure 1.7 – Checking installed Python packages and version

If you have a supported Python version installed, then proceed with the Ansible installation as explained in the next section.

Installing Ansible using the package manager

Depending on the OS, you need to add and enable the appropriate repositories before installing Ansible. For example, if you are using **Red Hat Enterprise Linux** (**RHEL**), then you need to ensure the Red Hat subscriptions are in place and repositories are enabled to install Ansible:

```
● ● ●

## on RHEL/CentOS/Fedora system
[ansible@ansible ~]$ sudo dnf install ansible

## For an Ubuntu system, you can use the apt command as follows:
$ sudo apt install ansible
```

Figure 1.8 – Installing Ansible package

Verify an installed Ansible version with the following command:

● ● ●

```
[ansible@ansible ~]$ ansible --version
ansible 2.9.27
config file = /etc/ansible/ansible.cfg
configured module search path = ['/home/ansible/.ansible/plugins/modules', '/usr/share/ansible/plugins/modules']
ansible python module location = /usr/lib/python3.6/site-packages/ansible
executable location = /usr/bin/ansible
python version = 3.6.8 (default, Mar 18 2021, 08:58:41) [GCC 8.4.1 20200928 (Red Hat 8.4.1-1)]
```

Figure 1.9 – Verifying Ansible installation

From this output, see the version of Ansible (*2.9.27*), the default configuration file used(config file = /etc/ansible/ansible.cfg), the module search path, the Python version in use, and other details.

The preceding Ansible version is coming from the default repository that you have configured on the OS. If you want to install the latest or different version of Ansible, then you need to follow different methods.

Installing Ansible using Python pip

You need to install pip if it is not already available on the system and then install Ansible using Python pip as follows:

● ● ●

```
## download and install Python pip
$ curl https://bootstrap.pypa.io/get-pip.py -o get-pip.py
$ python get-pip.py --user

## If pip is already installed, then make sure it is upgraded to the latest supported version.
$ python -m pip install --upgrade pip

## Then, install Ansible using pip:
$ python -m pip install --user ansible
```

Figure 1.10 – Installing Ansible using Python pip

Please note, when you execute pip install ansible, you are installing the Ansible package, which contains ansible-core and Ansible collections. I have already explained Ansible collections earlier in this chapter.

> **Pip-Based Ansible Installation and Support**
>
> It is best practice to follow the installation based on the OS package manager to get the appropriate support and updates automatically. Also, pip-based installations are hard to maintain and upgrade when there are newer versions of software available.

If you want to install a specific version of `ansible`, `ansible-base`, or `ansible-core`, then use the version information as follows:

```
## Installing old ansible version (ansible + modules)
$ python -m pip install ansible==2.9.25 --user

## Installing ansible package (ansible-core + Ansible collections)
$ python -m pip install ansible==4 --user

## Installing ansible base (ansible-base only; you need to install required collections separately)
$ python -m pip install ansible-base==2.10.13 --user

## Installing ansible core (ansible-core only; you need to install required collections separately)
$ python -m pip install ansible-core==2.11.4 --user
```

Figure 1.11 – Installing specific version of Ansible using pip

Let's check the Ansible version now:

```
[ansible@ansible ~]$ ansible --version
[DEPRECATION WARNING]: Ansible will require Python 3.8 or newer on the controller starting with Ansible
2.12. Current version: 3.6.8 (default, Mar 18 2021, 08:58:41) [GCC 8.4.1 20200928 (Red Hat 8.4.1-1)]. This
feature will be removed from ansible-core in version 2.12. Deprecation warnings can be disabled by setting
deprecation_warnings=False in ansible.cfg.
ansible [core 2.11.6]
config file = None
configured module search path = ['/home/ansible/.ansible/plugins/modules', '/usr/share/ansible/plugins/modules']
ansible python module location = /home/ansible/.local/lib/python3.6/site-packages/ansible
ansible collection location = /home/ansible/.ansible/collections:/usr/share/ansible/collections
executable location = /home/ansible/.local/bin/ansible
python version = 3.6.8 (default, Mar 18 2021, 08:58:41) [GCC 8.4.1 20200928 (Red Hat 8.4.1-1)]
jinja version = 3.0.3
libyaml = True
```

Figure 1.12 – Check Ansible version

See the DEPRECATION WARNING message and ignore that in a development environment. But in the production environment, you need to make sure you are using the supported Python version for Ansible.

> **Ansible Installation**
>
> Check the Ansible documentation for instructions for different OSs and methods at `https://docs.ansible.com/ansible/latest/installation_guide/intro_installation.html`.

Deploying Ansible

Before you start with automation jobs, you need to configure Ansible for your environment using the `ansible.cfg` file. Ansible will look for a configuration file in four places in order, as listed in the following:

- `$ANSIBLE_CONFIG`: Configuration file path in an environment variable
- `./ansible.cfg`: Configuration file in the current directory
- `~/.ansible.cfg`: Configuration file in the home directory
- `/etc/ansible/ansible.cfg`: Default configuration file

It is a best practice to keep the project-specific `ansible.cfg` file in the project directory:

```
[ansible@ansible ~]$ mkdir ansible-demo
[ansible@ansible ~]$ cd ansible-demo/
[ansible@ansible ansible-demo]$ vim ansible.cfg
```

Figure 1.13 – Creating the ansible.cfg file

Add some basic configurations inside the `ansible.cfg` file:

```
[Defaults]
inventory = ./hosts
remote_user = devops
ask_pass = false
```

Figure 1.14 – Content of the ansible.cfg file

It is not a best practice to keep `remote_user = devops` in `ansible.cfg`; instead, configure it inside the inventory for different hosts and host groups. You will learn about this later in this chapter, in the *Creating an Ansible inventory* section.

Now, check the Ansible version again to see the difference and see that Ansible is taking `/home/ansible/ansible-demo/ansible.cfg` as the configuration:

● ● ●

```
ansible@ansible ansible-demo]$ ansible --version
ansible [core 2.11.6]
config file = /home/ansible/ansible-demo/ansible.cfg
.
..<output omitted for brevity>..
```

Figure 1.15 – Checking which ansible.cfg is taken by Ansible

You have many other parameters to configure in `ansible.cfg` but not all of them are mandatory as if you have not specified parameters in your custom `ansible.cfg` file, then Ansible will load the default configurations as needed.

See another sample `ansible.cfg` file, which contains some of the important parameters, deciding the privilege escalation in Ansible. These parameters decide the `sudo` or `su` operation on the target nodes:

● ● ●

```
[ansible@ansible ansible-demo]$ cat ansible.cfg
[Defaults]
inventory = ./hosts
remote_user = devops
ask_pass = false

[privilege_escalation]
become = true
become_method = sudo
become_user = root
become_ask_pass = true
```

Figure 1.16 – Another ansible.cfg sample with privilege escalation parameters

In the preceding example, the following happens:

- Ansible will log in to the remote node as a `devops` user and without asking for a password (`ask_pass = false`).

- Ansible will automatically escalate privilege (`become = true`) to `root` (`become_user = root`) by using the `sudo` method (`become_method = sudo`) and it will also ask for a `sudo` password for the user. Turn this off with the `become_ask_pass = false` setting.

Configure different credentials for different nodes in the Ansible inventory. You will learn about that in a later section of this chapter.

Also note, in the preceding example I have not created any inventory files yet. I just mentioned the inventory filename there as `./hosts`.

Creating an Ansible inventory

Use any name for your inventory file, such as hosts, inventory, myhosts, or production-servers. Do not confuse this hosts file with the /etc/hosts file. When you have more and more managed nodes in inventory files, split the inventory into groups or into separate inventory files. For example, put production servers and devices in a file called production-hosts, and staging nodes into the staging-hosts file; there are no restrictions on this. Also note, if you are installing Ansible using the yum or apt utilities, then there will be a default inventory file called /etc/ansible/hosts with sample inventory content. Refer to that file to start with, but it is best practice to create your project-specific inventory file inside your project directory itself. It is possible to create inventory files in the ini or yaml formats as follows:

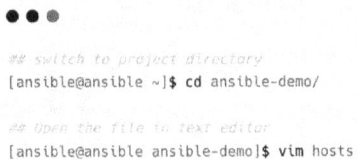

```
## switch to project directory
[ansible@ansible ~]$ cd ansible-demo/

## Open the file in text editor
[ansible@ansible ansible-demo]$ vim hosts
```

Figure 1.17 – Creating inventory file inside project directory

Add content inside the file as follows and save the file:

```
[local]
localhost

[dev]
192.168.100.4
```

Figure 1.18 – Sample inventory file

You do not need to add localhost as the localhost entry is implicit in Ansible. That means we can still call the localhost node and Ansible will create an implicit entry. In the preceding demonstration, I have added this to show an entry as a sample. Since localhost is the local machine, you will not be using an SSH connection and it is possible tell Ansible the connection type for localhost as local:

```
localhost ansible_connection=local
```

It is not easy to remember the nodes with IP addresses or nodes with long hostnames. In Ansible, use any name for your hosts, and mention the actual name or IP address using the ansible_host parameter.

Update your inventory with the correct entry name now. Please note, this is optional, but it is a best practice to use human-readable names rather than IP addresses and long **Fully Qualified Domain Names (FQDNs)**:

```
[ansible@ansible ansible-demo]$ cat hosts
[local]
localhost ansible_connection=local

[dev]
node01 ansible_host=192.168.100.4
```

Figure 1.19 – Ansible inventory with human-readable names and ansible_host

The default `ansible_connection` is `ssh` and you do not need to mention that in the inventory. If you are using any other connection types, then you need to mention them, such as `local`, `winrm`, or `paramiko`.

Add any number of nodes here with multiple host groups (as the `dev` host group in the preceding example).

Now, test your inventory to make sure Ansible is able to read and understand your inventory and hosts. Please note that you are not connecting to the hosts from Ansible using this command:

```
[ansible@ansible ansible-demo]$ ansible all --list-hosts
hosts (2):
  localhost
  node01
```

Figure 1.20 – List inventory hosts

Create another inventory file with some dummy managed nodes:

● ● ●

```
[ansible@ansible ansible-demo]$ cat myinventory
[myself]
localhost

[intranetweb]
servera.techbeatly.com
serverb.techbeatly.com

[database]
db101.techbeatly.com

[everyone:children]
myself
intranetweb
database
```

Figure 1.21 – Another Ansible inventory with more hosts and groups

In the preceding inventory, see the following:

- You have four host groups: myself, intranet, database, and everyone.

- everyone is the parent of the myself, intranet, and database host groups.

You have two inventory files here now:

● ● ●

```
[ansible@ansible ansible-demo]$ ls -l
total 12
-rw-rw-r--. 1 ansible ansible 181 Nov 19 15:40 ansible.cfg
-rw-rw-r--. 1 ansible ansible  90 Nov 19 15:33 hosts
-rw-rw-r--. 1 ansible ansible 162 Nov 19 15:44 myinventory
```

Figure 1.22 – Multiple inventory files in project directory

To use a different inventory file, you do not need to change the content of ansible.cfg; instead, use the -i switch to specify the inventory file dynamically. Ansible will take the mentioned inventory files instead of the one configured in the ansible.cfg file as follows:

● ● ●

```
[ansible@ansible ansible-demo]$ ansible all --list-hosts -i myinventory
hosts (4):
  localhost
  servera.techbeatly.com
  serverb.techbeatly.com
  db101.techbeatly.com
```

Figure 1.23 – List inventory hosts with different a inventory file

Use the --help options to see all the available switches with the Ansible command:

```
●●●
[ansible@ansible ansible-demo]$ ansible --help
.
.

-h, --help              show this help message and exit
-i INVENTORY, --inventory INVENTORY, --inventory-file INVENTORY
                        speciy inventory host path or comma separated host
                        list. --inventory-file is deprecated
-l SUBSET, --limit SUBSET
                        further imit selected hosts to an additional pattern
-m MODULE_NAME, --module-name MODULE_NAME
                        Name of the actionto execute (default=command)
-o, --one-line          condense output
-t TREE, --tree TREE  log output to this directory
-v, --verbose           verbose mode (-vvv for more, -vvvv to enable
                        connection debugging)
.
...<output omitted for brevity>...
```

Figure 1.24 – List inventory hosts with different a inventory file

It is also possible to use patterns to filter the managed hosts with supported patterns. For example, let us display only managed nodes with *techbeatly.com:

```
●●●
[ansible@ansible ansible-demo]$ ansible --list-hosts -i myinventory *techbeatly.com
hosts (3):
  servera.techbeatly.com
  serverb.techbeatly.com
  db101.techbeatly.com

** Print only db servers**
[ansible@ansible ansible-demo]$ ansible --list-hosts -i myinventory db*
hosts (1):
  Db101.techbeatly.com
```

Figure 1.25 – Host selection using patterns

> **Ansible host patterns**
>
> Look up more information about patterns, targeting hosts, and groups at https://docs.ansible.com/ansible/latest/user_guide/intro_patterns.html.
>
> Look up more information about the Ansible dynamic inventory at https://docs.ansible.com/ansible/latest/user_guide/intro_dynamic_inventory.html and https://github.com/ansible/ansible/tree/stable-2.9/contrib/inventory.

In the next section, we will learn how to configure managed nodes and connections.

Configuring your managed nodes

Use any supported authentication mechanisms to connect from the Ansible control node to managed node, such as SSH key-based authentication, username and password-based authentication, and SSL certificate authentication, for example.

Setting up SSH key-based authentication

It is possible to automate most of the following steps using Ansible ad hoc commands, but we will be using the manual approach to understand what backend configurations are needed.

The steps to create a user (for example devops) and enable SSH key based access on node-1 are as follows:

1. Create a dedicated user on the target node (e.g.: node01) as follows. (This is not mandatory, and it is possible to use any existing user accounts and configure that in Ansible.):

```
## create a new user - devops
[root@node01 ~]# useradd devops

## set password for the new user
[root@node01 ~]# passwd devops
Changing password for user devops.
New password:
Retype new password:
passwd: all authentication tokens updated successfully.
```

Figure 1.26 – Create new user and set password

If you do not have **Domain Name System** (**DNS**) servers in the network, then directly use the IP address with the ansible_host option, or add entries in /etc/hosts as local DNS resolution.

2. Enable the sudo access for the new user because, for any kind of privileged operation on the target node, you will be required to have this access:

```
[root@node01 ~]# echo "devops ALL=(ALL) NOPASSWD: ALL" > /etc/sudoers.d/devops
```

Figure 1.27 – Enabled privileged access for the new user

3. Create an SSH key pair on the **Ansible control node**. It is possible to create any supported type and size. Please note, if you have any existing key with the same name, please remember to use a different name or backup the original SSH key pairs:

```
[ansible@ansible ansible-demo]$ ssh-keygen -t rsa -b 4096 -C "ansible@ansible.lab.local"
Generating public/private rsa key pair.
Enter file in which to save the key (/home/ansible/.ssh/id_rsa):
Created directory '/home/ansible/.ssh'.
Enter passphrase (empty for no passphrase):
Enter same passphrase again:
Your idetification has been saved in /home/ansible/.ssh/id_rsa.
Your ublic key has been saved in /home/ansible/.ssh/id_rsa.pub.
..<output omitted>..
+----[SHA256]-----+
```

Figure 1.28 – Generating SSH key pair on Ansible control node

4.   Verify the SSH key permissions:

```
[ansible@ansible ansible-demo]$ ls -la ~/.ssh/
total 8
drwx------. 2 ansible ansible   38 Nov 19 16:14 .
drwx------. 7 ansible ansible  175 Nov 19 16:14 ..
-rw-------. 1 ansible ansible 3389 Nov 19 16:14 id_rsa
-rw-r--r--. 1 ansible ansible  751 Nov 19 16:14 id_rsa.pub
```

Figure 1.29 – Verify SSH key permission

5.   Copy the SSH public key from the Ansible control node to managed nodes under the devops user using the ssh-copy-id command:

```
[ansible@ansible ansible-demo]$ ssh-copy-id -i ~/.ssh/id_rsa devops@node01
/usr/bin/ssh-copy-id: INFO: Source of key(s) to be installed: "/home/ansible/.ssh/id_rsa.pub"
The authenticity of host 'node01 (192.168.100.4)' can't be established.
RSA key fingerprint is SHA256:UEQ72EtSvn+0/tuEDbeclQuhHNTtp/uPf+VVvKkuB6k.
Are you sure you want to continue connecting (yes/no/[fingerprint])? yes
/usr/bin/ssh-copy-id: INFO: attempting to log in with the new key(s), to filter out any that are already installed

/usr/bin/ssh-copy-id: INFO: 1 key(s) remain to be installed -- if you are prompted now it is to install the new
keys
devops@node01's password:

Number of key(s) added: 1

Now try logging into the machine, with:   "ssh 'devops@node01'"
and check to make sure that only the key(s) you wanted were added.
```

Figure 1.30 – Copy SSH public key to managed node

If you have issues with password authentication or copying, then manually copy the public key content from `/home/ansible/.ssh/id_rsa.pub` to the `/home/devops/.ssh./authorized_keys` file on the managed node.

6. Verify the passwordless SSH access from the Ansible control node to the managed node:

```
[ansible@ansible ansible-demo]$ ssh devops@node01node-1
Last login: Fri Nov 19 16:23:25 2021
[devops@node01node-1 ~]$

## check sudo access
[devops@node01node-1 ~]$ sudo -i
[root@node01node-1 ~]# hostname
Node01Node-1.lab.local
```

Figure 1.31 – Login to managed node without password

> **How to Set Up SSH Key-Based Authentication**
>
> Check out the steps on how to set up SSH key-based authentication at `https://www.techbeatly.com/2018/06/how-to-setup-ssh-key-based-authentication.html`.

In the next section, we will explore the option to use multiple credential for different managed nodes.

Multiple users and credentials

If you have different credentials for different managed nodes, then configure the remote username, SSH key to be used, and more in your inventory file. Let me show a sample for our `node01` managed node in our inventory:

```
[dev]
node01 ansible_host=192.168.100.4 ansible_ssh_private_key_file=/home/ansible/.ssh/id_rsa ansible_user=devops

## Or, you can configure the variable details
## separately in the inventory file:
[dev]
node01 ansible_host=192.168.100.4

[dev:vars]
ansible_ssh_private_key_file=/home/ansible/.ssh/id_rsa
ansible_user=devops
```

Figure 1.32 – Configuring SSH key information for managed nodes

In the latter example, we have used a variable section for the `dev` host group and mentioned the SSH key and remote user details.

Ansible ad hoc commands

The `ansible` command can be used to execute single jobs on managed nodes without a playbook; this is called an **Ansible ad hoc** command. It can be a simple connectivity check (`ping`) to the managed nodes, creating a user account, copying some files, or restarting a service, and execute these tasks without writing an Ansible playbook.

For example, it is possible to use the `ping` module to test the connection from Ansible to the managed node, `node01`, using this user and SSH key pair:

```
[ansible@ansible ansible-demo]$ ansible all -m ping
localhost | SUCCESS => {
  "ansible_facts": {
      "discovered_interpreter_python": "/usr/libexec/platform-python"
  },
  "changed": false,
  "ping": "pong"
}
node01 | SUCCESS => {
  "ansible_facts": {
      "discovered_interpreter_python": "/usr/libexec/platform-python"
  },
  "changed": false,
  "ping": "pong"
}
```

Figure 1.33 – Ansible ad hoc command using ping module

In the preceding snippet, as you have `localhost` also in the inventory (by implicit), the task will be executed on both localhost and node01 nodes when you mention `all`. The Ansible `ping` module is not just a regular network ping (ICMP); instead, it will log in to the managed node and return the result, `pong`.

Execute another Ansible ad hoc command using the `shell` module to check what remote user Ansible is using for connection:

```
[ansible@ansible ansible-demo]$ ansible all -m shell -a "whoami"
localhost | CHANGED | rc=0 >>
ansible
node01 | CHANGED | rc=0 >>
devops
```

Figure 1.34 – Ansible ad hoc command using shell module

From the preceding output, see that `localhost` is executed with the default `ansible` user and the `dev` node with the `devops` user.

Now, execute multiple commands using the `shell` module:

● ● ●

```
[ansible@ansible ansible-demo]$ ansible all -m shell -a "hostname;uptime;date;cat /etc/*release| grep ^NAME;uname
-a"
localhost | CHANGED | rc=0 >>
ansible
16:58:15 up  1:37,  1 user,  load average: 0.00, 0.00, 0.00
Fri Nov 19 16:58:15 UTC 2021
NAME="Red Hat Enterprise Linux"
Linux ansible 4.18.0-305.el8.x86_64 #1 SMP Thu Apr 29 08:54:30 EDT 2021 x86_64 x86_64 x86_64 GNU/Linux
node01 | CHANGED | rc=0 >>
node01.lab.local
16:58:15 up  1:43,  2 users,  load average: 0.24, 0.05, 0.02
Fri Nov 19 16:58:15 UTC 2021
NAME="Red Hat Enterprise Linux"
Linux node01.lab.local 4.18.0-305.el8.x86_64 #1 SMP Thu Apr 29 08:54:30 EDT 2021 x86_64 x86_64 x86_64 GNU/Linux
```

Figure 1.35 – Multiple commands in shell module

Please note that the preceding example was used to demonstrate the `shell` module, and similar details can be collected using `ansible_facts` without such tasks.

● ● ●

```
[ansible@ansible ansible-demo]$ ansible all -m setup -a "filter=ansible_distribution*"
```

Figure 1.36 – Ansible ad hoc command using setup module

You will learn more about `ansible_facts` in *Chapter 3*, *Automating Your Daily Jobs*.

Installing a package using Ansible

You need to ensure that you have package repositories (`yum` or `apt`) configured and enabled on the target machine.

Install the `vim` package on `node01`:

● ● ●

```
[ansible@ansible ansible-demo]$ ansible node01 -m dnf -a 'name=vim state=latest'
node01 | FAILED! => {
  "ansible_facts": {
      "discovered_interpreter_python": "/usr/libexec/platform-python"
  },
  "changed": false,
  "msg": "This command has to be run under the root user.",
  "results": []
}
```

Figure 1.37 – Ansible ad hoc command using dnf module

From the preceding output, see that you are using the devops user for connecting to managed nodes, which is a normal user. You do not need to add the become details in ansible.cfg; instead, pass this become switch while executing the ansible command, which is -b. (For Ansible playbooks, you can enable or disable the privilege escalation at the play level or tasks level):

● ● ●

```
[ansible@ansible ansible-demo]$ ansible node01 -m dnf -a 'name=vim state=latest' -b
node01 | CHANGED => {
  "ansible_facts": {
      "discovered_interpreter_python": "/usr/libexec/platform-python"
  },
  "changed": true,
  "msg": "",
  "rc": 0,
  "results": [
      "Installed: vim-common-2:8.0.1763-16.el8.x86_64",
      "Installed: vim-enhanced-2:8.0.1763-16.el8.x86_64",
      "Installed: gpm-libs-1.20.7-17.el8.x86_64"
  ]
}
```

Figure 1.38 – Installing package using dnf module and privileged mode

The package installation is successful as per the output.

The preceding ad hoc execution can be written in a playbook as follows:

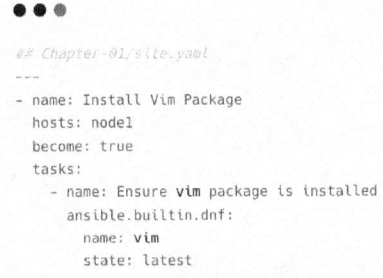

```
## Chapter-01/site.yaml
---
- name: Install Vim Package
  hosts: node1
  become: true
  tasks:
    - name: Ensure vim package is installed
      ansible.builtin.dnf:
        name: vim
        state: latest
```

Figure 1.39 – Package installation using an Ansible playbook

You will learn more about writing playbooks in *Chapter 2, Starting with Simple Automation.*

Now, remove the same package using an Ansible ad hoc command; instead of `state=latest`, use `state=absent`:

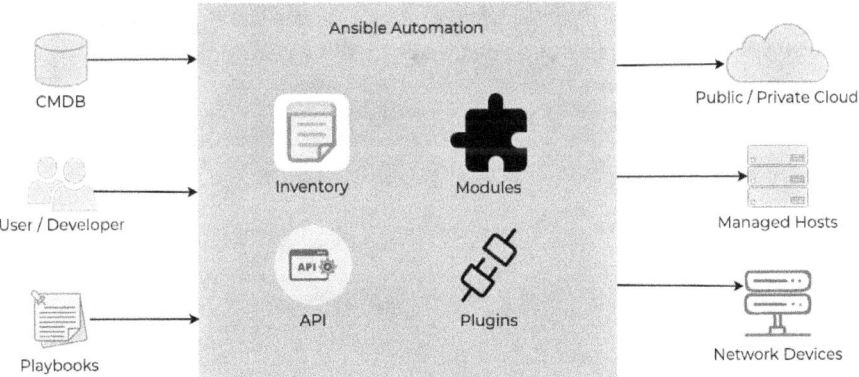

Figure 1.40 – Removing package using Ansible ad hoc command

We have now successfully installed and uninstalled a package using Ansible.

Summary

In this chapter, you have learned about the core concepts of Ansible, how Ansible works, and the key terminology, including Ansible inventory, playbooks, tasks, Ansible control node, managed nodes, and modules. You have also learned about basic Ansible installation, configuration, and deployment, and the importance of managed node configuration, SSH key credentials, and passwordless SSH.

In the next chapter, you will learn how to start with simple automation playbooks and execution. Then, you will learn how to find suitable Ansible modules for playbook development and remote node connection methods.

Further reading

- *Ansible documentation*: `https://docs.ansible.com/ansible/latest/index.html`

- *Ansible control node requirements*: `https://docs.ansible.com/ansible/latest/installation_guide/intro_installation.html#control-node-requirements`

- *Ansible dynamic inventory*: `https://docs.ansible.com/ansible/latest/user_guide/intro_dynamic_inventory.html` and `https://github.com/ansible/ansible/tree/stable-2.9/contrib/inventory`

- *Ansible Automation Platform*: `https://www.ansible.com/blog/introducing-red-hat-ansible-automation-platform`

- *Installing Ansible*: `https://docs.ansible.com/ansible/latest/installation_guide/intro_installation.html`

- *How network automation is different*: `https://docs.ansible.com/ansible/latest/network/getting_started/network_differences.html`

Starting with Simple Automation

When you start your automation journey, start with simple use cases instead of automating complex workflows. Find three small use cases that you can use to learn automation faster and implement it in your environment. Ansible has a smooth learning curve but it is also important to choose the right use cases for your first automation project. Three great examples for initial use cases for automation are simple tasks such as application deployment, asset information collection, and simple file manipulation such as copy operations.

In this chapter, we are going to cover the following topics:

- Identifying manual tasks to be automated
- Finding the Ansible modules to use
- Configuring your text editor for Ansible
- Connecting to remote nodes

You will start by creating basic automation tasks by finding suitable modules before learning how to use credentials and other parameters.

Technical requirements

You will need the following technical requirements for this chapter:

- A Linux machine for the Ansible control node
- One or more Linux machines with Red Hat repositories configured (if you are using other Linux operating systems instead of **Red Hat Enterprise Linux** (**RHEL**), then make sure you have the appropriate repositories configured to get packages and updates)

All the Ansible code and the Ansible playbooks, commands, and snippets for this chapter can be found in this book's GitHub repository at `https://github.com/PacktPublishing/Ansible-for-Real-life-Automation/tree/main/Chapter-02`.

Identifying manual tasks to be automated

In the previous chapter, you learned how to use Ansible ad hoc commands to manually execute tasks on remotely managed nodes using Ansible modules. Now, you will learn how to start with simple Ansible playbooks and tasks. Remember, you need to add your managed node details to your inventory file before you can execute any Ansible tasks.

We will start with a simple automation job to understand the basics of the Ansible playbook. For this example, we are assuming you have installed and configured the chronyd application. The chrony application is an implementation of the **Network Time Protocol** (**NTP**). chronyd is the default NTP client and server in Red Hat Enterprise Linux 8 and SUSE Linux Enterprise Server 15 and is available in many Linux distributions.

For our example Ansible playbook, we will do the following:

1. Install the chrony package on all nodes.

2. Adjust the chrony configurations.

3. Start the chronyd service and enable it.

The following diagram shows the Ansible to node1 connection for deploying and configuring the chrony application:

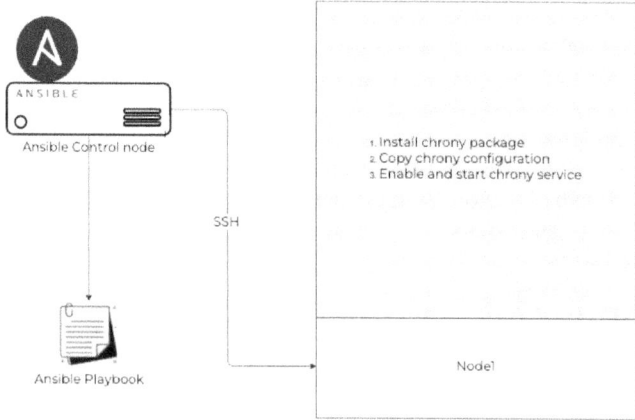

Figure 2.1 – Configurations for Ansible to automate chrony

Follow these steps to create Ansible artifacts for the chronyd installation, including ansible. cfg, hosts, and the Ansible playbook:

1. First, create a new directory called Chapter-02 and create the ansible.cfg file inside the directory, as you did in the previous chapter:

```
● ● ●

[defaults]
inventory = ./hosts
remote_user = devops
ask_pass = false

deprecation_warnings = False
[privilege_escalation]
become = false
become_method = sudo
become_user = root
become_ask_pass = false
```

Figure 2.2 – Ansible configuration file (ansible.cfg)

2. Next, create the `hosts` file in the same directory as the `node1` managed node:

```
● ● ●

.
.
[nodes]
node1 ansible_host=192.168.56.25

[nodes:vars]
ansible_ssh_private_key_file=/home/ansible/.ssh/id_rsa
ansible_user=devops
```

Figure 2.3 – Ansible inventory file with node1

3. Create a playbook file called `install-package.yaml` in the same directory with the following content:

```
● ● ●

---
- name: Install Chrony Package
  hosts: node1
  become: true
  tasks:
    - name: Ensure chrony package is installed
      ansible.builtin.dnf:
        name: chrony
        state: latest
```

Figure 2.4 – Ansible playbook for installing chrony

In the preceding playbook, you only have one *play* called `Install Chrony Package` and only one task called `Ensure Chronry package is installed`. Also, note the `hosts: node1` line as you are installing the package on your managed node – that is, `node1`.

Ansible Fully Qualified Collection Name (FQCN)

As we learned in *Chapter 1, Ansible Automation – Introduction,* in the *Ansible Content Collections* section, many plugins and modules were moved to content collections in Ansible 2.10. The playbooks will work without any issues but it is best practice to use **FQCN** for the modules, plugins, and roles to explicitly specify which items are to be used. In the preceding example, `ansible.builtin.dnf` is an FQCN in which dnf is the module name and part of `ansible-core`. As another example, there's `google.cloud.gcp_compute_disk`, where `google` is the author of the collection, `cloud` is the collection name, and `gcp_compute_disk` is the module name to create a Google Cloud disk. Read more about collections here: `https://docs.ansible.com/ansible/latest/user_guide/collections_using.html`.

---

4. Execute the playbook using the `ansible-playbook` command; you will see the following output:

● ● ●

```
[ansible@ansible Chapter-02]$ ansible-playbook install-package.yaml

PLAY [Install Chrony Package] ********************************************************************

TASK [Gathering Facts] ********************************************************************
ok: [node1]

TASK [Ensure Chronry package is installed] ********************************************************
changed: [node1]

PLAY RECAP ********************************************************************
dev-rhel8-55node1              : ok=2    changed=1    unreachable=0    failed=0    skipped=0    rescued=0
 ignored=0
```

Figure 2.5 – The chrony package installation playbook

With that, the playbook has been executed successfully. The following screenshot shows the success status:

```
● ● ●

TASK [Ensure Chronry package is installed] *********************************************************
changed: [node1]
```

Figure 2.6 – The chrony package installation message

5. Now, log in to the target machine, node1, and verify the installation:

```
● ● ●

[devops@node-1 ~]$ sudo yum list installed chrony
Updating Subscription Management repositories.
Installed Packages
chrony.x86_64          4.1-1.el8          @rhel-8-for-x86_64-baseos-rpms
```

Figure 2.7 – The chrony package installed on node1

Now, you need to create a chrony configuration that you can use for all of your servers.

6. Create the Chapter-02/chrony.conf.sample file in the same directory (Chapter-02). Add more details to the chrony configuration as per your organization's standards:

```
● ● ●

server 0.sg.pool.ntp.org
server 1.sg.pool.ntp.org
server 2.sg.pool.ntp.org
server 3.sg.pool.ntp.org
driftfile /var/lib/chrony/drift
makestep 1.0 3
rtcsync
keyfile /etc/chrony.keys
leapsectz right/UTC
logdir /var/log/chrony
```

Figure 2.8 – Sample chrony configuration

7. Add a task called Copy chrony configuration to node to the playbook to copy the sample chrony configuration to the node using the template module. Then, start the chronyd service using the service module.

The completed playbook will look as follows:

```
---
# Chapter-02/install-package.yaml
- name: Install Chrony Package
  hosts: node1
  become: true
  tasks:
    - name: Ensure chrony package is installed
      ansible.builtin.dnf:
        name: chrony
        state: latest

    - name: Copy chrony configuration to node
      ansible.builtin.copy:
        src: chrony.conf.sample
        dest: /etc/chrony.conf
        mode: 644
        owner: root
        group: root

    - name: Enable and start chrony Service
      ansible.builtin.systemd:
        name: chronyd
        state: started
        enabled: yes
        masked: no
```

Figure 2.9 – Playbook for installing and configuring chrony

8. Execute the playbook again and verify it, as shown in the following screenshot:

```
[ansible@ansible Chapter-02]$ ansible-playbook install-package.yaml

PLAY [Install Chrony Package] **********************************************************

TASK [Gathering Facts] *****************************************************************
ok: [dev-rhel8-55]

TASK [Ensure chrony package is installed] **********************************************
ok: [dev-rhel8-55]

TASK [Copy chrony configuration to node] **********************************************
changed: [node1]

TASK [Enable and start chrony Service] ************************************************
changed: [node1]

PLAY RECAP ****************************************************************************
node1              : ok=4    changed=2    unreachable=0    failed=0    skipped=0    rescued=0    ignored=0
```

Figure 2.10 – Expanded playbook with chrony configuration

In the preceding example, the status of the `Ensure chrony package is installed` task is ok. This means that the desired state has already been reached and that you do not need to install the `chrony` package again. Therefore, Ansible will not take any action for that task.

In Ansible, this feature is called **idempotency**, which means that if the result of performing an action is the same as the current state, then no further action is required for that task. Most of the Ansible modules are idempotent, which will help you run the same playbook multiple times on managed nodes, without any impact (`https://docs.ansible.com/ansible/latest/reference_appendices/glossary.html#term-Idempotency`).

Now, verify the details on the target node:

```
[devops@node-1 ~]$ cat /etc/chrony.conf
server 0.sg.pool.ntp.org
server 1.sg.pool.ntp.org
server 2.sg.pool.ntp.org
server 3.sg.pool.ntp.org
driftfile /var/lib/chrony/drift
makestep 1.0 3
rtcsync
keyfile /etc/chrony.keys
leapsectz right/UTC
logdir /var/log/chrony[devops@node-1 ~]$

[devops@node-1 ~]$ sudo systemctl status chronyd
● chronyd.service - NTP client/server
   Loaded: loaded (/usr/lib/systemd/system/chronyd.service; enabled; vendor preset: enabl>
   Active: active (running) since Sun 2022-07-24 07:58:06 UTC; 1h 8min ago
     Docs: man:chronyd(8)
           man:chrony.conf(5)
...output omitted...
```

Figure 2.11 – The chrony configuration and service status on node1

The preceding playbook can be used to automate your `chrony` configuration for thousands of servers and it will only take a few minutes to complete the tasks.

> **Ansible Module References**
>
> Refer to the following documentation for the modules you have used in the playbook:
>
> - `https://docs.ansible.com/ansible/latest/collections/ansible/builtin/dnf_module.html`
>
> - `https://docs.ansible.com/ansible/latest/collections/ansible/builtin/copy_module.html`
>
> - `https://docs.ansible.com/ansible/latest/collections/ansible/builtin/systemd_module.html`

Finding the Ansible modules to use

In this section, you will learn how to find suitable modules and documentation to use inside the Ansible playbook.

Find the available modules and details using the `ansible-doc` command:

```
[ansible@ansible Chapter-02]$ ansible-doc -l
add_host                                                          Add a host ...
amazon.aws.aws_az_facts                                          Gather info...
amazon.aws.aws_az_info                                           Gather info...
amazon.aws.aws_caller_facts                                      Get informa...
amazon.aws.aws_caller_info                                       Get informa...
amazon.aws.aws_s3                                                manage obje...
amazon.aws.cloudformation                                        Create or d...
amazon.aws.cloudformation_facts                                  Obtain info...
amazon.aws.cloudformation_info                                   Obtain info...
...output omitted...
```

Figure 2.12 – Ansible module list

It will be a long or short list, depending on your type of Ansible installation. (Recall the difference between `ansible`, `ansible-base`, and `ansible-core`, which was explained in the previous chapter.) You can check the total module count that's available as follows:

```
[ansible@ansible Chapter-02]$ ansible-doc -l |wc -l
6108
```

Check the module details by calling the module name with the `-s` (`--snippet`) argument, as follows:

```
[ansible@ansible Chapter-02]$ ansible-doc -s dnf
- name: Manages packages with the      package manager
  dnf:
      allow_downgrade:         # Specify if the named package and version is allowed to downgrade
                               a maybe already installed higher
                               version of that package. Note
                               that setting allow_downgrade=True
                               can make this module behave in a
                               non-idempotent way. The task
                               could end up with a set of
                               packages that does not match the
...output omitted...
```

Figure 2.13 – Ansible module snippet for the dnf module

Alternatively, Check the full details of the module as follows:

```
● ● ●
[ansible@ansible Chapter-02]$ ansible-doc dnf
> ANSIBLE.BUILTIN.DNF    (/home/ansible/.local/lib/python3.6/site-packages/ansible/modules/dnf.>

        Installs, upgrade, removes, and lists packages and groups with the
        dnf` package manager.

OPTIONS (= is mandatory):

- allow_downgrade
        Specify if the named package and version is allowed to downgrade a
        maybe already installed higher version of that package. Note that
        setting allow_downgrade=True can make this module behave in a non-
        idempotent way. The task could end up with a set of packages that
...output omitted...
VERSION_ADDED_COLLECTION: ansible.builtin

EXAMPLES:

- name: Install the latest version of Apache
  dnf:
    name: httpd
    state: latest

- name: Install Apache >= 2.4
  dnf:
    name: httpd>=2.4
    state: present
...output omitted...
```

Figure 2.14 – Ansible module details for the dnf module

The preceding output shows the example usages and all the arguments for the module. This is like an offline copy of the module documentation. If you want to search for a specific module interactively inside the module list, execute the `ansible-doc -l` command, then press the / key and type the module's name to search, as shown in the following screenshot:

```
● ● ●

add_host                                           Add a host (a...
ansible.netcommon.cli_command                      Run a cli com...
ansible.netcommon.cli_config                       Push text bas...
ansible.netcommon.cli_parse                        Parse cli out...
ansible.netcommon.net_banner                       (deprecated, ...
ansible.netcommon.net_get                          Copy a file f...
ansible.netcommon.net_interface                    (deprecated, ...
ansible.netcommon.net_l2_interface                 (deprecated, ...
ansible.netcommon.net_l3_interface                 (deprecated, ...
ansible.netcommon.net_linkagg                      (deprecated, ...
ansible.netcommon.net_lldp                         (deprecated, ...
ansible.netcommon.net_lldp_interface               (deprecated, ...
ansible.netcommon.net_logging                      (deprecated, ...
ansible.netcommon.net_ping                         Tests reachab...
ansible.netcommon.net_put                          Copy a file f...
ansible.netcommon.net_static_route                 (deprecated, ...
ansible.netcommon.net_system                       (deprecated, ...
ansible.netcommon.net_user                         (deprecated, ...
/dnf
```

Figure 2.15 – Searching for the module in ansible-doc

When you hit the *Enter* key, the searched item will be highlighted if available. This can be seen in the following screenshot. Press the *N* key to find the next item with the same text:

```
● ● ●

dnf                                                       Manages packa...
dpkg_selections                                           Dpkg package ...
expect                                                    Executes a co...
f5networks.f5_modules.bigip_apm_acl                       Manage user-d...
f5networks.f5_modules.bigip_apm_network_access            Manage APM Ne...
f5networks.f5_modules.bigip_apm_policy_fetch              Exports the A...
f5networks.f5_modules.bigip_apm_policy_import             Manage BIG-IP...
f5networks.f5_modules.bigip_asm_advanced_settings         Manage BIG-IP...
f5networks.f5_modules.bigip_asm_dos_application           Manage applic...
f5networks.f5_modules.bigip_asm_policy_fetch              Exports the A...
f5networks.f5_modules.bigip_asm_policy_import             Manage BIG-IP...
f5networks.f5_modules.bigip_asm_policy_manage             Manage BIG-IP...
f5networks.f5_modules.bigip_asm_policy_server_technology  Manages Serve...
f5networks.f5_modules.bigip_asm_policy_signature_set      Manages Signa...
f5networks.f5_modules.bigip_cgnat_lsn_pool                Manage CGNAT ...
f5networks.f5_modules.bigip_cli_alias                     Manage CLI al...
f5networks.f5_modules.bigip_cli_script                    Manage CLI sc...
f5networks.f5_modules.bigip_command                       Run TMSH and ...
/dnf
```

Figure 2.16 – Finding modules in the ansible-doc list

Use `ansible-doc` to list all the other plugins, including the `connection` plugins, `become` plugins, `lookup`, `filters`, and so on.

The following screenshot shows the available `become` plugins:

```
[ansible@ansible Chapter-02]$ ansible-doc -t become -l
ansible.netcommon.enable      Switch to elevated permissions on a network device
community.general.doas        Do As user
...output omitted...
runas                         Run As user
su                            Substitute User
sudo                          Substitute User DO
```

Figure 2.17 – Ansible become plugins

The following screenshot shows the available `connection` plugins:

```
[ansible@ansible Chapter-02]$ ansible-doc -t connection -l
ansible.netcommon.httpapi       Use httpapi to run command on network appliances
ansible.netcommon.libssh        (Tech preview) Run tasks using libssh for ssh connection
...output omitted...
community.docker.docker         Run tasks in docker containers
community.docker.docker_api     Run tasks in docker containers
...output omitted...
community.kubernetes.kubectl    Execute tasks in pods running on Kubernetes
community.libvirt.libvirt_lxc   Run tasks in lxc containers via libvirt
community.libvirt.libvirt_qemu  Run tasks on libvirt/qemu virtual machines
community.okd.oc                Execute tasks in pods running on OpenShift
community.vmware.vmware_tools   Execute tasks inside a VM via VMware Tools
containers.podman.buildah       Interact with an existing buildah container
containers.podman.podman        Interact with an existing podman container ...output omitted...
```

Figure 2.18 – Ansible connection plugins

With that, you have learned how to use the `ansible-doc` command to find the necessary modules and module documentation. Next, you will learn how to configure your text editor for editing Ansible playbooks.

Configuring your text editor for Ansible

Since **YAML** is highly sensitive to indentation, you need to take extra care while developing and editing playbooks using your text editor. You can use any text editor of your choice to edit Ansible playbooks and configure the editor as needed.

If you can use a GUI editor such as Visual Studio Code or Atom, skip this section as GUI editors can easily be configured with multiple plugins to perform Ansible content development more efficiently. Please refer to `https://docs.ansible.com/ansible/latest/community/other_tools_and_programs.html` to find details about tools and programs for Ansible content development. Now, let's learn how to configure the **Vim** editor for Ansible YAML files. Use Vim variables to enable or disable the features in the Vim editor:

```
[ansible@ansible Chapter-02]$ vim install-package.yaml
```

Now, press *Esc* followed by : and type `set nu` to enable line numbers, as shown in the following screenshot:

```
  1 ---
  2 - name: Install Chrony Package
  3   hosts: node1
  4   become: true
  5   tasks:
  6     - name: Ensure chrony package is installed
  7       ansible.builtin.dnf:
  8         name: chrony
  9         state: latest
 10
 11     - name: Copy chrony configuration to node
 12       ansible.builtin.copy:
 13         src: chrony.conf.sample
 14         dest: /etc/chrony.conf
 15         mode: 644
 16         owner: root
 17         group: root
 18
 19     - name: Enable and start chrony Service
 20       ansible.builtin.systemd:
 21         name: chronyd
 22         state: started
 23         enabled: yes
:set nu
```

Figure 2.19 – Configuring the Vim editor with line numbers

You will see the line numbers visible on the left-hand side of your editor. But when you close the Vim editor session and reopen it, all these variables will be reset. Instead of enabling the features one by one, you can configure these Vim variables in a file called `.vimrc` under your home directory.

Create a `~/.vimrc` file and add the following content to configure Vim for YAML files:

```
[ansible@ansible Chapter-02]$ cat ~/.vimrc
autocmd FileType yaml setlocal et ts=2 ai sw=2 nu sts=0
colorscheme desert
```

Figure 2.20 – Configuring the ~/.vimrc file

The following table lists some of the available Vim variables:

Variable	Description	Example
ts	`tabstop` is the number of spaces that a `<Tab>` in the file counts for.	`tabstop=2`
et	`expandtab` uses the appropriate number of spaces to insert a `<Tab>`.	`et`
sw	`shiftwidth` is the number of spaces to use for each step of (auto)indent.	`shiftwidth=2`
sts	`softtabstop` is the number of spaces that a `<Tab>` counts for while performing editing operations, such as inserting a `<Tab>` or using `<BS>`.	`softtabstop=2`
nu	`number` shows the line number in the file.	`nu`

Table 2.1 – List of Vim variables

Once you have completed the `~/.vimrc` file, check the Vim editor by editing any YAML files:

```
[ansible@ansible Chapter-02]$ vim install-package.yaml
```

You will see line numbers and other syntax highlighted. Try to edit the file; the indentation will be created automatically when you edit lines. You can still use *Tab* as Vim will replace *Tab* with two spaces based on your `vimrc` configuration. The following screenshot shows a sample Vim editor screen after enabling the `vimrc` configuration:

```
 1 ---
 2 - name: Install Chrony Package
 3   hosts: dev-rhel8-55
 4   become: true
 5   tasks:
 6     - name: Ensure chrony package is installed
 7       ansible.builtin.dnf:
 8         name: chrony
 9         state: latest
10
11     - name: Copy chrony configuration to node
12       ansible.builtin.copy:
13         src: chrony.conf.sample
14         dest: /etc/chrony.conf
15         mode: 644
16         owner: root
17         group: root
18
19     - name: Enable and start chrony Service
20       ansible.builtin.systemd:
21         name: chronyd
22         state: started
23         enabled: yes
24         masked: no
```

Figure 2.21 – The Vim editor configured for YAML files

Please refer to *Setup Your Vim editor for Ansible Playbook* at `https://www.techbeatly.com/setup-your-vim-editor-for-ansible-playbook/` to learn more about Vim editor configuration for YAML.

> **What is Vim editor?**
>
> Vim is a well-known text editor available for Linux platforms. The Vim editor is highly configurable and useful for developing and editing any complex files in Linux. Refer to `https://www.vim.org` for more details.

Connecting to remote nodes

It is the best practice to use dynamic inventories to avoid frequent changes in static inventory files. However, this depends on your environment. It is also a best practice to separate inventory files based on environment, criticality, or other parameters. The following screenshot shows sample inventory files based on the workload environment. As you can see, there are different directories and files for production, development, and staging devices:

```
[ansible@ansible Chapter-02]$ tree inventories/
inventories/
├── development
│   └── hosts
├── production
│   └── hosts
└── staging
    └── hosts

3 directories, 3 files
```

Figure 2.22 – Ansible inventory separation based on environment

You need to ensure that the Ansible control node to managed nodes connection is safe and secure. For Linux/Unix managed nodes, use the `ssh` connection (which is the default connection method) with key-based authentication, as explained in *Chapter 1's, Configuring Your Managed Nodes* section. There might be cases where you cannot use SSH keys. In that case, you can use encrypted passwords with a username for authentication; this will be explained in *Chapter 13, Using Ansible for Secret Management*.

For Windows machines, use the `WinRM` protocol, which Ansible can use to connect to and execute tasks on Windows machines. However, you need to configure a few items on the Windows machine, such as enabling a WinRM listener, opening the port for WinRM, and so on.

View the `WinRM` connection plugin using the `ansible-doc` command, as shown here:

```
[ansible@ansible Chapter-02]$ ansible-doc -t connection -l |grep winrm
winrm                          Run tasks over Microsoft's WinRM
```

Figure 2.23 – Ansible WinRM connection plugin

It is also possible to configure the connection method for the managed host in your inventory or playbook. You can also do this while executing the playbook. The following screenshot shows the `ansible_connection=winrm` method for the `win2019` inventory group. All the hosts under that group will use `winrm` as the connection method:

```
[ansible@ansible Chapter-02]$ cat inventories/production/hosts
[win2019]
prod-db-101 ansible_host=192.168.110.10

[win2019:vars]
ansible_connection=winrm
```

Figure 2.24 – WinRM configured in the Ansible inventory

You can also mention the connection type in your Ansible playbook, as follows:

```
---
- name: Install Package
  hosts: win2019
  become: true
  connection: local
```

Figure 2.25 – WinRM configured in the Ansible playbook

If you want to dynamically provide the connection method, pass this information while executing the `ansible-playbook` command:

```
[ansible@ansible Chapter-02]$ ansible-playbook playbook.yml --connection=local
```

Figure 2.26 – Passing the WinRM connection while executing the Ansible playbook

For network and firewall devices, use the supported connection protocols based on the device's type and compatibility. You will learn more about this in *Chapter 6, Automating Microsoft Windows and Network Devices*.

Please refer to the Ansible connection plugin documentation at `https://docs.ansible.com/ansible/latest/plugins/connection.html` to learn more.

Summary

In this chapter, you learned how to develop a simple playbook to automate the `chrony` package's deployment and service management. You learned how to use the `ansible-doc` command to find the modules and details, including module examples and arguments to use. You also explored how to use the `ansible-doc` command to find the connection plugins and become plugins. After that, you learned how to configure your Vim editor to edit Ansible YAML files easily. Finally, you learned how to configure managed node connection methods.

In the next chapter, you will learn how to automate your daily tasks using Ansible, develop real use cases, and apply them to your workplace.

Further reading

For more information on the topics covered in this chapter, please visit the following links:

- *Tools and programs for Ansible*: `https://docs.ansible.com/ansible/latest/community/other_tools_and_programs.html`

- *Ansible idempotency*: `https://docs.ansible.com/ansible/latest/reference_appendices/glossary.html#term-Idempotency`

- *Ansible Automation for Windows*: `https://www.techbeatly.com/ansible-windows/`

- *Introduction to Ansible modules*: `https://docs.ansible.com/ansible/latest/user_guide/modules_intro.html`

- *Ansible connection plugins*: `https://docs.ansible.com/ansible/latest/plugins/connection.html`

3
Automating Your Daily Jobs

Are you struggling to find automation use cases to start with Ansible automation? Your workplace is a great place to start your search for automation use cases. Track the most repeated jobs that you or your team are doing every day and you will see the opportunity to automate these tasks. This can be simple server information gathering, collecting operating system versions, or a simple weekly reboot job.

In this chapter, you will learn how to use the Jinja2 template to create reports and emails with the help of Ansible. You will also learn how to develop Ansible artifacts in a modular way and include tasks and variables dynamically.

In this chapter, we will cover the following topics:

- Using Ansible to collect server details
- Collecting system information
- System scanning and remediation using Ansible
- Automated weekly system reboot using Ansible
- Automating notifications

We will start with `ansible_facts` and learn how to extract the data for your system reports. The chapter will explain how to collect data, insert it into an HTML report, and store it on web servers. Then, you will learn how to configure standard system files and automate a server reboot with an email notification.

Technical requirements

The following are the technical requirements for this chapter:

- A Linux machine for the Ansible control node
- Two or more Linux machines with Red Hat repositories configured (if you are using other Linux operating systems instead of **Red Hat Enterprise Linux** (**RHEL**) machines, then make sure you have the appropriate repositories configured to get packages and updates)

All the Ansible configurations, playbooks, commands, and snippets for this chapter can be found in this book's GitHub repository at `https://github.com/PacktPublishing/Ansible-for-Real-life-Automation/tree/main/Chapter-03`.

Using Ansible to collect server details

In the previous chapter, you learned how to use Ansible for basic automation by using simple playbooks and tasks. In this chapter, you will learn more by automating the simple day-to-day jobs in your workplace.

An up-to-date system inventory with easy access is the dream of every system engineer and IT team. In large enterprises, it is common to use **configuration management database** (**CMDB**) software. However, engineers must maintain their spreadsheets to keep the server and device information they are managing. When you have software-defined infrastructures such as virtual machines and virtual appliances, verifying and updating these local spreadsheets will become a tedious task.

Maintaining such information can be automated using Ansible, as shown in the following diagram:

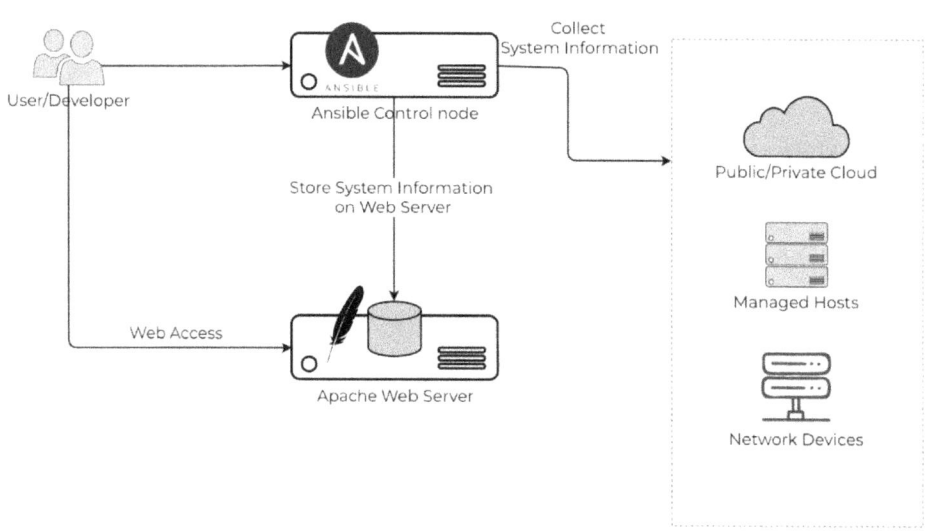

Figure 3.1 – Maintaining a system information database using Ansible

Ansible and `ansible_facts` can be used to create and update your system inventory database or your own CMDB. `ansible_facts` provides detailed and informative data that's gathered from the target nodes (managed nodes) and stored in **JSON** format. By default, Ansible will execute the `setup` module at the beginning of each play in the playbook. If you remember the `chrony` package installation playbook, check the `TASK [Gathering Facts]` line, which shows the `setup` module running in the background:

● ● ●

```
[ansible@ansible Chapter-02]$ ansible-playbook install-package.yaml

PLAY [Install Chrony Package] ****************************************************************

TASK [Gathering Facts] **********************************************************************
ok: [dev-rhel8-55]
...output omitted...
```

Figure 3.2 – An Ansible fact-gathering task by the setup module

ansible_facts can be used to make decisions inside the playbook, such as skipping the task if the system memory is less than the required value or installing specific packages, depending on the operating system version of the target node.

Ansible roles

An Ansible role is a collection of tasks, handlers, templates, and variables for configuring the target system so that it meets the desired state. Ansible roles enable content sharing without much trouble as we can include the required variables, tasks, templates, and other files in the role directory itself. Ansible roles are meant for reusability and collaborative support in such a way that the same roles can be distributed to teams or the public. Other users can use the roles to achieve the same task:

- With Ansible roles, we can easily share a playbook's content with other teams by sharing the entire role directory. For example, it is possible to write a role for install_dbserver or setup_webserver and later share it with the public/other teams.

- Larger projects can be created in a modular way.

- Different users can create different roles in parallel and share the same project. For example, developer 1 writes a role for install_dbserver and developer 2 focuses on the setup_webserver role.

Let us learn about the directory structure of Ansible roles in the next section.

Roles directory structure

The content of Ansible roles is arranged in an organized way. The top-level directory defines the name of the role itself, as shown in the following screenshot (we will create these roles and their content later in this chapter):

```
[ansible@ansible Chapter-03]$ tree ./
./
├── ansible.cfg
├── deploy-web.yml
├── hosts
├── node1-ansible-facts
├── README.md
├── roles
│   ├── deploy-web-server
│   │   ├── defaults
│   │   │   └── main.yml
│   │   ├── handlers
│   │   │   └── main.yml
│   │   ├── meta
│   │   │   └── main.yml
│   │   ├── README.md
│   │   ├── tasks
│   │   │   └── main.yml
│   │   ├── tests
│   │   │   ├── inventory
│   │   │   └── test.yml
│   │   └── vars
│   │       └── main.yml
│   ├── security-baseline-rhel8
│   │   ├── defaults
│   │   │   └── main.yml
│   │   ├── files
...<output omitted>...
```

Figure 3.3 – Sample project directory with roles

In the preceding screenshot, I have two roles named deploy-web-server and security-baseline-rhel8 under my roles directory. Some of the directories contain main.yml, which contains tasks, variables, or handlers:

- Defaults/main.yml: This contains variables for the role that can be overwritten when the role is used.

- tasks/main.yml: This contains the main list of tasks to be executed when using the role.

- vars/main.yml: This contains internal variables for the role.

- files: This contains static files that can be referenced from this role.

- templates: This is the jinja2 templates that can be used via this role.

- handlers/main.yml: This contains the handler definitions.

- meta/main.yml: This defines some metadata for this role, such as the author, license, platform, and dependencies.

- tests: This is an inventory. test.yml file that can be used to test this role.

The default variables can be defined under `defaults/main.yml`; additional role variables can be defined inside `vars/main.yml`. However, depending on the location of your variable and variable precedence, Ansible will apply the appropriate value for the variable. Refer to `https://docs.ansible.com/ansible/latest/user_guide/playbooks_variables.html#understanding-variable-precedence` to learn and understand more about variable precedence.

> **Ansible Roles**
>
> There are over 10,000 roles available in Ansible Galaxy (`https://galaxy.ansible.com`) that have been contributed by the community. Users can freely download and use them for their Ansible automation process. If the organization is looking for certified content collections (collections and roles), then we can find them by going to Red Hat Automation Hub (at the `console.redhat.com` portal). Read `https://docs.ansible.com/ansible/latest/user_guide/playbooks_reuse_roles.html` to learn more about Ansible roles. Find the official Red Hat roles at `https://galaxy.ansible.com/RedHatOfficial`.

Ansible Jinja2 templates

Jinja2 is an extensive templating engine that dynamically generates content using variables and facts. It is possible to create any kind of complex template file that contains variables, loops, and other controls. Inside the playbook, use the `template` module (or the Ansible `template` filter) to convert the Jinja2 template into actual content or a file. Ansible will replace the variable with values as needed.

For example, a standard `/etc/motd` file can be deployed to your servers by using the appropriate values dynamically. Your Jinja2 template will look as follows (it may look different, based on your customization):

```
Welcome to {{ ansible_facts.hostname }}
(IP Address: {{ ansible_facts.default_ipv4.address }})

Access is restricted; if you are not authorized to use it
please logout from this system

If you have any issues, please contact {{ system_admin_email }}.
Phone: {{ system_admin_phone | default('1800 1111 2222') }}

-----------------------------------------
This message is configured by Ansible
-----------------------------------------
```

Figure 3.4 – Jinja2 template for the motd file

Inside the playbook, you will `template` the file and transfer it to the target machine using the `template` module (do not use the `copy` module as the templating will not work):

● ● ●

```
tasks:
  - name: Deploy motd
    template:
      dest: /etc/motd
      src: motd.j2
```

Figure 3.5 – Using template module in playbook

On the target machine, the variables inside the Jinja2 template will be replaced with values, as follows:

● ● ●

```
Welcome to node1
(IP Address: 10.1.10.25)

Access is restricted; if you are not authorized to use it
please logout from this system

If you have any issues, please contact sysops@lab.local.
Phone: 1800 1111 2222

------------------------------------
This message is configured by Ansible
------------------------------------
```

Figure 3.6 – The /etc/motd file that was created using the Jinja2 template

Any level of complexity and loops can be used inside the template to create dynamic output files. Some of the use cases generate reports, web server configuration (HTTPS, Nginx), system configuration files (/etc/hosts, /etc/motd, and /etc/fstab), and more.

> **Ansible Jinja2 Templating**
>
> Read more about Ansible Jinja2 templating at https://docs.ansible.com/ansible/latest/user_guide/playbooks_templating.html.

In this use case, you will extract the required information using `ansible_facts` from `node1` (and other machines if you have some) and store it as an HTML report inside a web server (`node2`), as shown in the following diagram:

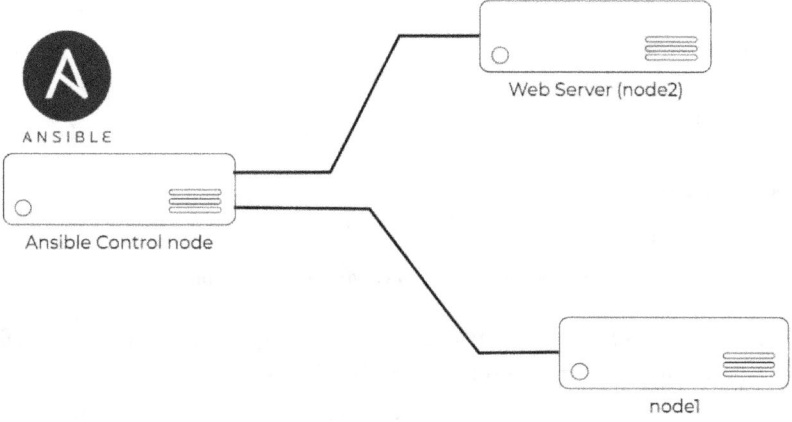

Figure 3.7 – Machines for the Ansible CMDB use case

For the following use case, we need to deploy a web server. We will use Ansible to deploy and configure the web server on node2.

Follow these steps to deploy a web server using an Ansible role:

1. On your Ansible control node, create a new directory called Chapter-03 and create an ansible.cfg file, as follows:

Figure 3.8 – The ansible.cfg file

2. Create a hosts file in the same directory as the node-1 managed node (use the correct IP address as per your lab or environment):

```
● ● ●

[nodes]
node1 ansible_host=192.168.56.25

[web]
webserver1 ansible_host=192.168.56.24

[all:vars]
ansible_ssh_private_key_file=/home/ansible/.ssh/id_rsa
```

Figure 3.9 – The Ansible inventory (hosts) file

In the preceding screenshot, [nodes:vars] is the group variable. The variables will be available for all managed nodes under the nodes host group.

3. Now, create an Ansible role to deploy a web server on node2 using Apache:

    ```
    [ansible@ansible Chapter-03]$ mkdir roles
    [ansible@ansible Chapter-03]$ cd roles/
    ```

4. Once you have created the roles directory, initiate a new Ansible role using the ansible-galaxy command. The ansible-galaxy command will initiate and create the skeleton directory structure for the role:

```
● ● ●

[ansible@ansible roles]$ ansible-galaxy role init deploy-web-server
- Role deploy-web-server was created successfully
```

Figure 3.10 – Initializing the Ansible role with the ansible-galaxy command

5. Verify the content of the roles directory:

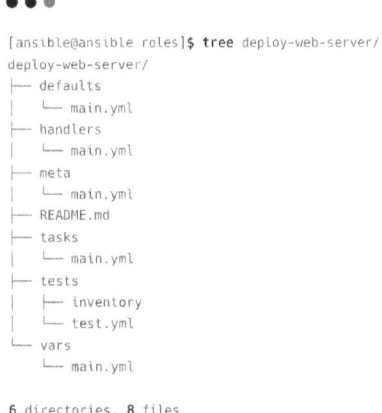

```
[ansible@ansible roles]$ tree deploy-web-server/
deploy-web-server/
├── defaults
│   └── main.yml
├── handlers
│   └── main.yml
├── meta
│   └── main.yml
├── README.md
├── tasks
│   └── main.yml
├── tests
│   ├── inventory
│   └── test.yml
└── vars
    └── main.yml

6 directories, 8 files
```

Figure 3.11 – The roles directory's content

6. Add the following tasks to the `roles/deploy-web-server/tasks/main.yml` file to install the `firewalld` and `httpd` packages:

```
---
# tasks file for deploy-web-server

- name: Create directory
  ansible.builtin.file:
    state: directory
    path: /var/www/html
    mode: '0755'

- name: Install httpd and firewalld
  ansible.builtin.yum:
    name:
      - httpd
      - firewalld
    state: latest
```

Figure 3.12 – Installing httpd and firewalld

7. Add more tasks to the role to enable the firewall service and permit the `httpd` service in the firewall, as follows:

```
---
# tasks file for deploy-web-server
.
.
- name: Enable and Run Firewalld
  ansible.builtin.service:
    name: firewalld
    enabled: true
    state: started

- name: Firewalld permit httpd service
  ansible.posix.firewalld:
    service: http
    permanent: true
    state: enabled
    immediate: yes
```

Figure 3.13 – Ansible web deploy role – enabling the firewalld service and permitting httpd in the firewall

8. Finally, add the task for enabling and starting the `httpd` service, as follows:

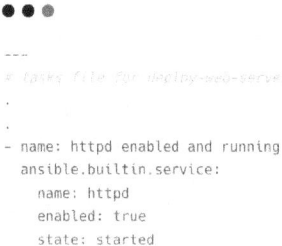

```
---
# tasks file for deploy-web-server

.

.

- name: httpd enabled and running
  ansible.builtin.service:
    name: httpd
    enabled: true
    state: started
```

Figure 3.14 – Ansible web deploy role – enabling and starting httpd

The `tasks/main.yml` file will create the `/var/www/html` directory and install the `httpd` and `firewalld` packages. Then, it will start the `httpd` and `firewalld` services and open `httpd` ports in the firewall.

Remove other unwanted directories and files (automatically generated) inside the `deploy-web-server` role directory. In this case, leave it as-is and proceed with the next step.

9. Create a playbook called `Chapter-03/deploy-web.yml` so that you can deploy the web server (remember to go back to the main directory – that is, `Chapter-03`):

```
# Chapter-03/deploy-web.yml

- name: Deploy Webserver using apache
  hosts: web
  become: true
  tasks:
    - name: Deploy Web service
      include_role:
        name: deploy-web-server
```

Figure 3.15 – Ansible playbook to call the role

Now, your directory will contain the following contents:

```
[ansible@ansible Chapter-03]$ ls -l
total 16
-rw-rw-r--. 1 ansible ansible  209 Jan  8 14:16 ansible.cfg
-rw-rw-r--. 1 ansible ansible  158 Jan  9 09:41 deploy-web.yml
-rw-rw-r--. 1 ansible ansible  159 Jan  8 14:17 hosts
-rw-rw-r--. 1 ansible ansible 1249 Jan  8 13:45 README.md
drwxrwxr-x. 3 ansible ansible   31 Jan  9 09:24 roles
```

Figure 3.16 – The project directory's content

10. Execute the `deploy-web.yml` playbook to deploy the web server on `node2`:

```
[ansible@ansible Chapter-03]$ ansible-playbook deploy-web.yml

PLAY [Deploy Webserver using apache] *********************************************

TASK [Gathering Facts] ***********************************************************
ok: [webserver1]

TASK [Deploy Web service] ********************************************************

TASK [deploy-web-server : Create directory] **************************************
changed: [webserver1]

TASK [deploy-web-server : Install httpd and firewalld] ***************************
changed: [webserver1]

TASK [deploy-web-server : Enable and Run Firewalld] ******************************
changed: [webserver1]

TASK [deploy-web-server : Firewalld permit httpd service] ************************
ok: [webserver1]

TASK [deploy-web-server : httpd enabled and running] *****************************
changed: [webserver1]

PLAY RECAP **********************************************************************
webserver1              : ok=6    changed=4    unreachable=0    failed=0    skipped=0    rescued=0    ignored=0
```

Figure 3.17 – Executing the web server deployment playbook

The same playbook can be executed multiple times and Ansible will execute or skip the operation in the backend based on the desired status. This feature is called **idempotency** in Ansible, by which the Ansible module will check the desired status (for example, installing packages or copying files) and execute the operation only if required.

> **Ansible Idempotency**
>
> *"An operation is idempotent if the result of performing it once is the same as the result of performing it repeatedly without any intervening actions."* – Ansible Glossary (`https://docs.ansible.com/ansible/latest/reference_appendices/glossary.html`)

Let's execute the same playbook again and check the difference. Here, we can see the `ok` status instead of `changed` (*Figure 3.17*) since Ansible has not executed the operation. This is because the `firewalld` and `httpd` packages have been installed and the services have been started already (the desired state has already been met):

```
[ansible@ansible Chapter-03]$ ansible-playbook deploy-web.yml

PLAY [Deploy Webserver using apache] ***************************************************************

TASK [Gathering Facts] *****************************************************************************
ok: [webserver1]

...<output omitted>....

TASK [deploy-web-server : Firewalld permit httpd service] ******************************************
ok: [webserver1]

TASK [deploy-web-server : httpd enabled and running] **********************************************
ok: [webserver1]

PLAY RECAP *****************************************************************************************
webserver1                 : ok=6    changed=0    unreachable=0    failed=0    skipped=0    rescued=0    ignored=0
```

Figure 3.18 – Executing the deployment playbook again and noticing the ok status instead of changed

Now, verify the web server by visiting the server IP or **Fully Qualified Domain Name** (**FQDN**) in a web browser (you may see different or similar pages, depending on the operating system and version):

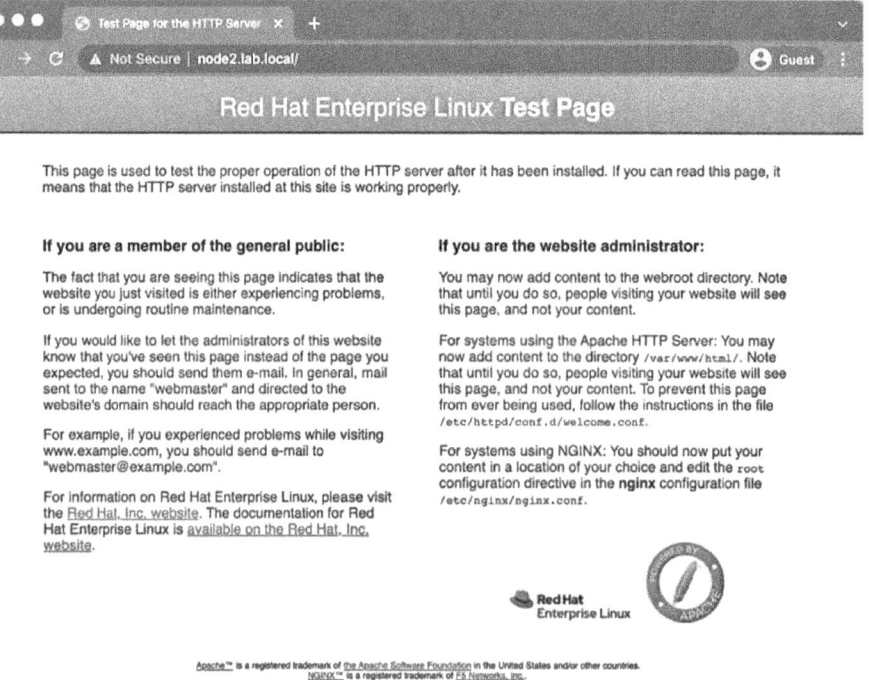

Figure 3.19 – Default Apache web page

Log in to `node2` and remove the default `welcome.conf` file so that you can see the directory content of `/var/www/html`:

```
[root@node-2 ~]# mv /etc/httpd/conf.d/welcome.conf .
```

This is not the best practice in production, and you need to configure your web server with adequate permissions and directory listing options. The preceding step was mentioned to explain the demo and web server functionality.

With that, you have deployed a web server using an Ansible role to keep the systems information reports or CMDB.

> **Note**
>
> There are many open source projects that can be used to implement CMDB, such as using Ansible facts such as Ansible-CMDB (`https://ansible-cmdb.readthedocs.io/en/latest/`). Refer to the project documentation and repository for detailed information.

Collecting system information

In this section, you will extract the required information from `ansible_facts` and generate HTML reports inside the web server that you created in the previous section.

`ansible_facts` contains a lot of information about nested dictionaries and lists. Search and go through the content and find the important information you need for your report.

To see the content of `ansible_facts`, execute the following ad hoc command:

```
● ● ●
[ansible@ansible Chapter-03]$ ansible node1 -m setup |less
node1 | SUCCESS => {
    "ansible_facts": {
        "ansible_all_ipv4_addresses": [
            "192.168.100.101",
            "192.168.56.25",
            "10.0.2.15"
        ],
        ...output omitted...

        "ansible_date_time": {
            "date": "2022-01-10",
            "day": "10",
            ...output omitted...
        },
        "module_setup": true
    },
    "changed": false
}
```

Figure 3.20 – Ansible facts output after using an ad hoc command

Using the `less` or `more` commands after the pipe (|) symbol will keep the output on top without you having to scroll to the bottom. It is possible to scroll down or up using the arrow keys or find the text by searching for it (/ + <text>).

Find sample `ansible_facts` details for a Linux machine at `https://github.com/ PacktPublishing/Ansible-for-Real-life-Automation/tree/main/Chapter- 03/node1-ansible-facts`.

Follow these steps to use some of the variables from the preceding `setup` output:

1. Create a role called `Chapter-03/roles/system-report` to generate the HTML reports, as follows:

```
[ansible@ansible Chapter-03]$ cd roles/
[ansible@ansible roles]$ ansible-galaxy role init system-report
- Role system-report was created successfully
```

Figure 3.21 – Creating a new Ansible role for the system report

2. Create a Jinja2 template file called `roles/system-report/templates/system- information.html.j2` and add the HTML header and other details inside the template file:

```
<!DOCTYPE html PUBLIC "-//W3C//DTD XHTML 1.0 Transitional//EN" "https://www.w3.org/TR/xhtml1/DTD/xhtml1-
transitional.dtd">
<html xmlns="https://www.w3.org/1999/xhtml">
<head>
<title>{{ ansible_hostname }} - System Information | Ansible Automation</title>

...<output omitted>...

<body style="margin:0px; padding:0px; width: 700px; text-align: center;" >
  <table valign="top" width="100%" cellspacing="0" cellpadding="0" border="0" align="center">
    <tbody><tr>
    ...<output omitted>...
    <tr>
      <td valign="top" align="center"><h2>System Information for {{ ansible_hostname }}</h2></td>
    </tr>
    <tr>
      <td style="min-width:700px;background-color:#ffffff; text-align: center;" text-align="center">
        <table valign="top" width="100%" cellspacing="0" cellpadding="0" border="1" align="center" >
          <tr>
            <td valign="top" align="left">System Name</td>
            <td valign="top" align="left">{{ ansible_hostname }}</td>
          </tr>
          <tr>
            <td valign="top" align="left">IP Address</td>
            <td valign="top" align="left">{{ ansible_all_ipv4_addresses }}</td>
          </tr>

...<output omitted>...

        </table>
      </td>
    </tr>
    <tr>
      <td style="font-size:12px; line-height:18px; color:#999999; padding: 20px;">
        If you find any mismatch in report, please report to <a href="mailto:{{ report_admin_email }}"
target="_blank" style="text-decoration:none; color:#999999;">{{ report_admin_email }}</a>
      </td>
    </tr>
    </tbody>
  </table>
</body></html>
```

Figure 3.22 – Jinja2 template for HTML report

Find the full Jinja2 template at `https://github.com/PacktPublishing/Ansible-for-Real-life-Automation/blob/main/Chapter-03/roles/system-report/templates/system-information.html.j2`. You may have already noticed the variables that are being used inside the Jinja2 template, as follows:

- `{{ ansible_hostname }}`: The hostname of the managed node.

- `{{ ansible_all_ipv4_addresses }}`: The IP address list.

- `{{ ansible_architecture }}`: The architecture of the target machine.

- {{ ansible_distribution }} and {{ ansible_distribution_version }}: The operating system's distribution and version.

- {{ report_admin_email }}: This is a custom variable you need to define in the playbook.

3. Create the roles/system-report/tasks/main.yml file with the following content:

● ● ●

```
---
# task file for system-report
- name: Generate and save system report in html format
  template:
    dest: /var/www/html/{{ inventory_hostname }}.html
    src: system-information.html.j2
  delegate_to: node2
```

Figure 3.23 – The task file for the system report role

In the preceding example, the template module will convert the template into the target HTML file with variables replaced with values. Since the report needs to be saved in the web server path, the task will be delegated to node2 (the web server).

4. Create the Chapter-03/system-info.yml playbook and include the system-report role:

● ● ●

```
# Chapter-03/system-info.yml
- name: Collect System Information
  hosts: nodes
  become: true
  vars:
    report_admin_email: admin@lab.local
  tasks:
    - name: Generate System Report
      include_role:
        name: system-report
```

Figure 3.24 – The Ansible playbook for collecting system information

5. Execute the playbook and verify its output:

```
[ansible@ansible Chapter-03]$ ansible-playbook system-info.yml

PLAY [Collect System Information] ***************************************

TASK [Gathering Facts] *************************************************
ok: [node1]

TASK [Generate System Report] *****************************************

TASK [system-report : Generate and save system report in html format] *************
changed: [node1 -> node2]

PLAY RECAP ************************************************************
node1                     : ok=2    changed=1    unreachable=0    failed=0    skipped=0    rescued=0    ignored=0
```

Figure 3.25 – Collecting system information

6. Check the web server to view the report, as follows:

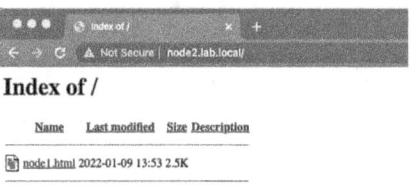

Figure 3.26 – The report that was generated on the web server

7. Click on node1.html to see its content, as follows:

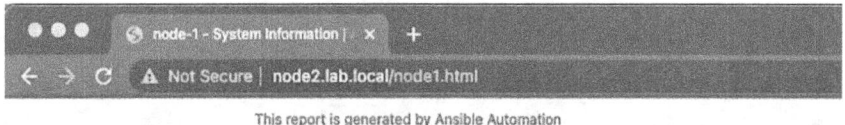

This report is generated by Ansible Automation

System Information for node-1

System Name	node-1
IP Address	['192.168.100.101', '192.168.56.25', '10.0.2.15']
Network Interfaces	['eth0', 'eth2', 'eth1', 'lo']
Architecture	x86_64
Operating System	RedHat 8.4

If you find any mismatch in report, please report to admin@lab.local

Figure 3.27 – System information report generated by Ansible

This is a very basic HTML report template that explains the capability of the Jinja2 template and `ansible_facts`. Expand the template with additional items, CSS styles, or even different formats such as Markdown, CSV, or JSON. Also, it is possible to keep the report in alternate locations such as a GitHub server or web server with authentication.

System scanning and remediation using Ansible

Security scanning and remediation are critical, and organizations are spending more time and money on this area every year. When there are new features and changes in the operating system and applications, you will have more configurations to check and validate to ensure the best security practices are in place. With the help of Ansible, it is possible to automate the security scanning and remediation tasks for your systems and devices.

In this section, you will automate a few basic security and compliance configurations based on the CIS Red Hat Enterprise Linux 8 Benchmark.

> **CIS Benchmark**
>
> CIS provides the best practices and configurations for systems and platforms to ensure security and compliance. Refer to `https://www.cisecurity.org/cis-benchmarks` to learn more about CIS Benchmarks.

When we have several tasks in a playbook or role, then we can split the tasks into multiple files and call them using the `include_tasks` module dynamically. For example, different parts of security remediation tasks can be split into different tasks files so that they can be called from `main.yaml`, as shown in the following diagram:

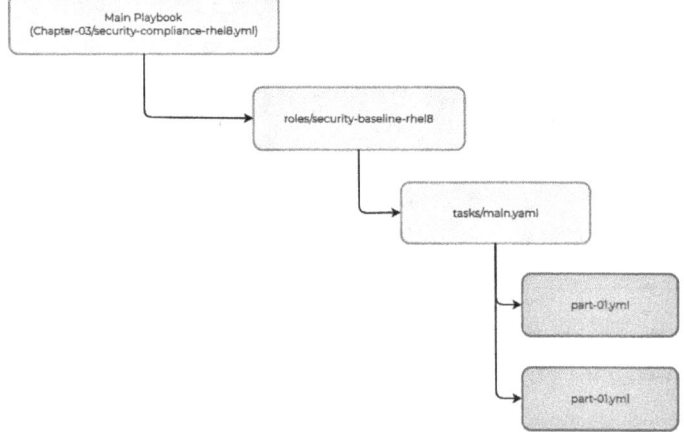

Figure 3.28 – Splitting tasks into multiple files

Such methods can help us develop, test, and execute the tasks dynamically and in a modular way.

Follow these steps to develop a security remediation use case:

1. Create a new role called `Chapter-03/roles/security-baseline-rhel8`:

```
[ansible@ansible Chapter-03]$ cd roles/
[ansible@ansible roles]$ ansible-galaxy role init security-baseline-rhel8
- Role security-baseline-rhel8 was created successfully
```

Figure 3.29 – Creating a new role for security baselining

2. Add the necessary security baseline configurations to `roles/security-baseline-rhel8/tasks/main.yml`.

   As I mentioned at the beginning of this section (*Figure 3.28*), in this example, you will learn how to split the tasks into multiple files and develop playbooks in a modular way. Add the following content to `roles/security-baseline-rhel8/tasks/main.yml`:

```
---
# tasks file for security-baseline-rhel8

- name: "Running Part 01 checks"
  include_tasks: "part-01.yml"
  when: "'01.01' not in baseline_exclusions"

- name: "Running Part 02 checks"
  include_tasks: "part-02.yml"
  when: "'02.01' not in baseline_exclusions"
```

Figure 3.30 – The main task file for the security baselining role

You have two tasks and both are calling other tasks files via the `include_tasks` module. There is an important line you need to take note of, as follows:

- `when: "'01.01' not in baseline_exclusions"`: This is a mechanism we are adding to the task to control the execution of specific baseline rule using the `when` statement. Ansible will check the condition and execute or skip the tasks based on this condition. In this case, you need to define a list variable called `baseline_exclusions` and add the specific item to exclude from execution. (Use any other string or numbering system; this is just a sample list for this demonstration.)

3. Create a file called `roles/security-baseline-rhel8/tasks/part-01.yml` that contains the following content to install `sudo` and enable `sudo` logging:

```
• • •

# part-01.yml

- name: "Ensure sudo is installed"
  dnf:
    name: sudo
    state: present

- name: "Ensure sudo log file exists"
  lineinfile:
    path: /etc/sudoers
    regexp: '^Defaults\s*logfile="{{ sudo_log }}"'
    line: 'Defaults logfile="{{ sudo_log }}"'
    insertafter: '^# Defaults specification'
    validate: /usr/sbin/visudo -cf %s
```

Figure 3.31 – part-01.yml for the sudo configuration

The first task will install the sudo package, while the second will enable sudo logging. There is an important line you need to take note of, as follows:

- `line: 'Defaults logfile="{{ sudo_log }}"'`: You need to define this sudo_log variable in the playbook.

4. Create another task file called `roles/security-baseline-rhel8/tasks/part-02.yml` with the following content to deploy the default /etc/motd and /etc/issue files:

```
• • •

# part-02.yml

- name: "Ensure message of the day is configured properly"
  copy:
    src: "{{ motd_file }}"
    dest: /etc/motd
    owner: root
    group: root
    mode: 0644

- name: "Ensure local login warning banner is configured properly"
  copy:
    src: "{{ issue_file }}"
    dest: /etc/issue
    owner: root
    group: root
    mode: 0644
```

Figure 3.32 – part-02.yml for the motd configuration

The first task will deploy default content to /etc/motd, while the second will deploy content to the /etc/issue file.

5.   Create the following default content for the `motd` and `issue` files under the `roles/security-baseline-rhel8/files/` directory:

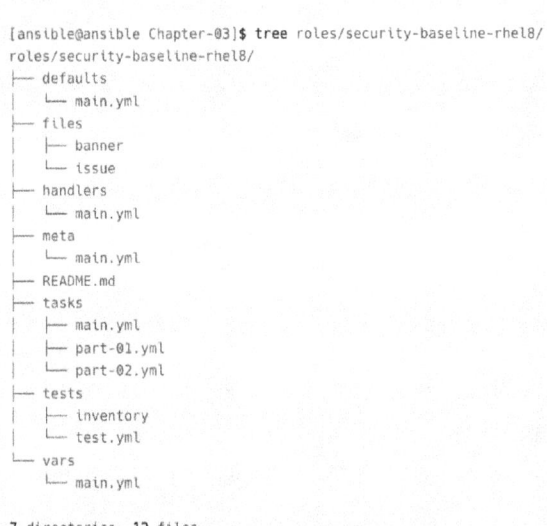

```
[ansible@ansible Chapter-03]$ cat roles/security-baseline-rhel8/files/banner
Authorized uses only. All activities will be monitored and reported.

[ansible@ansible Chapter-03]$ cat roles/security-baseline-rhel8/files/issue
Authorized uses only. All activities will be monitored and reported.
```

Figure 3.33 – The default motd and issue files

6.   Verify the files of your Ansible role to ensure all the content is in place, as shown in the following screenshot:

```
[ansible@ansible Chapter-03]$ tree roles/security-baseline-rhel8/
roles/security-baseline-rhel8/
├── defaults
│   └── main.yml
├── files
│   ├── banner
│   └── issue
├── handlers
│   └── main.yml
├── meta
│   └── main.yml
├── README.md
├── tasks
│   ├── main.yml
│   ├── part-01.yml
│   └── part-02.yml
├── tests
│   ├── inventory
│   └── test.yml
└── vars
    └── main.yml

7 directories, 12 files
```

Figure 3.34 – The content of the security baseline role

You can remove other unwanted directories that are not in use. However, in this case, keep everything as-is.

7.   The variables can be kept in your playbook or even inside the inventory file, but it will not be easy to manage the content when you have more variables to maintain. Create a directory to keep the variables in, as follows:

```
[ansible@ansible Chapter-03]$ mkdir vars
```

8. Create a variable file called `vars/common.yml` to keep the common variables in and add `sudo_log` and other variables there (remember to add the variables and values as needed inside the file):

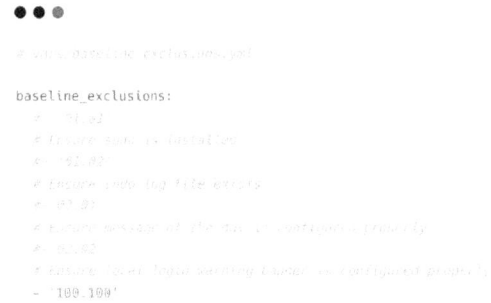

```
[ansible@ansible Chapter-03]$ cat vars/common.yml
sudo_log: "/var/log/sudoers"
motd_file: "banner"
issue_file: "issue"
```

Figure 3.35 – Creating a variable file

9. Create another variable file called `vars/baseline_exclusions.yml` to keep the `baseline_exclusions` variable in:

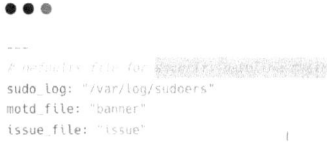

```
# vars/baseline_exclusions.yml

baseline_exclusions:
  #  - '01.01'
  # Ensure auth is installed
  #  - '02.02'
  # Ensure sudo log file exists
  #  - '03.03'
  # Ensure message of the day is configured properly
  #  - '04.04'
  # Ensure local login warning banner is configured properly
  - '100.100'
```

Figure 3.36 – Creating a variable file for baseline exclusions

We can disable the security check by uncommenting the line (for example, `'01.01'`) so that Ansible will check before executing the task.

10. As a best practice, you need to keep the default values for all variables that are used inside the Ansible role in a file. The `roles/security-baseline-rhel8/defaults/main.yml` file can be used for this purpose:

```
---
# defaults file for security-baseline-rhel8
sudo_log: "/var/log/sudoers"
motd_file: "banner"
issue_file: "issue"
```

Figure 3.37 – Default variables for the security-baseline-rhel8 role

11. Now, create the main playbook, `security-compliance-rhel8.yml`, and ensure it contains the following content:

```
---
# Chapter-03/security-compliance-rhel8.yml

- name: Performing Security Scanning and Configuration - RHEL8
  hosts: "{{ NODES }}"                # give NODES during playbook.
                                      # eg: -e "NODES=webservers"

  become: true
  vars_files:
    - vars/common.yml                # common variables
    - vars/baseline_exclusions.yml   # exclusion list

  tasks:
    - name: 'Starting Scanning'
      include_role:
        name: security-baseline-rhel8
```

Figure 3.38 – The main playbook – security-compliance-rhel8.yml

Note that in the preceding playbook, we are not hardcoding the hosts. Instead, we are using a variable called NODES. This variable will be passed to the playbook using the **extra variables** option while executing the playbook (see the next step). It is a best practice to not hardcode hosts to avoid the playbook being executed accidentally on the incorrect servers. (More about extra-vars will be covered in the next section.)

Also, note vars_files where we included the two variable files we created earlier. Here, the playbook is calling the security-baseline-rhel8 role.

12. Execute the playbook and pass the NODES details as an extra variable:

```
[ansible@ansible Chapter-03]$ ansible-playbook security-compliance-rhel8.yml -e "NODES=nodes"

PLAY [Performing Security Scanning and Configuration - RHEL8] ********************************

...<output omitted>...

TASK [security-baseline-rhel8 : Running Part 01 checks] ********************************
included: /home/ansible/ansible-book-packt/Chapter-03/roles/security-baseline-rhel8/tasks/part-01.yml for node1

TASK [security-baseline-rhel8 : Ensure sudo is installed] ********************************
ok: [node1]

TASK [security-baseline-rhel8 : Ensure sudo log file exists] ********************************

...<output omitted>...

PLAY RECAP ********************************************************************************
node1                      : ok=6    changed=2    unreachable=0    failed=0    skipped=1    rescued=0    ignored=0
```

Figure 3.39 – Executing the security baseline playbook and ensuring the subtasks are executed

13. Log in to node1 and verify the implemented items (note the login prompt):

```
● ● ●
[ansible@ansible Chapter-03]$ ssh devops@node1
Authorized uses only. All activities will be monitored and reported.
Last login: Mon Jan 10 08:09:50 2022 from 192.168.56.23
[devops@node-1 ~]$
```

Figure 3.40 – Verifying the content of motd on the login screen for node1

Enhance your playbook by adding more validations and verifications to the tasks. Also, create reports while executing the job and send or save them for later auditing purposes.

> **Ansible Integration with Third-Party Security Tools**
>
> It is possible to integrate Ansible with other third-party platforms and tools such as OpenSCAP or Red Hat Insight. In such cases, you need to develop playbooks to control scanning and remediation instead of manually scanning and fixing the configurations on the systems directly. Refer to `https://www.ansible.com/use-cases/security-and-compliance` and `https://www.open-scap.org/` for more information.

Ansible --extra-vars

`extra-vars` contains the variables that will override all other variables. Using a dynamic extra variable will help you control the playbook based on the values and is also useful when you are using survey forms in Ansible AWX or Ansible Automation Controller, where variables can be defined in the GUI method (survey forms). `--extra-vars` can be passed as a single value, multiple key-value pairs, in JSON format, or read from a variable file, as follows:

```
● ● ●
$ ansible-playbook site.yml --extra-vars "version=1.23.45 other_variable=foo"

$ ansible-playbook site.yml --extra-vars '{"version":"1.23.45","other_variable":"foo"}'

$ ansible-playbook site.yml --extra-vars "@vars_file.json"
```

Figure 3.41 – Ansible extra-vars examples

> **Ansible --extra-vars**
>
> Read more about runtime variables at `https://docs.ansible.com/ansible/latest/user_guide/playbooks_variables.html#defining-variables-at-runtime`. Also check out the automation controller survey forms at `https://docs.ansible.com/automation-controller/latest/html/userguide/job_templates.html#surveys`.

In the next section, you will learn how to automate scheduled reboot jobs using Ansible.

Automated weekly system reboot using Ansible

A scheduled and planned system reboot is a standard process in an IT environment to ensure the servers and applications are working well and the environment is stable with service restart operations. The `reboot` command might be simple when it executes but the reboot process and its formalities are not straightforward.

A generic server reboot activity involves multiple steps, as shown in the following diagram:

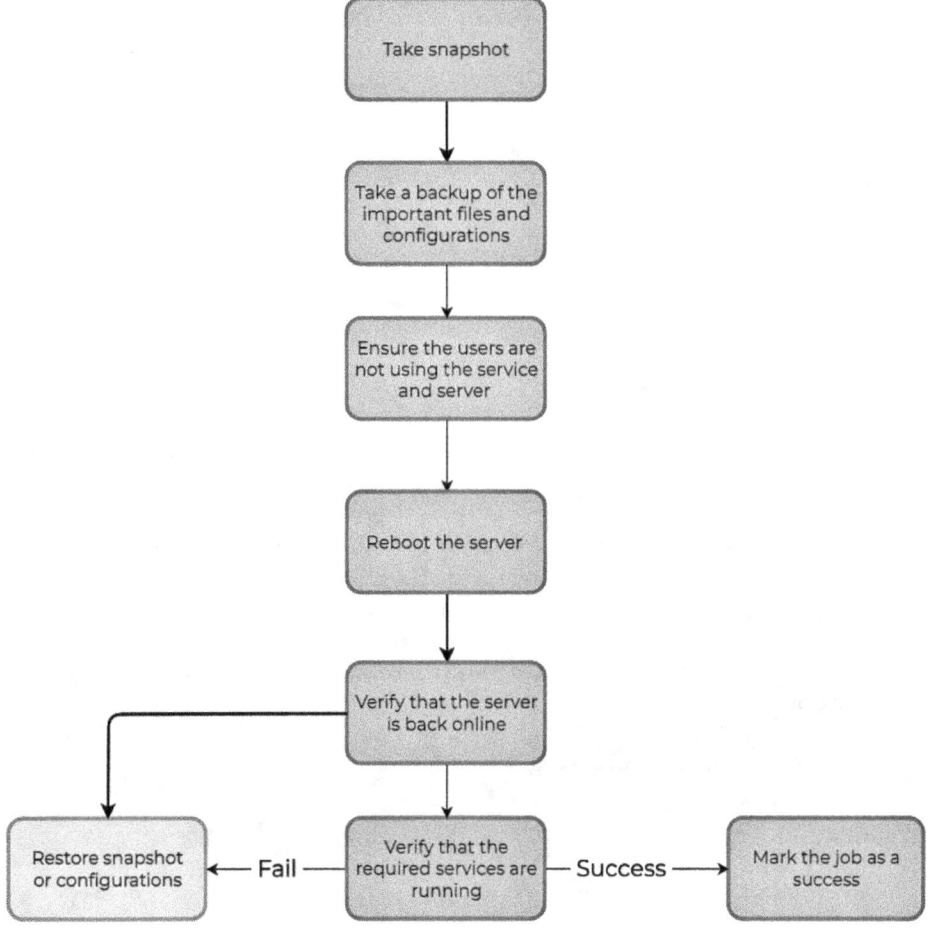

Figure 3.42 – Typical system reboot job workflow

Imagine that you have hundreds of servers to reboot every week and your team is too small to handle such critical operations on weekends. It is possible to automate the entire workflow using Ansible by using backup operations before reboot and service verifications after reboot.

The Ansible `reboot` module was introduced in Ansible 2.7 (2018). At the time of writing, this module is part of `ansible-core` and included in all Ansible installations.

Create an Ansible playbook to reboot the machine as follows:

1. Create the `Chapter-03/system-reboot.yml` playbook with the following content:

```
---
# Chapter-03/system-reboot.yml

- name: System Reboot Linux
  hosts: "{{ NODES }}"
  gather_facts: no
  become: true
  tasks:
    - name: Running Pre-reboot tasks
      debug:
        msg: "Taking backup and snapshot"
      # you may include your backup and other jobs here

    - name: Reboot node and wait for 5 min
      reboot:
        reboot_timeout: 300

    - name: Running Post-reboot tasks
      debug:
        msg: "Verifying services and filesystem"
      # you may include your verification tasks here
```

Figure 3.43 – Ansible playbook for the reboot job

2. Execute the playbook:

```
[ansible@ansible Chapter-03]$ ansible-playbook system-
reboot.yml -e "NODES=nodes"
```

3. Verify the reboot status on `node1`, as follows:

```
[devops@node-1 ~]$ uptime
 09:03:22 up 0 min,  1 user,  load average: 0.76, 0.24, 0.08
```

Figure 3.44 – Verifying the reboot status on node1

Enhance the playbook with the snapshot jobs (for example, a VMWare or OpenStack backup) or file backup (for example, `/etc/hosts`, `/etc/motd`, or `/etc/fstab`) before rebooting the system. Also, create additional tasks to verify the required services are running on the server, such as the `httpd` or `mysql` services. If you are using an automation controller or Ansible AWS, then schedule these automation tasks as weekly or daily jobs; the tasks will be executed based on the schedule without any user interaction (refer to *Chapter 12*, *Integrating Ansible with Your Tools*, for more details).

> **Ansible reboot Module**
>
> For more information about the Ansible `reboot` module, refer to `https://docs.ansible.com/ansible/latest/collections/ansible/builtin/reboot_module.html`.

Automating notifications

It is very important to notify the administrators and end users about the changes you are making in the environment. Whether it's a simple package installation or a system reboot, the end user should be aware of the downtime and changes that are occurring. Instead of sending emails manually, the Ansible `mail` module can be used to automate email notifications. The Ansible `mail` module is powerful and can support most email features, including custom headers, attachments, and security.

Encrypting sensitive data using Ansible Vault

If the email server (**SMTP**) is not **open** (configured to send email without authentication), then you need to authenticate the SMTP server with a username and password (app password or secret key). Keeping such sensitive information in plain text is not a best practice, so you need to store it in a safe method. To store such sensitive information, use key vault tools, in which the information will be saved in an encrypted format. Fortunately, there's an built-in feature in Ansible for storing and managing encrypted content called **Ansible Vault**.

Ansible Vault will encrypt the files with the password (vault secret) you are providing and make the sensitive data unreadable to normal users. When Ansible wants to read the data, Ansible will ask for the vault password; you need to provide the password by keystrokes or via a secret file.

Create a vault file using the `ansible-vault create` command. Do not forget the vault password as you will not be able to decrypt the content otherwise:

```
● ● ●
[ansible@ansible Chapter-03]$ ansible-vault create vars/secrets
New Vault password:
Confirm New Vault password:
```

Figure 3.45 – Creating a secret variable file using Ansible Vault

Once you've done this, a text editor (for example, `vim`) will open so that you can enter the content of your sensitive file. Add your content and save the file:

```
mysecretusername: username
mysecretpassword: password
```

If you try to read the file, you will see the encrypted content, as follows:

```
[ansible@ansible Chapter-03]$ cat vars/secrets
$ANSIBLE_VAULT;1.1;AES256
38393063373031356638353866353937306462663565366266323166363130356435326564343735
30616638313262237356430353361646235396661663538310a373337376339383561353762356265
39363830316465346166303666373064353061343536361373434333665363065653393739643238
31363061306633761610a646138326130333435373836303832343335373373730353535365616430
32323537303765356366383930623631666561396361626535666313531636232613462306662332
31373138616137346132626223062646434303430663731663663353966353030333338396163666131
38323762616262263343761336630393366331
```

Figure 3.46 – The Ansible Vault file after being encrypted

View the content of the file using the `ansible-vault view` command; Ansible will ask for the vault password, as follows:

```
[ansible@ansible Chapter-03]$ ansible-vault view vars/secrets
Vault password:
mysecretusername: username
mysecretpassword: password
```

Figure 3.47 – Viewing the content of the Ansible Vault file

You will be using this Ansible Vault in the upcoming exercise to keep the SMTP server username and password safe.

We will discuss Ansible Vault in more detail in *Chapter 13*, *Using Ansible for Secret Management*.

> **Encrypting Content with Ansible Vault**
>
> You have more options with Ansible Vault, such as edit, encrypt, decrypt, rekey, and more. Refer to *Chapter 13*, *Using Ansible for Secret Management*, for more details. Check out the documentation at `https://docs.ansible.com/ansible/latest/user_guide/vault.html` for more details about Ansible Vault.

In this exercise, you will enhance the previous **weekly reboot** playbook by adding email notifications before and after the system reboot:

1. Create a new role called `Chapter-03/roles/send-email`:

```
[ansible@ansible Chapter-03]$ cd roles/
[ansible@ansible roles]$ ansible-galaxy role init send-email
- Role send-email was created successfully
```

Figure 3.48 – Creating an Ansible role for sending emails

2. Add a task inside `roles/send-email/tasks/main.yml`:

```
---
# roles/send-email/tasks/main.yml

- name: Sending notification email
  mail:
    host: "{{ email_smtp_server }}"
    port: "{{ email_smtp_server_port }}"
    secure: try
    from: "{{ email_smtp_from_address }}"
    to: "{{ email_smtp_to_address }}"
    #cc: "{{ email_smtp_cc_address }}"
    subject: "{{ email_smtp_subject }}"
    body: "{{ email_report_body }}"
    #attach:
    #  - "{{ report_file_name }}"
    headers:
      - Reply-To="{{ email_smtp_replyto_address }}"
    username: "{{ email_smtp_username }}"
    password: "{{ email_smtp_password }}"
  delegate_to: localhost
```

Figure 3.49 – The task file for the send-email role

> **Note**
>
> The `delegate_to: localhost` line of the `mail` module needs to be executed on the localhost (the Ansible control node here) rather than the managed node.

Skip the `username` and `password` variables if your SMTP server has been configured as open and no authentication is required. In this case, you need to create a secret file to keep the username and password in. (The `cc` and `attach` options have been kept here as comments for demonstration purposes. It is possible to enhance the use case by adding those features.)

3. Create a new variable file called `vars/smtp_secrets.yml` using Ansible Vault (remember the vault password):

```
[ansible@ansible Chapter-03]$ ansible-vault create vars/smtp_secrets.yml
New Vault password:
Confirm New Vault password:
```

Figure 3.50 – Creating a secret variable file using Ansible Vault

4. Add the content of the secret file and save it:

```
email_smtp_username: 'ansible-automation@lab.local'
email_smtp_password: 'secretpassword'
~
~
~
~
~
~
~
:wq
```

Figure 3.51 – Adding a variable to the vault file and saving it (:wq)

5. Create a new playbook called `Chapter-03/system-reboot-with-email.yml` with the following content (alternatively, copy the previous `system-reboot.yml` file and rename it):

```
---
# Chapter-03/system-reboot-with-email.yml

- name: System Reboot - Linux with email notification
  hosts: "{{ NODES }}"
  gather_facts: no
  become: true
  vars_files:
    vars/smtp_secrets.yml
  vars:
    email_smtp_server: 'smtp.mail.com'
    email_smtp_server_port: '587'
    email_smtp_from_address: 'ansible@lab.local (Ansible Automation)'
    email_smtp_to_address:
      #  - 'John Doe <john@lab.local>'
      #  - 'Linda <linda@lab.local>'
    # email_smtp_cc_address:
    #  - 'John Doe <john@gmail.com>'
    email_smtp_replyto_address: 'no-reply@lab.local'
```

Figure 3.52 – Ansible playbook for rebooting with an email notification (1)

6. Add tasks for the reboot and email, as follows:

```
---
# Chapter-03/system-reboot-with-email.yml

#...<playbook continues>...
  tasks:
    - name: Email notification before reboot
      include_role:
        name: send-email
      vars:
        email_report_body: "Alert: {{ inventory_hostname }} is rebooting as per schedule. Please do not use the
server. Notification will be sent after the reboot activity is completed."
        email_smtp_subject: "Weekly System Reboot - {{ inventory_hostname }} - Initiated"

    - name: Running Pre-reboot tasks
      debug:
        msg: "Taking backup and snapshot"        # include your backup and other jobs here.

    - name: Reboot node and wait for 5 min
      reboot:
        reboot_timeout: 300

    - name: Running Post-reboot tasks
      debug:
        msg: "Verifying services and filesystem" # include your verification tasks here.

    - name: Email notification after reboot
      include_role:
        name: send-email
      vars:
        email_report_body: "Alert: {{ inventory_hostname }} reboot activity has been completed."
        email_smtp_subject: "Weekly System Reboot - {{ inventory_hostname }} - completed"
```

Figure 3.53 – Ansible playbook for rebooting with an email notification (2)

Note that different values should be used for email_report_body and email_smtp_ subject based on the job (pre-reboot or post-reboot).

With that, you have all required variables in the playbook for the send-email role except email_smtp_username and email_smtp_password. This is because they are sensitive and you cannot store them as plain text here; instead, you should include them in your secret variable file (vars/smtp_secrets.yml) that's being encrypted by Ansible Vault.

7. Execute the playbook:

```
[ansible@ansible Chapter-03]$ ansible-playbook system-reboot-with-email.yml -e "NODES=nodes"
ERROR! Attempting to decrypt but no vault secrets found
```

Figure 3.54 – Executing the Ansible playbook without the vault password

Ansible is trying to decrypt the secret file, but no vault secret (vault password) has been provided. Execute the same playbook but add the `--ask-vault-password` switch at the end (Ansible will not ask or prompt for vault secrets by default):

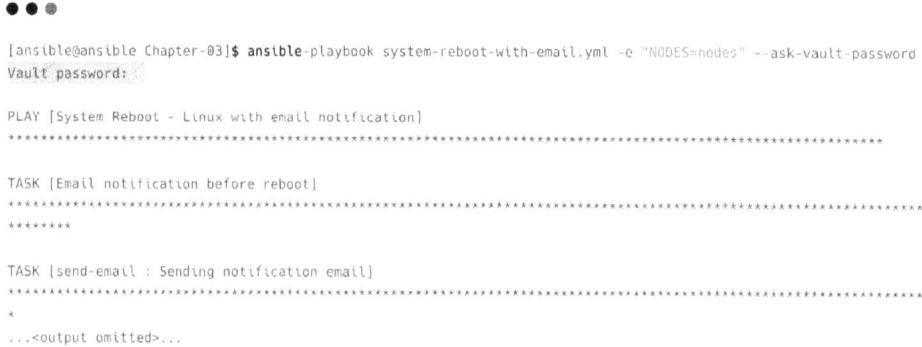

Figure 3.55 – Executing the Ansible playbook with the vault password

8. Check your inbox (your `email_smtp_to_address`) for an automated email from Ansible:

Figure 3.56 – Email notification

As an exercise, enhance the playbook by sending consolidated emails for all servers rather than sending an individual email or attaching a job summary report in the email.

Send an Email Using Ansible

For more information about the Ansible mail module, refer to `https://docs.ansible.com/ansible/latest/collections/community/general/mail_module.html`.

Summary

In this chapter, you learned how to create Ansible roles, Jinja2 templates, and Ansible Vault secrets. You also learned how to collect system information from Ansible facts and use the Jinja2 template to create reports and configurations. The use cases you have practiced were very generic, such as collecting system information, configuring standard system files, rebooting servers, and sending email notifications. As an exercise, enhance the use cases by adding more tasks and validation (such as validating the reboot activity before sending an email and so on).

In the next chapter, you will learn about the importance of **version control systems** (**VCSs**) in Ansible, the best practices to keep your Ansible artifacts safe, and how to enable collaboration and sharing to improve the quality of Ansible artifacts.

Further reading

To learn more about the topics that were covered in this chapter, take a look at the following resources:

- *How to pass extra variables to an Ansible playbook*: `https://www.redhat.com/sysadmin/extra-variables-ansible-playbook`

- *Ansible for Security and Compliance*: `https://www.ansible.com/use-cases/security-and-compliance`

Exploring Collaboration in Automation Development

When you work as a team, collaboration is the key to your team's harmony. Instead of keeping your automation content and knowledge to yourself, you can share it with your team, or even other departments. By doing that, the content will be useful to many others and also, they can contribute with their own ideas and tips. Compared to custom scripts, Ansible content is human-readable and easy for others to understand. Hence, they can modify it and later contribute to the content by fixing bugs or adding features. It is possible to use any standard methods to keep and distribute your Ansible automation content, such as a Git server, Subversion, or any other **Version Control System** (**VCS**).

In this chapter, you will learn about the following topics:

- The importance of version control in IT automation
- Where should I keep automation artifacts?
- Managing automation content in a Git server
- Collaboration is the key to automation

You will start by looking at GitHub account configuration and access management to share content with your team. You will also learn how to manage contributions from other users without losing the original content.

This chapter focuses on Git and how to use Git for Ansible content. If you are familiar with Git and GitHub, then you may skip the sections about those topics.

Technical requirements

The following are the technical requirements to proceed with this chapter:

- One Linux machine for the Ansible control node.

- One or more Linux machines with Red Hat repositories configured (if you are using other Linux operating systems instead of RHEL machines, then make sure you have appropriate repositories configured to get packages and updates).

- An email ID to create a new GitHub account (if you don't already have a GitHub account).

- Basic knowledge about source control server and version control systems.

All the Ansible code, Ansible playbooks, commands, and snippets for this chapter can be found in the GitHub repository at `https://github.com/PacktPublishing/Ansible-for-Real-life-Automation/tree/main/Chapter-04`.

The importance of version control in IT automation

Like any other software, configurations, or scripts, it is not a best practice to keep your Ansible playbooks and configurations on the local machine, which is the Ansible control node. There are many reasons for not keeping the automation content on the local Ansible control node. A few of them are listed here:

- If something happens to the Ansible control node, you will lose all your automation content, which is not desirable.

- If someone accidentally deletes any files or changes any configurations, you will not have the opportunity to restore the original content.

- If you want to make any changes to configurations or playbooks, then you need to make a backup of files and configurations. This is general practice in case something goes wrong and you want to restore an old version of your files.

You need to consider the Ansible automation content as software code, which should keep track of every change and have the option to use old versions at any point in time. Keeping multiple versions of the content will give you the freedom and confidence to make continuous changes to your automation playbooks and configurations. This will also implement the *single-source-of-truth* practice where your Ansible playbooks and variables reside in a central place and track all the changes.

VCSs, such as Git and Subversion, will help you to keep a track of the changes for your Ansible automation content and configurations. Depending on the technology, all these tools will create new versions of your content whenever there are changes made to it.

You will keep your content in Git servers as repositories, which are collections of files and directories. It is possible to create and maintain multiple Git repositories based on the content, such as one Git repository for keeping package installation playbooks and tasks, another Git repository for Linux remediation automation, and so on. You will practice creating and managing Git repositories in the following sections of this chapter.

> **Git Documentation**
>
> Check `https://git-scm.com/doc` for the reference manual, books, and videos to learn Git. Learn the important terminologies in Git, such as branch, commit, push, pull, clone, and staging. Since this book is focused on Ansible, we will not go into detail about these topics.
>
> Please note that **Software Configuration Management (SCM)** is not the same as VCS but VCS is a part or subset of SCM.

Selecting a Git server

Choose any type of VCS, but in this chapter, we will focus on how to use Git servers for storing your Ansible automation content. Organizations use enterprise Git solutions such as **GitHub Enterprise**, GitLab **Enterprise Edition (EE)**, and **Atlassian Bitbucket**. If you do not have a Git server in your environment, then it is possible to easily set up one using any of the free and open source solutions available, such as GitLab **Community Edition (CE)**, Gogs, and Gitea. Most of them are included a web **Graphical User Interface (GUI)** and you can use them for any general Git use cases, including Ansible automation content.

GitHub

GitHub is a hosted service provider for version control repositories and other **Continuous Integration/Continuous Delivery (CI/CD)** operations. It is possible to create a personal GitHub account and create unlimited public or private repositories to store your software or configurations. One of the main limitations of GitHub is that you cannot host a private GitHub server for your environment.

GitLab

GitLab (`gitlab.com`) is another hosted VCS provider similar to GitHub but with different features and services. GitLab offers public-hosted servers (`gitlab.com`) and also private hosting via GitLab EE or CE.

> **How to Install a Git Server**
>
> Refer to `https://www.techbeatly.com/build-your-own-git-server-using-gogs/` to learn how to install a Git server using Gogs inside a simple Docker container. If you want to install GitLab CE, then refer to the documentation at `https://about.gitlab.com/install/` for detailed instructions.

We have learned the importance of storing Ansible automation content in Git servers and about the different Git servers available to use. In the next section, we will learn the best practices for storing Ansible content in Git and repository guidelines.

Where should I keep automation artifacts?

Keep your playbooks and configurations in multiple Git repositories based on the automation and content type.

Ansible and Git repositories – best practices

There are many best practices for keeping your Ansible automation content in a VCS.

Repository for Ansible roles

If you are creating Ansible roles alone (it is no longer common to create individual roles for distribution without a collection), then create one Git repository per role so that the development and collaboration will be easy without depending on other tasks and configurations. See the sample ansible-role repositories in *Figure 4.1*.

```
$ ls -l
total 0
drwxr-xr-x  14 gini  staff  448 21 Jan 12:46 ansible-role-pgsql-replication
drwxr-xr-x  12 gini  staff  384 21 Jan 12:45 ansible-role-repo-epel
drwxr-xr-x  11 gini  staff  352 21 Jan 12:42 ansible-role-setup-user
drwxr-xr-x  14 gini  staff  448 21 Jan 12:45 ansible-role-system-facts-report
drwxr-xr-x  15 gini  staff  480 21 Jan 12:41 ansible-role-tower-setup
```

Figure 4.1 – Separate repositories for Ansible roles

Repositories for Ansible collections

If you are creating Ansible collections, then create one Git repository per collection to make the development and management easy. Move your existing Ansible roles, libraries, modules, and other plugins to an Ansible collection repository and distribute them in a standard way. See the sample ansible-collection repositories in *Figure 4.2*.

```
[ansible@ansible ansible-collections]$ ls -l
total 0
drwxr-xr-x  14 gini  staff  448 21 Jan 13:22 ansible-collection-custom-modules
drwxr-xr-x   5 gini  staff  160 21 Jan 13:21 ansible-collection-kubernetes_home_lab
```

Figure 4.2 – Ansible collection repositories

Dedicated repositories for teams

It is a best practice to create different repositories for different teams if they are working on different items as it will make repository management easy and transparent.

Dedicated repositories for inventories

Keeping inventories for different environments and groups in separate Git repositories will help you to use them efficiently in playbooks. Also, this practice will avoid the accidental execution of automation jobs in wrong inventories. For example, you can keep production servers in one repository and development servers in another repository, as shown in *Figure 4.3*:

```
├── ansible-inventory-development
│   ├── group_vars
│   │   └── mysqlservers
│   ├── host_vars
│   └── inventory
├── ansible-inventory-production
│   ├── group_vars
│   │   ├── mysqlservers
│   │   └── webservers
│   ├── host_vars
│   └── inventory
└── ansible-inventory-staging
    ├── group_vars
    │   ├── mysqlservers
    │   └── webservers
    ├── host_vars
    └── inventory
```

Figure 4.3 – Ansible inventory directory structure for different environments

By separating inventories for different environments into different Git repositories, it is possible to control the access to these inventories' data as well, such as who can modify or use this content.

If you use an Ansible controller, this practice will also help you as it is possible to import inventories from project repositories. Read `https://docs.ansible.com/automation-controller/latest/html/userguide/inventories.html#sourced-from-a-project` to find out more.

Managing automation content in a Git server

In this section, you will learn how to create a GitHub (`github.com`) account, create, install, and configure the repositories, and keep Ansible automation content inside the repositories.

Setting up a GitHub account

If you already have a GitHub personal or enterprise account, then you can skip the account creation steps:

1. Open your web browser and go to `github.com`, then click the **Signup** button in the top-right corner of the page.

2. Enter your email address and a password and username on the next screen, as shown in *Figure 4.4*. GitHub will tell you whether the username is available as usernames in GitHub must be unique:

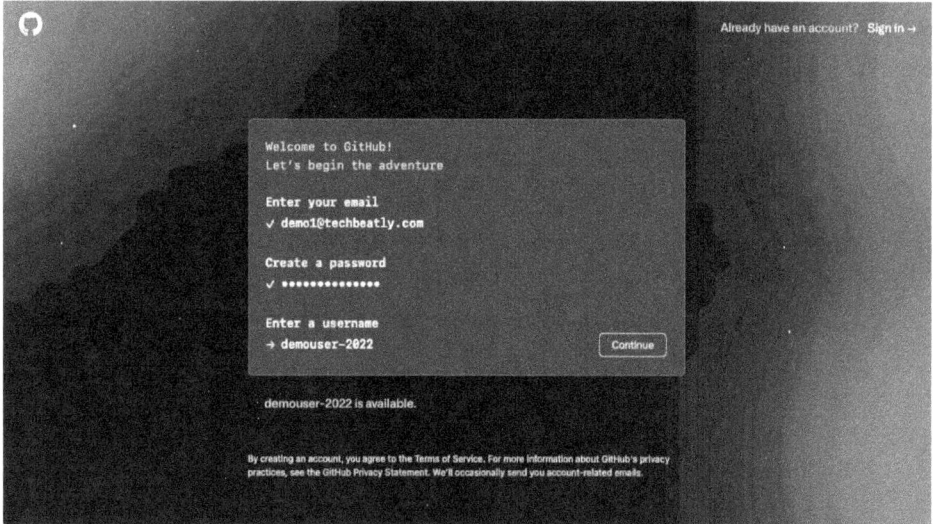

Figure 4.4 – Creating a GitHub account

3. Click **Continue** and finish the simple puzzle (CAPTCHA) on the next screen to verify your identity. Once done, create your account.

4. On the next screen, GitHub will ask you for the one-time code that you will receive on your registered email. Check your email inbox for the code (*Figure 4.5*).

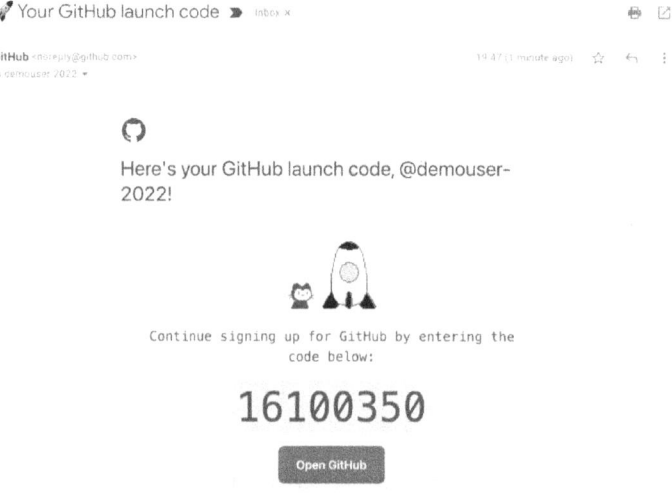

Figure 4.5 – Fetching the activation token from your mailbox

5. Enter the code into the GitHub window and complete the signup process (*Figure 4.6*).

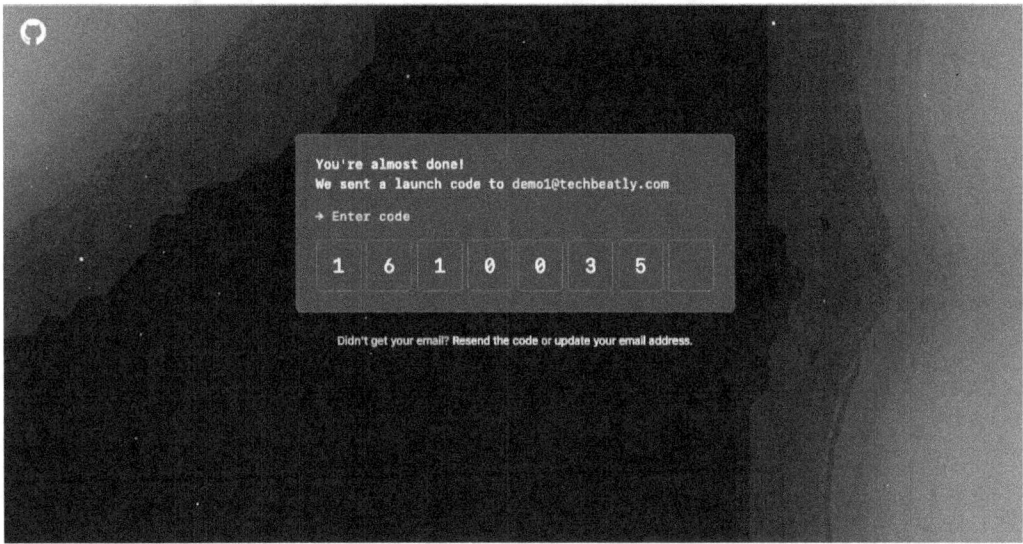

Figure 4.6 – Inputting the one-time token in GitHub

6. Now, you will be taken to the GitHub home page with your activated account (*Figure 4.7*).

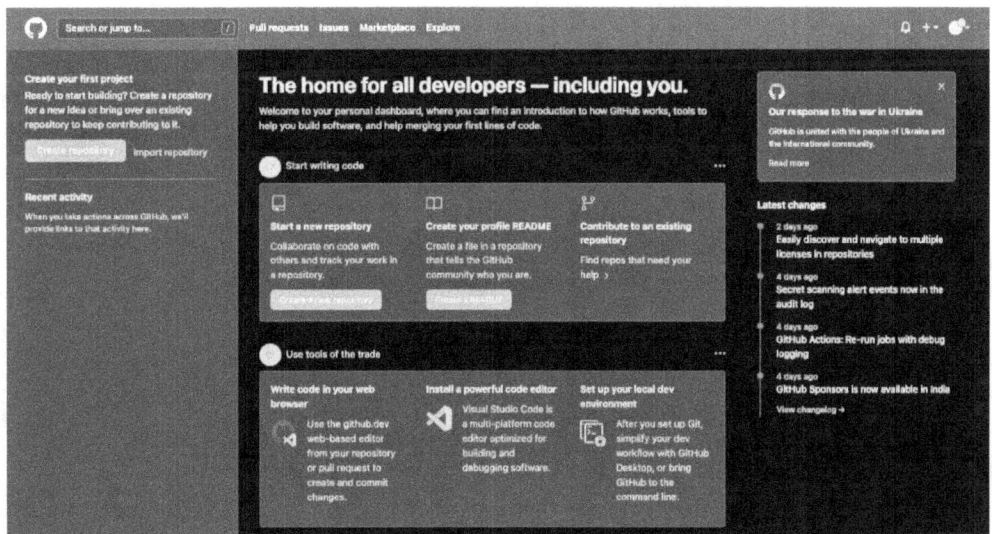

Figure 4.7 – GitHub account home page

> **Other Git Servers**
>
> Use any other Git services, such as GitLab (`https://gitlab.com`) or Bitbucket (`https://bitbucket.org`), and the sign-up process will be more or less the same.

Creating your first Git repository

In this exercise, you will create a new Git repository to store your playbooks and configurations:

1. From the GitHub home page, click the + icon in the upper-right corner and select **New repository** from the menu.

Figure 4.8 – Creating a new repository

2. Enter the details of the new repository, such as the repository name and description. Also select the visibility of the repository as a public or private repository. Also, tick the **Add a README file** option as shown in *Figure 4.9*.

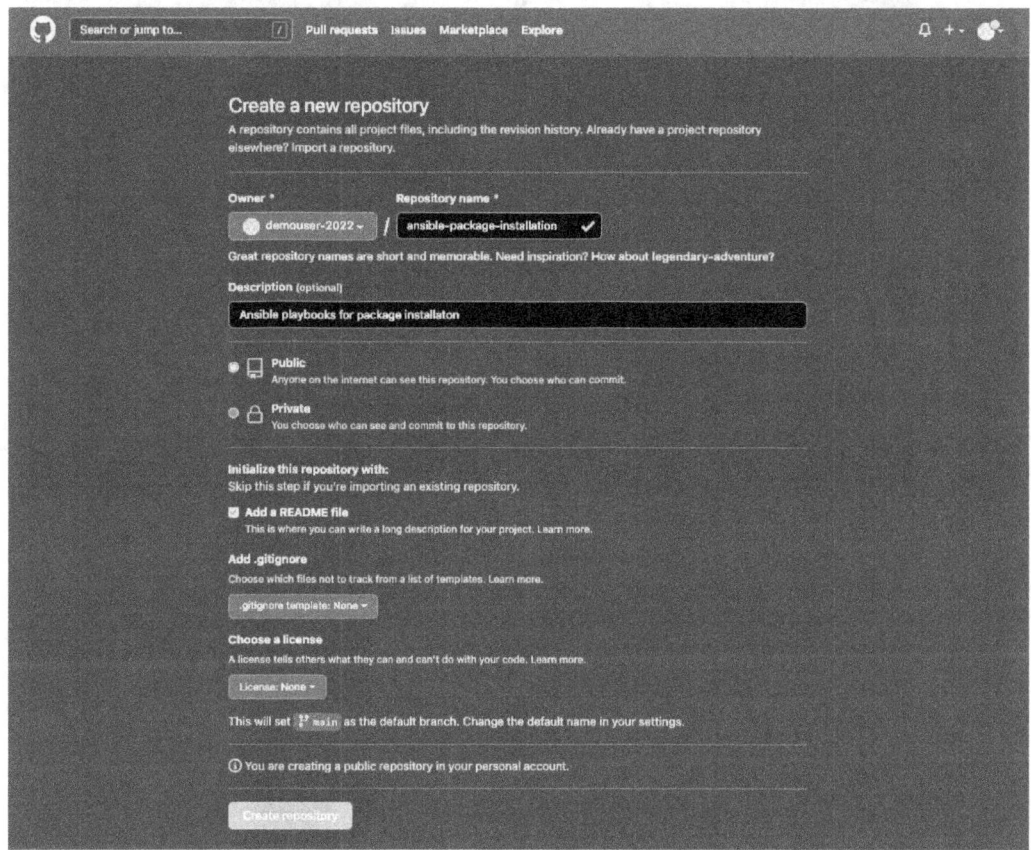

Figure 4.9 – Entering a new repository name and details

> **Note**
> It is possible to change all these configurations at any time, but it is a best practice not to change the repository name as it may break your integrations and paths.

Once you have entered all the details, click on the **Create repository** button.

3. GitHub will show the repository with default README.md file content (*Figure 4.10*). This file was created automatically because we ticked the **Add a README file** option in the previous step.

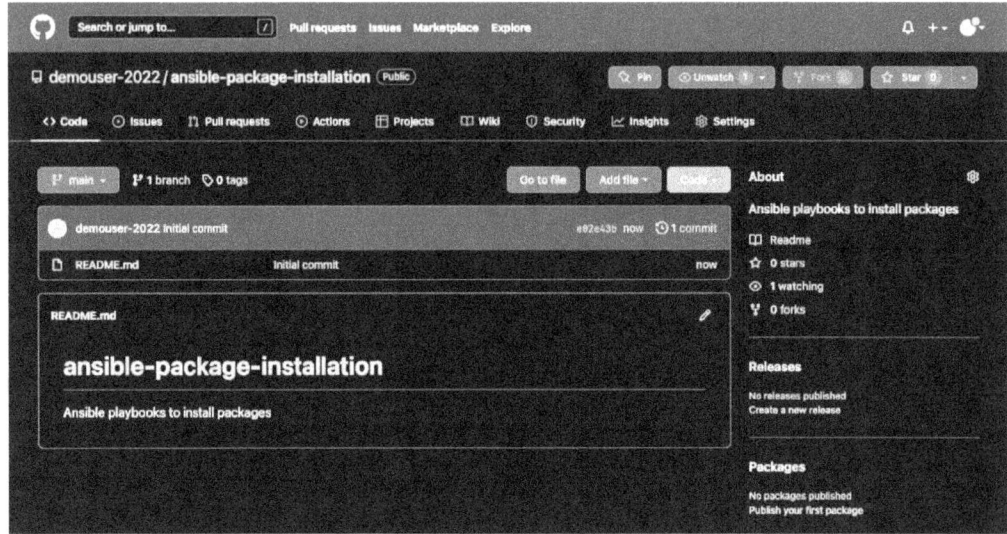

Figure 4.10 – GitHub repository default view with README.md file

README.md is a special file that will be used to communicate important information about the repository or directory inside a repository. GitHub will render and show the content of the README. md file as an HTML-equivalent format, which is a great way to add your repository information, documentation, and more.

> **Markdown Format**
>
> The .md extension is used for Markdown files, which is a lightweight markup language. Markdown language is used for creating formatted text using plain-text editors that is later rendered to HTML or other formats. Read https://www.markdownguide.org/getting-started for more details.

Installing Git on a local machine

Create and manage your Git repository and content from the GitHub web UI itself, but this is limited as you cannot do any bulk operations, such as changing multiple files in a single commit. You can manage your repository and content from any compatible Git CLI or GUI tools, such as the default Git CLI, GitHub Desktop, Sourcetree, TortoiseGit, SmartGit, and Tower.

In this exercise, you will install and use the Git command-line utility to access and manage your Git repositories:

1. To install Git on the Ansible control node, use the following:

    ```
    [ansible@ansible ~]$ sudo yum install git
    ```

2. Next, verify the Git version:

```
[ansible@ansible ~]$ git version
git version 2.27.0
```

3. Configure the Git username and email address. This step will update the global username and email address for your Git environment. You need to use the username and email address that you used during the GitHub account creation:

```
[ansible@ansible ~]$ git config --global user.name
"demouser-2022"
[ansible@ansible ~]$ git config --global user.email "M
demo1@techbeatly.com"
```

Note that it is possible to use any other dummy username or email address as this information is not used as credentials for GitHub access but the identity for local users.

> **Git CLI and GUI Clients**
>
> Check out `https://git-scm.com/book/en/v2/Getting-Started-Installing-Git` to find out how to install Git on different operating systems. Git GUI clients are used for better management of Git repositories without much command-line execution. Check out `https://git-scm.com/downloads/guis` for available Git GUI clients.

Configuring SSH keys in GitHub

It is possible to access and manage your GitHub repositories using your username and password, but it is not desirable to enter the username and password every time you want to update something in the Git server. This is where we can utilize SSH keys, and you can use the same or different SSH keys to configure GitHub access.

In this exercise, you will configure the SSH public keys to GitHub for seamless and password-less access from the Git client:

1. Fetch the SSH public key content. Use the same SSH key pairs that were created in the *Setting up SSH key-based authentication* section in *Chapter 1, Ansible Automation – Introduction*. Copy the public key content as shown in *Figure 4.11*:

```
[ansible@ansible ~]$ cat ~/.ssh/id_rsa.pub
ssh-rsa
AAAAB3NzaC1yc2EAAAADAQABAAABgQDgzrPJQ4Vp6FGO4XVGUpQNzpTOyO1+pS/9whfBqjvY8OOgfJM2eg/rpcubMsMAamCPzeFmy0RKXIHixAno5Snm9
VcENfobknHb4IQmRq0ATOiG1niyWDJB9fUIm/3YOPt+ZxPiiUa/iQvc8B4FqLGvBGSWB9GZE4OPPFk+sfCrmDrlI+2kgBeRJ3xKqMxoj70aReHDdO/jVN
9VcUiHQ+WrTqBSHyHObb1SCxWFScj7VKR2BnayyKrS1EDOluPKLwfcEM5scms6tLBcwnyCvko4W2afIQqSbEdhOesoGh/fQl4c7ycFnkIxaicnReEEDEX
nBso9Ndp3PCTojoT86RyqDUgpazjMsZkmL52YPcq2aX6RGOrE8eWIeATHNNM4nH5tTMf/35j3+3WXA/9NSdvsikGet5FKL21tIy2qo5hKHgMnL9Dipdoa
i3cnlCD/t4A/Z0bNsAMWDgzSPsmVjdDCBealRJYiLJimj8sTjleruah5DlZqfZoTymuMloInxsM= ansible@ansible-controlnode
```

Figure 4.11 – Copy SSH public key content

2.   Go to **GitHub | Settings**.

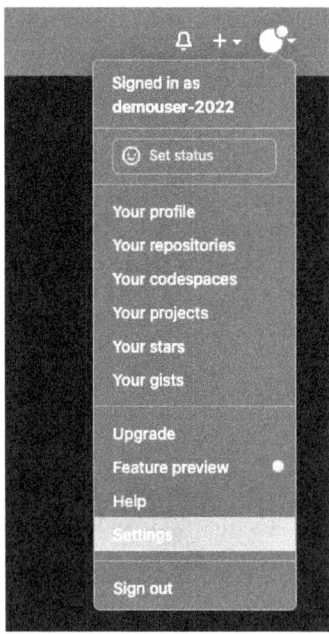

Figure 4.12 – Opening the GitHub settings

3.   Select the **SSH and GPG keys** tab on the left (*Figure 4.13*).

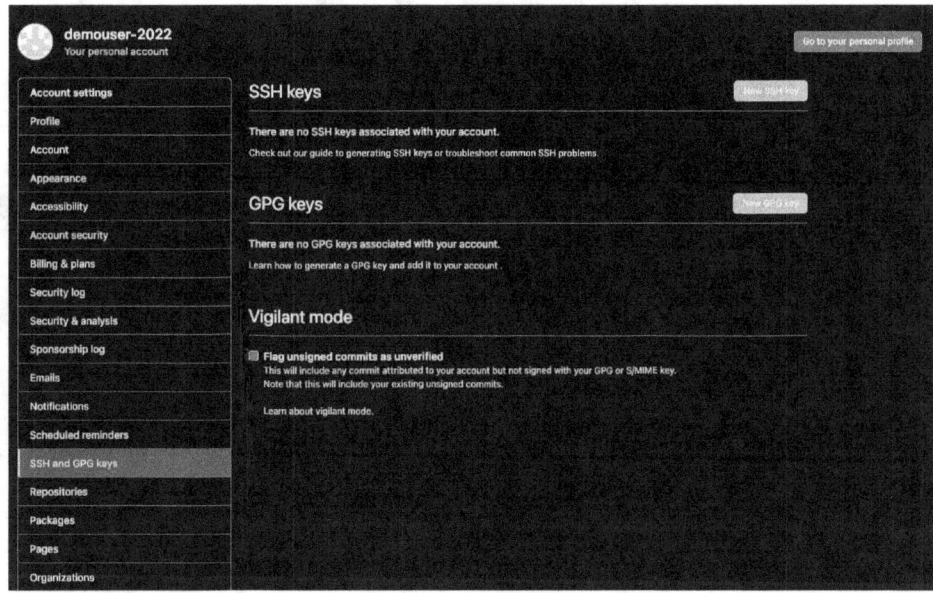

Figure 4.13 – GitHub account SSH and GPG keys configuration

4. Click on the **New SSH key** button and enter a title (any identifiable name) and the SSH public
    key content that we copied in *Step 1*. Click on **Add SSH key**.

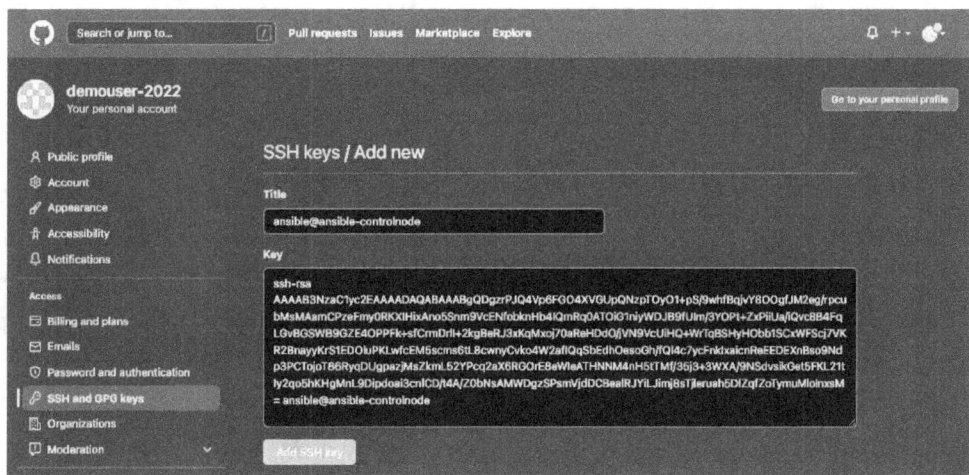

Figure 4.14 – Adding SSH public key content

GitHub will ask for your password to confirm this SSH key addition task.

5. Verify that the SSH key has been added (*Figure 4.15*).

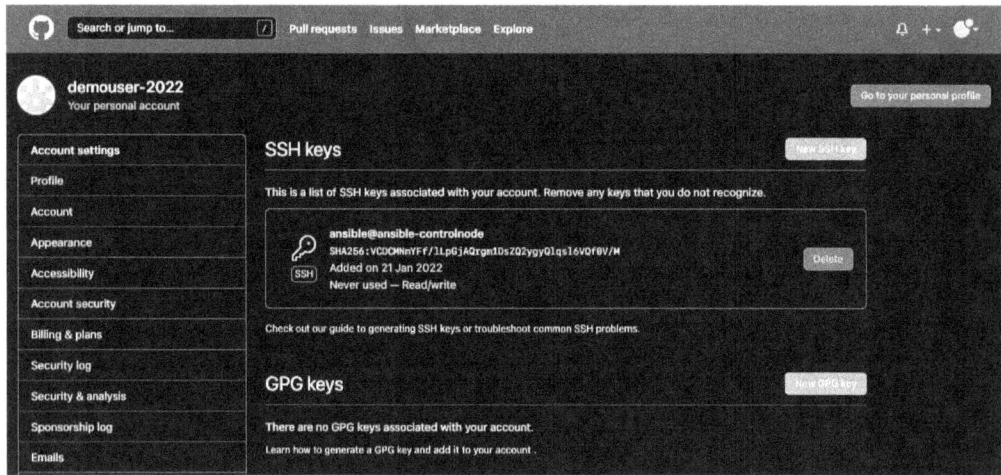

Figure 4.15 – Verifying SSH keys

Add any number of keys here; for example, if you want to manage your GitHub repository from another workstation (or your laptop), then add the SSH public key from that workstation here. Also, note that if you find that any of the SSH keys are not in use or have been compromised, you can remove them from this page and deny access.

> **Adding SSH Keys to GitHub**
>
> Refer to the documentation at `https://docs.github.com/en/authentication/connecting-to-github-with-ssh/adding-a-new-ssh-key-to-your-github-account` for more details.

Adding content to the Git repository

In this exercise, you will add your previously created automation playbooks (in *Chapter 2*, *Starting with Simple Automation*) and configurations to the newly created GitHub repository. To achieve that, you need to **clone** the remote repository (on `github.com`) to a local machine:

1. Go to GitHub and access your repository.

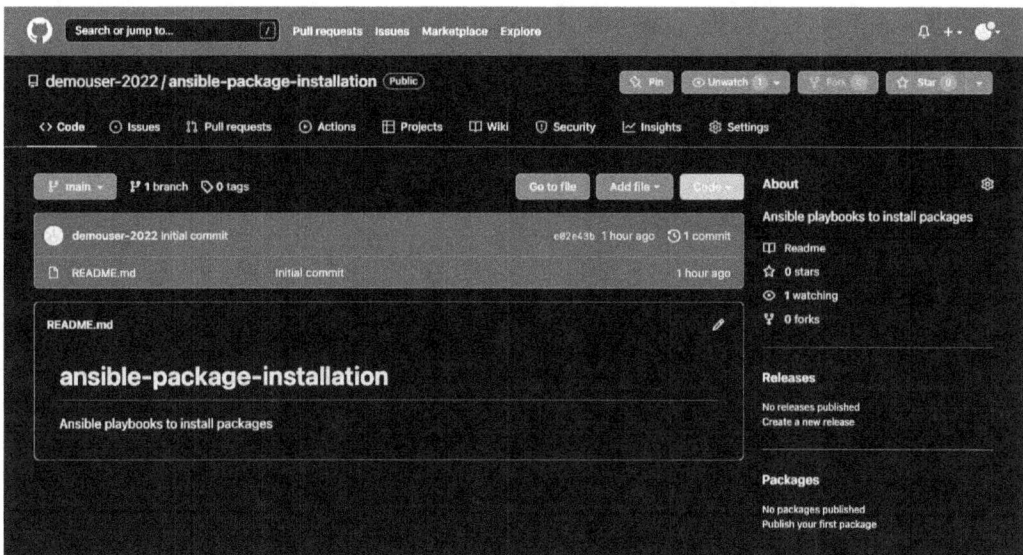

Figure 4.16 – GitHub repository details

2. Find the **Code** button, click on the drop-down arrow, switch to the **SSH** option, and copy the command to clone the repository to your local machine.

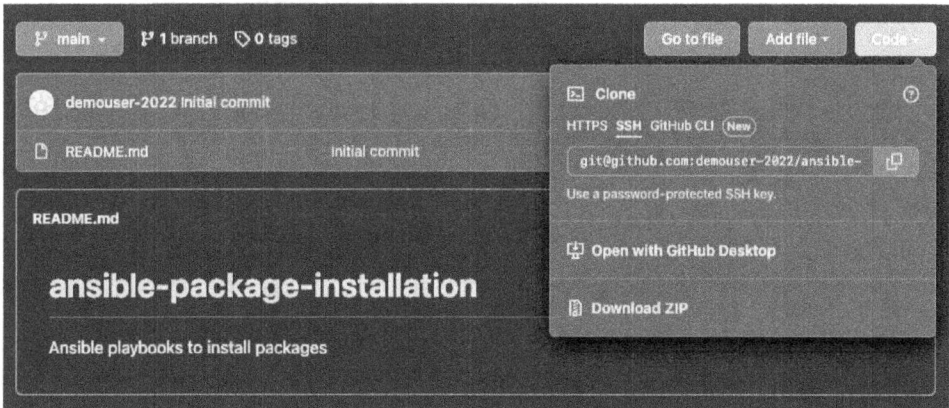

Figure 4.17 – Getting the GitHub repository URL

Since you have configured the SSH keys, you can use the SSH-based clone and access the repository. If you are using HTTPS-based cloning, GitHub will ask for your username and password every time you update the content back to the remote repository.

3. On your Ansible control node, execute the `git clone` command:

```
[ansible@ansible ~]$ git clone git@github.com:demouser-2022/ansible-package-installation.git
Cloning into 'ansible-package-installation'...
remote: Enumerating objects: 3, done.
remote: Counting objects: 100% (3/3), done.
remote: Compressing objects: 100% (2/2), done.
remote: Total 3 (delta 0), reused 0 (delta 0), pack-reused 0
Receiving objects: 100% (3/3), done.
```

Figure 4.18 – Clone Git repository to local machine

4. Check the content of the cloned Git repository:

```
[ansible@ansible ~]$ cd ansible-package-installation/
[ansible@ansible ansible-package-installation]$ ls -la
total 4
drwxrwxr-x. 3 ansible ansible  35 Jan 21 14:25 .
drwxrwxrwt. 9 root    root    208 Jan 21 14:25 ..
drwxrwxr-x. 8 ansible ansible 163 Jan 21 14:25 .git
-rw-rw-r--. 1 ansible ansible  69 Jan 21 14:25 README.md
```

Figure 4.19 – Listing content of cloned Git repository

See the `README.md` file that was created automatically when you created the Git repository. The `.git` directory contains all the information about this repository, including remote repository and commit details.

5. Move/copy the files you created in the `Chapter-02` exercise to this directory and verify the files are inside the directory as shown in *Figure 4.20*:

```
[ansible@ansible ansible-package-installation]$ ls -la
total 24
drwxrwxr-x.  3 ansible ansible  121 Jan 21 14:24 .
drwx------. 13 ansible ansible 4096 May 28 03:26 ..
-rw-rw-r--.  1 ansible ansible  209 Jan 21 14:24 ansible.cfg
-rw-rw-r--.  1 ansible ansible  222 Jan 21 14:24 chrony.conf.sample
drwxrwxr-x.  8 ansible ansible  185 Jan 21 14:32 .git
-rw-rw-r--.  1 ansible ansible  135 Jan 21 14:24 hosts
-rw-rw-r--.  1 ansible ansible  558 Jan 21 14:24 install-package.yaml
-rw-rw-r--.  1 ansible ansible   69 Jan 21 14:21 README.md
```

Figure 4.20 – Content of Git local repository after files moved

6. Check `git status` and notice the changes as shown in *Figure 4.21*:

```
● ● ●

[ansible@ansible ansible-package-installation]$ git status
On branch main
Your branch is up to date with 'origin/main'.

Untracked files:
  (use "git add <file>..." to include in what will be committed)
        ansible.cfg
        chrony.conf.sample
        hosts
        install-package.yaml

nothing added to commit but untracked files present (use "git add" to track)
```

Figure 4.21 – git status output for untracked files

From the preceding output, you can understand the following facts:

- The `ansible.cfg`, `chrony.conf.sample`, `hosts`, and `install-package.yaml` files are not in the Git database and are called **untracked files**.

- If you want to add them to Git, you need to use the `git add` command.

7. Add untracked files to Git (you can add them one by one or all at once):

    ```
    [ansible@ansible ansible-package-installation]$ git add *
    ```

    Check the `git status` again as shown in *Figure 4.22*.

```
● ● ●

[ansible@ansible ansible-package-installation]$ git status
On branch main
Your branch is up to date with 'origin/main'.

Changes to be committed:
  (use "git restore --staged <file>..." to unstage)
        new file:   ansible.cfg
        new file:   chrony.conf.sample
        new file:   hosts
        new file:   install-package.yaml
```

Figure 4.22 – git status after adding file to the Git repository

Also note that the files are not transferred to the remote repository (GitHub) yet.

8. Commit the changes to Git using the `git commit` command. Use appropriate comments to identify the change in the repository:

```
● ● ●

[ansible@ansible ansible-package-installation]$ git commit -m "First commit with Ansible files"
[main 302dfcc] First commit with Ansible files
4 files changed, 51 insertions(+)
create mode 100644 ansible.cfg
create mode 100644 chrony.conf.sample
create mode 100644 hosts
create mode 100644 install-package.yaml
```

Figure 4.23 – git commit output

9. Now, push the changes to the remote repository using the `git push` command as shown in *Figure 4.24*:

```
● ● ●

[ansible@ansible ansible-package-installation]$ git push
Enumerating objects: 7, done.
Counting objects: 100% (7/7), done.
Compressing objects: 100% (6/6), done.
Writing objects: 100% (6/6), 1.04 KiB | 1.04 MiB/s, done.
Total 6 (delta 0), reused 0 (delta 0), pack-reused 0
To github.com:demouser-2022/ansible-package-installation.git
   e02e43b..302dfcc  main -> main
```

Figure 4.24 – Push changes to the remote Git repository

This will transfer all the files and changes to the remote repository.

10. Verify the content on GitHub.

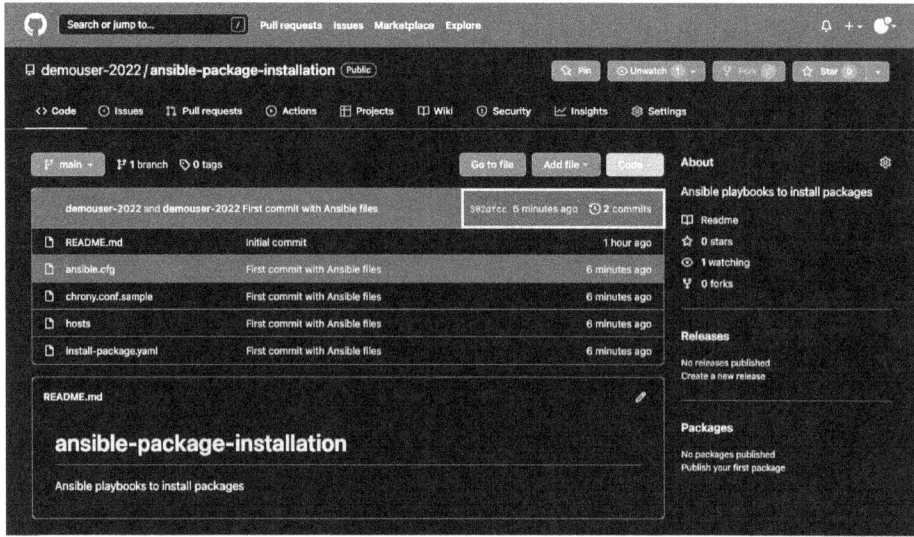

Figure 4.25 – Verifying pushed content on the remote repository

As we can see, the files are already available on the Git repository.

11. Verify the commit history on GitHub. Click on the **Commits** link (below the **Code** button as shown in *Figure 4.25*) and check the commits.

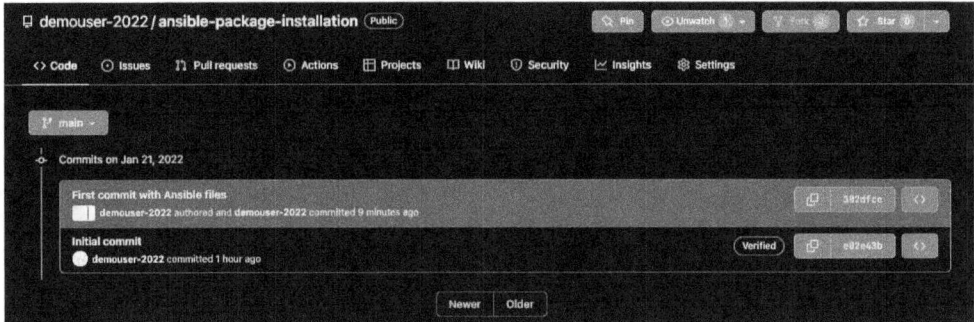

Figure 4.26 – Verifying Git commits in GitHub

You can explore commits and history by referring to the Git documentation here: `https://git-scm.com/docs/gittutorial`.

Now we have learned how to add Ansible content to a Git repository and see the version history from the GitHub UI. In the next section, we will learn how to manage contributions and collaborative development in a Git repository.

Collaboration is the key to automation

Now you have your Ansible automation content in your GitHub repository. There are several advantages to this:

- You do not need to take a backup of your files before you make changes (once you make the changes, remember to test, commit, and push the changes to a remote GitHub repository).

- Pull the content to any of the machines whenever needed and test it. For example, you can download the code to your local workstation and develop it further. Once you make the changes, push it back to the remote repository; a new version of the code will be stored there.

- Other users and developers can test and contribute to your code without having access to your Ansible control node. You just need to allow appropriate access to other users.

- If any of the code is not working after an update, you can revert to an old version of the code at any point in time.

Let's learn how to use Git branching in the next session.

Using Git branching

Git provides a feature called branching, which will help you to create multiple versions (or branches) of code in the same Git repository. A **branch** is a pointer to the latest commit in your Git repository. By default, Git will create a branch called main (formerly master) and all your Git commits go to the main branch.

You can create multiple branches on the Git repository to take advantage of the Git workflow:

- Create multiple branches for development and staging to track the changes. Once the content of development and staging branches are tested and confirmed as good content, then you can merge the content from these development and staging branches to the main branch of your Git repository. With this practice, the main branch will contain only tested and clean code.

- Create different branches for patching or bug fixes and merge them to the main branch once tested.

- Create a branch for users to contribute their code and merge to the main branch once tested.

You can choose any type of branching strategy based on your development workflow and the organization's requirements. For our exercise, we will be using a simple Git branching strategy, as shown in *Figure 4.27*:

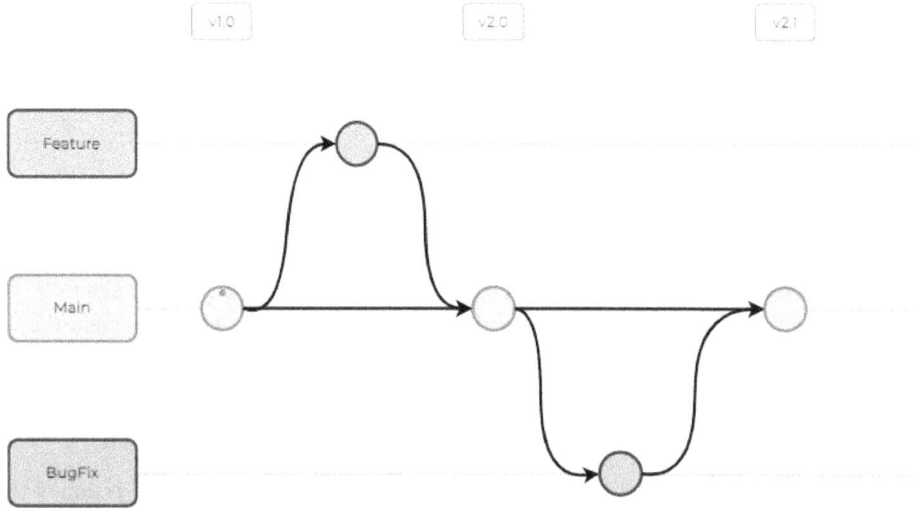

Figure 4.27 – Git branching

Git branching and merging may happen for different reasons. The following are the typical tasks involved in the Git workflow:

- The `main` branch will contain the code, for example, version 1.0 of your automation content.

- Create new feature branches as needed. Once the feature branch is tested and good to use, you will raise a merge request (**PR**, or **pull request**). This is the stage where a contributor informs the maintainer of the Git repository to review code in the feature branch and merge it into the main branch of the project. You can call this version of code **2.0**, for example.

- If you find any issues in the code, duplicate a bugfix branch (same as the feature branch) and raise another PR to merge the changes into the main branch. You can call this version **2.1** here.

Let's look at an example to understand this concept better.

Implementing Git branching

In this exercise, you will learn how to create multiple branches and contribute as different users.

In this exercise, I am using another workstation to clone the previously created repository as a different GitHub user (for example, `ginigangadharan`). Create another GitHub account or ask your friend if you can use their account to test this:

1. Fork the original repository from the web browser as a different user (eg: `ginigangadharan`). Click on the **Fork** button as shown in *Figure 4.28*.

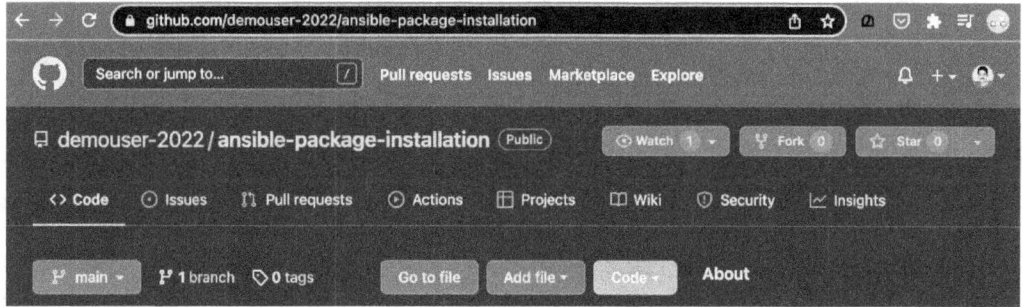

Figure 4.28 – Opening the repository as a different user

2. GitHub will ask for the target account as shown in the *Figure 4.29* (if you have other organization accounts) to fork the repository and will create a copy of the original repository in the new user account.

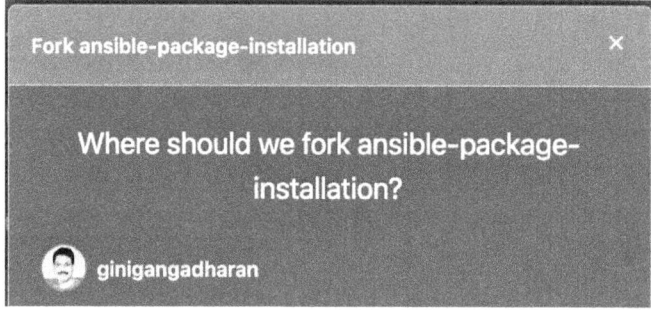

Figure 4.29 – Forking the repository

Now you can see a new repository created under the new account, which is forked from the original repository.

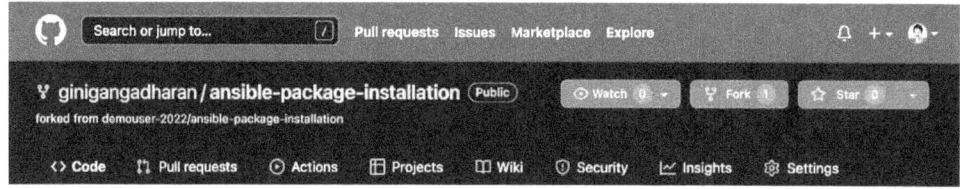

Figure 4.30 – Forked repository details

3.   Clone this repository to your workstation and check the content; remember to clone with the new repository URL:

```
$ git clone git@github.com:ginigangadharan/ansible-package-installation.git
Cloning into 'ansible-package-installation'...
remote: Enumerating objects: 9, done.
remote: Counting objects: 100% (9/9), done.
remote: Compressing objects: 100% (8/8), done.
Receiving objects: 100% (9/9), done.
remote: Total 9 (delta 0), reused 6 (delta 0), pack-reused 0

$ cd ansible-package-installation
$ ls -l
total 40
-rw-r--r--  1 gini  staff    69 22 Jan 12:33 README.md
-rw-r--r--  1 gini  staff   209 22 Jan 12:33 ansible.cfg
-rw-r--r--  1 gini  staff   222 22 Jan 12:33 chrony.conf.sample
-rw-r--r--  1 gini  staff   135 22 Jan 12:33 hosts
-rw-r--r--  1 gini  staff   558 22 Jan 12:33 install-package.yaml
```

Figure 4.31 – Clone the Git repository from new user account

4. Create a new Git branch called `feature-1`:

    ```
    $ git branch feature-1
    ```

5. Switch to the new branch:

    ```
    $ git checkout feature-1
    Switched to branch 'feature-1'
    ```

6. Check the Git branches:

    ```
    $ git branch
    * feature-1
      Main
    ```

    Here, you can see the `*` symbol, which denotes the current branch.

7. Now, you can update your code, for example, change some lines of code or add some tasks to the playbook.

8. Check git `status` to see the changes:

    ```
    $ git status
    On branch feature-1
    Changes not staged for commit:
      (use "git add <file>..." to update what will be committed)
      (use "git restore <file>..." to discard changes in working directory)
            modified:   install-package.yaml

    no changes added to commit (use "git add" and/or "git commit -a")
    ```

    Figure 4.32 – Git status after updating repositry content

    You can see that `install-package.yaml` (or whichever file you have changed) is highlighted there.

9. Add the changed file and commit the changes to Git; remember to use appropriate commit messages to identify the changes:

    ```
    $ git add *
    $ git commit -m "updated install-package.yaml"
    [feature-1 6e7004b] updated install-package.yaml
    1 file changed, 2 insertions(+), 2 deletions(-)
    ```

    Figure 4.33 – Add update to Git and commit changes

10. Check git `log` to review the commit history as shown in *Figure 4.34*:

```
● ● ●
$ git log
commit 898e5dfde4d90805feb579d245efdce5a18738c7 (HEAD -> feature-1)
Author: ginigangadharan <net.gini@gmail.com>
Date:    Sat Jan 22 13:04:26 2022 +0800

    updated install-package.yaml

commit 302dfccd4cc5b018e17619d8fb1a107b9f230350 (origin/main, origin/HEAD, main)
Author: demouser-2022 <M demo1@techbeatly.com>
Date:    Fri Jan 21 14:32:14 2022 +0000

    First commit with Ansible files

commit e02e43be5e66504e6c129443b38c228245a6444a
Author: demouser-2022 <98160880+demouser-2022@users.noreply.github.com>
Date:    Fri Jan 21 21:23:13 2022 +0800

    Initial commit
```

Figure 4.34 – Details of commits in Git logs

11. Push the new branch and changes to the remote repository:

```
● ● ●
$ git push -u origin feature-1
Enumerating objects: 5, done.
Counting objects: 100% (5/5), done.
...<output omitted>...
* [new branch]        feature-1 -> feature-1
Branch 'feature-1' set up to track remote branch 'feature-1' from 'origin'.
```

Figure 4.35 – Push the changes to remote GIt repository

Now the updated code is available in the new user's GitHub repository, which is a forked copy of the original repository.

12. Go to the new user's GitHub repository and select **Pull requests**.

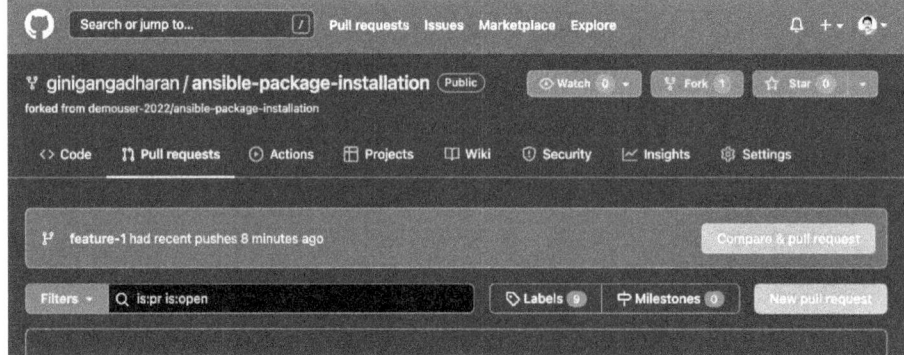

Figure 4.36 – Pull requests in the GitHub repository

13. Click on the **New pull request** button. Select the repository and branches and click on the **Create pull request** button to submit a PR (provide the PR comment as required):

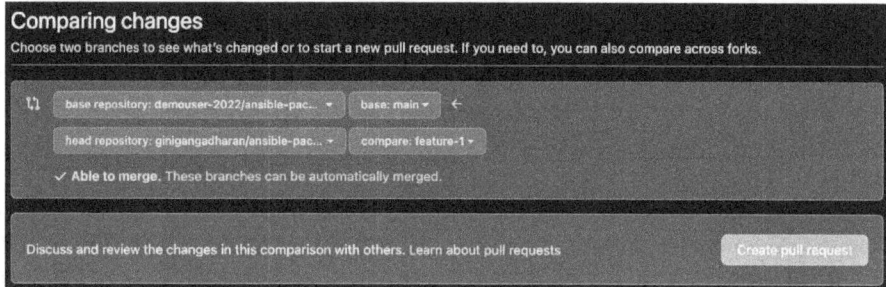

Figure 4.37 – Create a pull request in the GitHub repository

14. Now, go back to the `demouser-2022` user's GitHub account and check the PRs. You will find the PR from the other user (that is, `ginigangadharan`) and can open the PR. Verify the merge, check for any conflicts, and click on the **Merge pull request** button to accept the changes from this user.

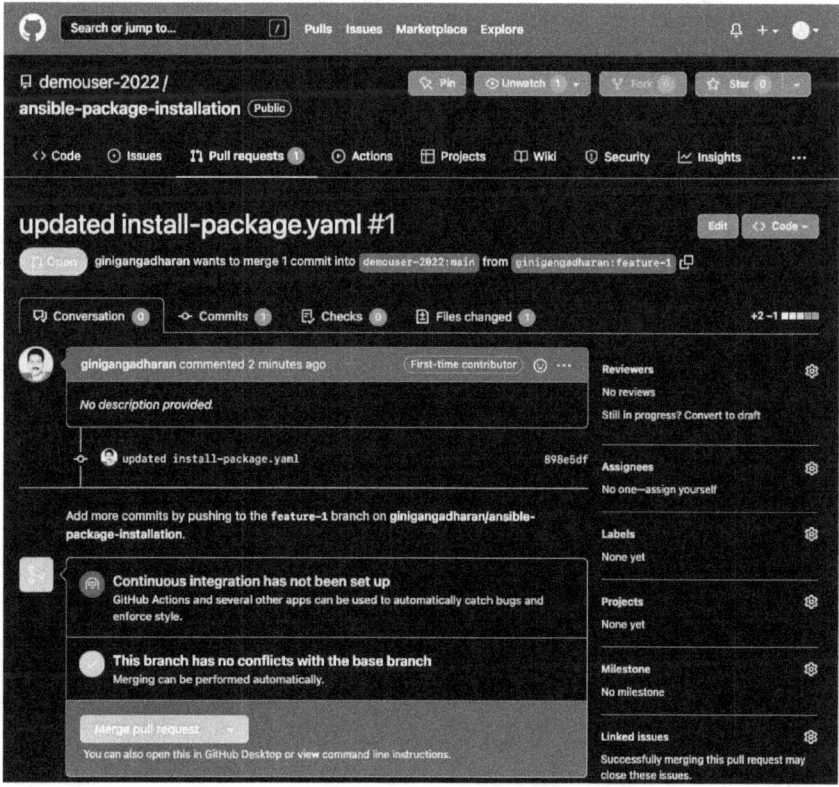

Figure 4.38 – Pull request details in GitHub repository

Now you have the latest contribution from another user in your Git repository.

Please note that this is a very basic Git workflow, and you need to include additional steps, such as adding PR approvals, creating tests before merging to the `main` branch, and other policies as required.

Accept contributions from other users and teams using the preceding workflow without giving full access to your repository, which is a common method used for any other open source software.

Summary

In this chapter, you have learned the importance of version control in an IT automation environment. You have learned the basics of Git and different Git servers and providers. You have practiced creating a GitHub account, Git repository, and other steps. You also learned how to accept contributions to your Git content, branching methods, and PR processes.

In the next chapter, you will learn how to find more automation use cases from your workplace and personal projects. You will also learn more about inventory management and different strategies to keep your managed nodes' information.

Further reading

To learn more about the topics covered in this chapter, please visit the following links:

- *What is Git?*: `https://git-scm.com/docs/gittutorial`

- *What is version control?*: `https://en.wikipedia.org/wiki/Version_control`

- *What are pull requests?*: `https://docs.github.com/en/pull-requests/collaborating-with-pull-requests/proposing-changes-to-your-work-with-pull-requests/about-pull-requests`

- *Syncing your branch*: `https://docs.github.com/en/desktop/contributing-and-collaborating-using-github-desktop/keeping-your-local-repository-in-sync-with-github/syncing-your-branch`

Part 2

Finding Use Cases and Integrations

This part of the book will explain how to use Ansible in real-world use cases in an IT environment. It will also cover most of the general items, such as infrastructure, platforms, and applications.

This part of the book comprises the following chapters:

5

Expanding Your Automation Landscape

It is important to find and apply automation at the right place and at the right level. Usually, when searching for automation use cases, people make mistakes by automating the inappropriate tasks and end up wasting money, effort, and time. This is the reason why we need to analyze the environment and day-to-day tasks and find the best tasks as automation candidates. We can use your statistical data such as event tickets, customer change requests, and project tasks to find this information.

In this chapter, you will learn about the following topics:

- Finding your automation use cases in your day-to-day work

- Automation feasibility and usability

- Involving teams in the automation journey

- The Ansible dynamic inventory

We will start with standard methods to analyze the tasks to find the highest number of tasks and check the feasibility of automation. We will explore the integration opportunities between **IT service management** (**ITSM**) tools and Ansible Automation Platform. We will also practice the Ansible dynamic inventory with public clouds such as **Amazon Web Services** (**AWS**) and explore the importance of Ansible host variables and group variables.

Technical requirements

The following are the technical requirements to proceed with this chapter:

- A Linux machine as an Ansible control node (with internet access)

- One or more Linux machines as managed nodes with Red Hat repositories configured (if you are using non-RHEL machines, then make sure you have appropriate repositories configured to get packages and updates)

- Basic knowledge of the AWS platform and an AWS account with one or more test machines created

All the Ansible artifacts, playbooks, commands, and snippets for this chapter can be found in the GitHub repository at `https://github.com/PacktPublishing/Ansible-for-Real-life-Automation/tree/main/Chapter-05`.

Finding your automation use cases in your day-to-day work

We all know that every member of staff working in an IT environment is executing some tasks and most of the time, they repeat the same job every day. Looking around, we can see many examples, as follows:

- A system engineer is building servers and virtual machines, installing packages, patching old systems, and more.

- A network engineer is configuring the new network device and firewall devices, configuring ports and **virtual local area networks** (**VLANs**) based on requests, patching the device firmware, and many other things.

- A developer is struggling to build his coding environment every time there is a new version of the programming language or software library. They are also spending a lot of time testing the code and waiting for test results.

- A storage administrator is spending their valuable time provisioning the disk space and configuring the storage devices.

- A database administrator is complaining about the provisioning delay of a new database server and issues with network or system readiness.

- A database user is struggling with the delay on a simple database password resetting task.

- The operation team is struggling with the flood of events and alerts and spending their time filtering out the false alerts, resulting in unproductive work.

- The security team is working hard to fix the violations and make sure the systems are compliant with the security standards.

The list is not comprehensive; you will be able to find many additional tasks and scenarios in your workplace as well.

Assessing the tasks

You can take your daily or weekly task reports or details from your ITSM systems such as ServiceNow, Remedy, and Jira. For this scenario, we will use the very common tasks undertaken by a platform team, but you can use any tasks, team, or methods for this assessment. We have extracted a report for a week, as shown in *Figure 5.1*:

Date	Task	Minutes	Source
01-01-2021	LDAP Password reset	10	Jira
01-01-2021	Package installation	120	ServiceNow
01-01-2021	Package installation	100	ServiceNow
01-01-2021	Package installation	100	ServiceNow
02-01-2021	OS Patching - Linux	120	ServiceNow
02-01-2021	OS Patching - Linux	120	ServiceNow
02-01-2021	Weekly system reboot - Linux	60	Email
02-01-2021	Weekly system reboot - Linux	60	Email
02-01-2021	Weekly system reboot - Linux	60	Email
02-01-2021	Weekly system reboot - Linux	60	Email
03-01-2021	OS Patching - Linux	120	ServiceNow
03-01-2021	OS Patching - Linux	120	ServiceNow
03-01-2021	OS Patching - Linux	120	ServiceNow
03-01-2021	OS Patching - Linux	120	ServiceNow
04-01-2021	New Virtual Machine Configuration	240	ServiceNow
04-01-2021	New Virtual Machine Configuration	240	ServiceNow
04-01-2021	New Virtual Machine Configuration	240	ServiceNow
04-01-2021	New Virtual Machine Configuration	240	ServiceNow
04-01-2021	Additional disk configuration	120	Email
04-01-2021	Additional disk configuration	120	Email
04-01-2021	Additional disk configuration	120	Email
04-01-2021	Additional disk configuration	120	Email
06-01-2021	LDAP Password reset	10	Jira
06-01-2021	System configurations	60	Email
06-01-2021	System configurations	60	Email
06-01-2021	System configurations	60	Email
07-01-2021	LDAP Password reset	10	Jira
07-01-2021	LDAP Password reset	10	Jira
07-01-2021	Weekly System Report	60	Email
07-01-2021	Weekly System Report	60	Email

Figure 5.1 – Sample task report

We can see on the chart (*Figure 5.2*) the most critical tasks that are taking the most time and effort of systems engineers:

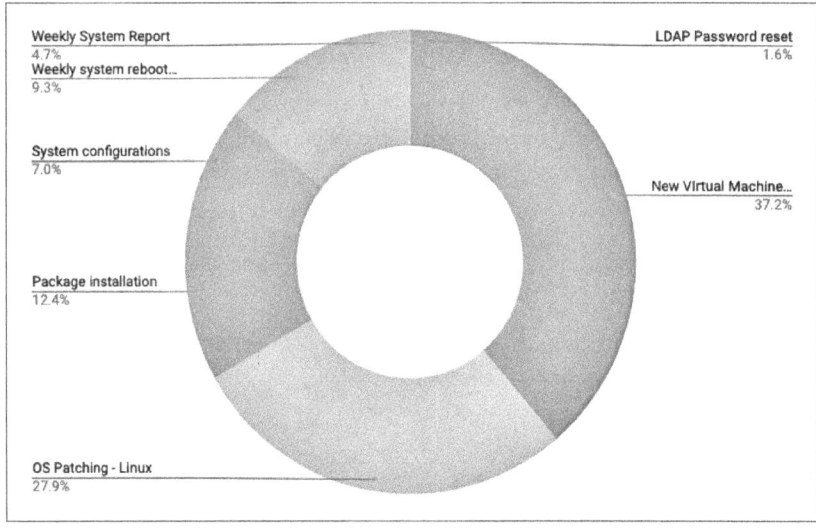

Figure 5.2 – Weekly task summary

Based on this very small chart, we can almost identify the tasks needed to automate and save time.

If we check some more details, we can see the source of the task requests (*Figure 5.3*) and further scope and opportunities for automation:

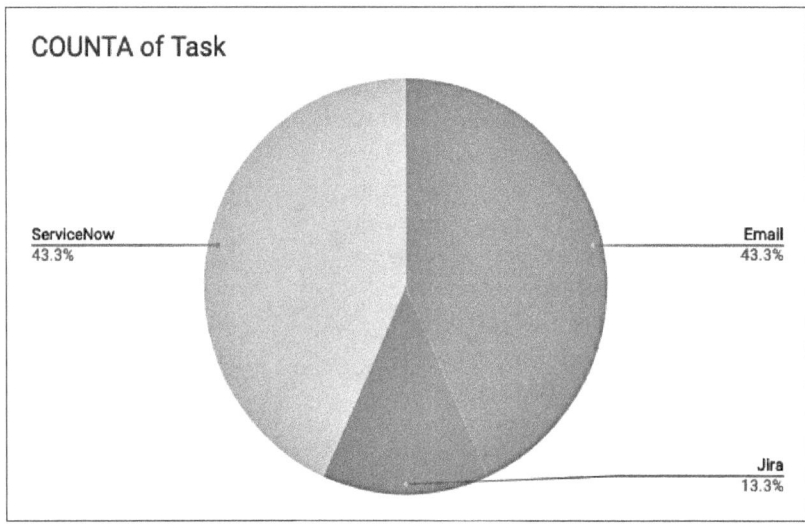

Figure 5.3 – Weekly task source summary

Most of the ITSM tools can integrate with automation tools that will help us to automate the execution trigger from the tools themselves.

Ansible and ITSM

For example, when a user requests a package deployment in a Linux server via the ServiceNow portal, you can configure your Ansible Automation Platform to trigger the package installation job based on the input details from ServiceNow. At the end of the automation job, Ansible will return the results to the ServiceNow ticket and the user who created the ticket can see the details without waiting for the engineer to update the ticket manually.

Please note that you need either Ansible AWX or Red Hat **Ansible Automation Platform** (**AAP**) for this integration, and this is not possible with a simple Ansible control node alone.

Figure 5.4 demonstrates a simple integration with ITSM tools and Ansible AWX or Red Hat AAP. You will learn more about Ansible and ITSM tools integration in *Chapter 12, Integrating Ansible with Your Tools*.

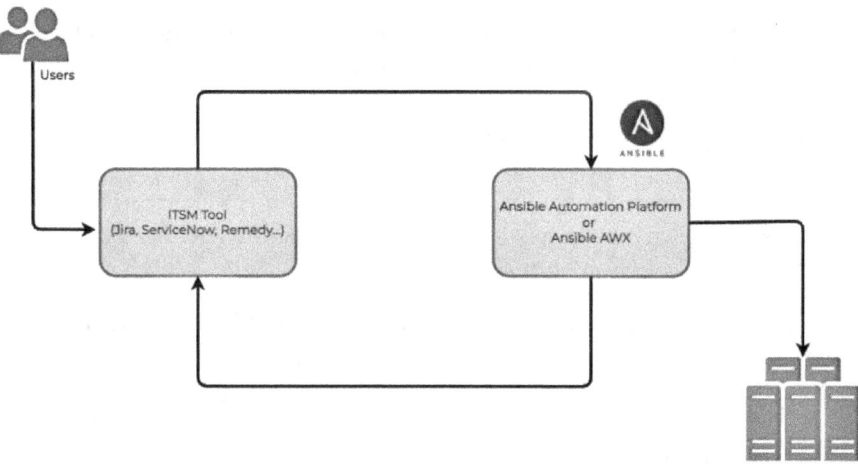

Figure 5.4 – Ansible integration with ITSM tools

ITSM Software

ITSM software will help organizations track and manage their IT infrastructure operations, resources, and other related usages: https://www.gartner.com/reviews/market/it-service-management-tools.

Automation and Information Technology Infrastructure Library compliance

When we talk about automated operations using ITSM and automation tools, it is normal for the stakeholders to raise concerns about **Information Technology Infrastructure Library** (ITIL) practices and compliances. IT automation does not mean that you need to bypass your organization's best practices or ITIL processes; instead, you are automating the tasks that a human needs to execute manually and repeatedly. For example, you will still follow the standard change approval process in the ITSM system, and the tool will trigger automation only when the request is approved to execute.

In this section, we have learned some simple methods to find automation use cases from your workspaces. You can break complex operations into small tasks and start with simple automation use cases. We will learn how to study those use cases and convert them into automation candidates in the next section.

Automation feasibility and usability

We all need to understand that not all use cases or repeated jobs are suitable for automation. We need to study and confirm the feasibility of implementing the use case using Ansible automation.

Once you find the use cases, you can ask yourself several questions to understand the feasibility of implementation and usability of your automation use cases.

How complex is the use case?

You need to Consider the complexity of the use case or workflow you are trying to automate. If the use case is too complex, then you can split it into smaller use cases. You need to start with smaller use cases to avoid any possible delay or obstruction.

For example, if you want to automate the Linux **operating system** (OS) patching task, then split the job into multiple use cases as follows:

1. Take a virtual machine snapshot.
2. Back up the configuration.
3. OS patching tasks.
4. Verification of the OS after patching and reboot.
5. Restore snapshot in case of any failure.

By doing this, you will get more confidence to start the automation journey and also, different people on your team can contribute to their own domain of expertise.

Can I reduce human error?

Consider how many human mistakes we can avoid by implementing automation for a use case. If the task contains multiple steps, there is a high chance that the engineer will miss some steps or execute typos while implementing the job. Such jobs are good candidates for automation, as you will not need to worry about errors and typos when using automated workflows.

For example, imagine you have a task to deploy a database cluster with hundreds of steps, a sample of which is as follows:

1. Configure the IP address for the cluster nodes.
2. Install multiple packages on the cluster nodes.
3. Configure multiple files on the nodes.
4. Configure clustering and heartbeats.
5. Configure the virtual IP and virtual interface, and so on.

In such cases, you can develop Ansible playbooks to cover each and every task and save enormous amounts of time and effort.

Can I reduce the deployment time and speed up my tasks?

It is a proven fact that automation can improve the speed of your tasks, deployment, and deliveries. Automation can significantly reduce the time required for an operation, as a single trigger can complete the entire workflow and the engineer doesn't need to observe the monitor or console for errors and status updates.

So, you should ask yourself about the use case and whether you can save time or not by automating this task.

For example, if you are automating the OS patching task using Ansible, you can trigger the patching for hundreds of servers with a single command and just wait for the tasks to be completed. You do not need to log in to different servers, switch between consoles, collect logs, or keep the time of events. You can include each and every task in Ansible playbooks and collect the summary or report for your later auditing purposes.

How frequently am I doing this task?

If the tasks are not executed frequently (for example, once a year), then it will not be much use to spend time developing the automation content for that task. But, if the task is to be executed for several servers, then that automation use case is valid and a good candidate, as during the execution time, you will need fewer resources for that job as the automation will take care of the complex workflows.

Always compare the time required for the manual task with the time required for the development of Ansible playbook content. You also have to bear in mind that the task might currently be executing less frequently due to resource unavailability or task complexity. By automating such tasks, you can increase the frequency of the tasks as you need less effort to execute them using your Ansible playbooks.

For example, assume you have thousands of servers to manage and the patching is possible only once every 6 months due to having a small team, and we all know that engineers need to spend a few hours to complete patching for a single server. Due to the complexity and criticality of OS patching tasks, you can do it only during non-business hours (usually weekends) and you need to rotate your engineers on weekend activities to accomplish this task. If you can automate this task using Ansible, then the engineers need to spend much less time and you can do the OS patching almost every month or whenever necessary.

How much time can I save by automating this task?

Consider whether the automation can help to save some time and help the engineers to focus on improving their workflows. For some tasks, engineers need to sit in front of their workstations until the task reaches a particular stage and they need to interactively complete the task regardless of their work hours.

For example, for a weekly system reboot job, you need to wait for the system to be up and running before focusing on the next machine in the workflow. Ansible can help you to automate this reboot, validation, and system restore process in case of an emergency. You can schedule the jobs in a parallel or serial workflow without waiting for one machine to complete the reboot activity.

Can I save some money?

It is not a major reason from the technical point of view, but reducing cost is one of the most common reasons organizations are looking for automation of their IT and application infrastructure. You can assess the cost savings together with the time saving, as the engineers will be spending more time on improvements and better practices rather than working on the same repeated jobs. Also, the reduction of human errors will help to reduce service outages, which, in turn, reduces the cost to the organization.

You can add more facts and questions in the use case selection criteria and assessment as per your organization's requirements.

In the next section, we will learn how to organize and store managed node information in the Ansible inventory, and different best practices to follow.

Involving teams in the automation journey

It is a common misunderstanding that the responsibility for finding use cases and implementing automation only falls to the systems team, platforms team, or infrastructure team. When we explore our work environment and day-to-day tasks, we will find thousands of tasks that we can automate using Ansible. It could be the database team managing database servers and instances, the network team handling network operations, or the application team who wants to deploy their application updates more effectively. Implementing automation in the environment is a collaborative journey, and we need support and guidance from different teams.

For example, typical database provisioning steps can be seen in *Figure 5.5*:

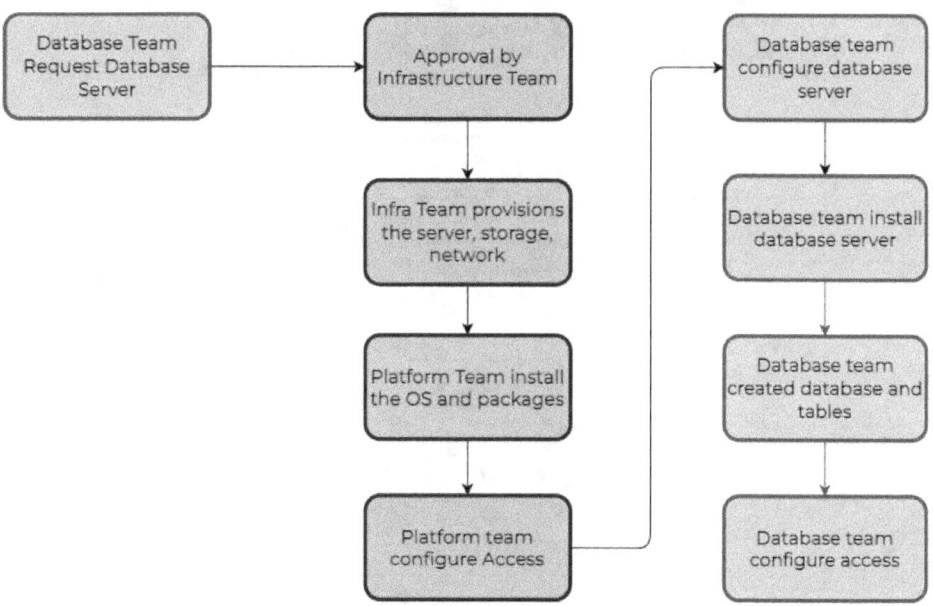

Figure 5.5 – Typical database provisioning tasks

Many of the tasks listed in *Figure 5.5* can be automated using Ansible, and the workflow can be completed in minutes instead of days and weeks. The database team needs to share more insight about the database operations and the automation opportunities, as they are the **subject matter experts (SMEs)** for database-related topics.

We have detailed sections on learning about database automation using Ansible, which you can find in *Chapter 8, Helping the Database Team with Automation*.

We can take Windows server automation as another example, as there is always a misconception that Ansible is only available for Linux and cannot be used for automation on Windows servers. This is incorrect as Ansible can be used to automate most of your Windows management and administration operations. There are about 100 Ansible modules available in the Ansible Windows collection, which can be downloaded from Ansible Galaxy as follows:

- `community.windows` – `https://galaxy.ansible.com/community/windows`

- `Ansible.windows` – `https://galaxy.ansible.com/ansible/windows`

Refer to *Figure 5.6* for a typical user creation job in Windows:

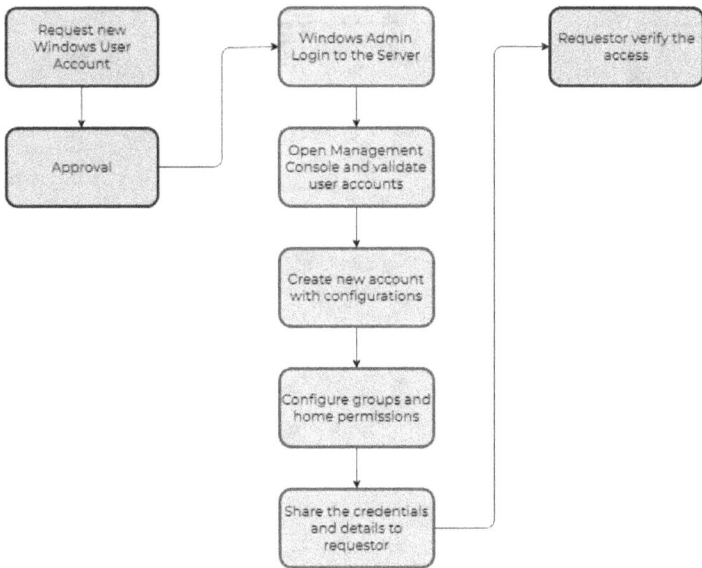

Figure 5.6 – Windows user creation workflow

The entire workflow can be automated using a few tasks inside an Ansible playbook and the Windows team can reuse the same automation artifacts for thousands of servers in the environment. As an Ansible content developer, you can collaborate with the Windows platform team and find more such use cases for implementing automation. You will learn more about Windows automation in *Chapter 6, Automating Microsoft Windows and Network Devices*.

Let's explore one more use case opportunity and challenge that the cloud platform team faces. When you introduce Ansible automation to the cloud platform team that manages the public or private cloud, they are always faced with a dilemma – how to update the managed node information in the Ansible inventory every time when there are frequent changes in terms of virtual machines, disks, network, and so on. In the next section, we will learn how to handle a large number of dynamically managed nodes in the Ansible inventory using the dynamic inventory plugins.

Ansible dynamic inventory

It is easy to manage your managed node information inside static inventory files when you have a smaller number of nodes or an almost fixed set of assets, such as bare-metal servers or virtual machines that are not frequently recreated. But, if your environment contains many dynamic nodes, such as virtual machines on multiple public or private cloud platforms, Kubernetes, or OpenShift platforms, then keeping your managed node information inside static files will be difficult, as you need to keep track of the changes and update your inventory files with them, including IP addresses, login credentials, and more. In such cases, you can use the **dynamic inventory** features in Ansible, which are basically some custom scripts and inventory plugins that collect inventory information from these virtualization or container platforms.

When you pass the dynamic inventories to Ansible, the inventory plugins will be executed and will collect the details of managed nodes from your virtualization platforms. This information will be passed to Ansible as regular inventory data and Ansible will execute the automation tasks for those managed nodes based on the node selection.

You can use dynamic inventory plugins for any supported platforms, such as VMware, OpenStack, AWS, Azure, GCP, or from other container platforms such as Kubernetes, OpenShift, and so on, as depicted in *Figure 5.7*:

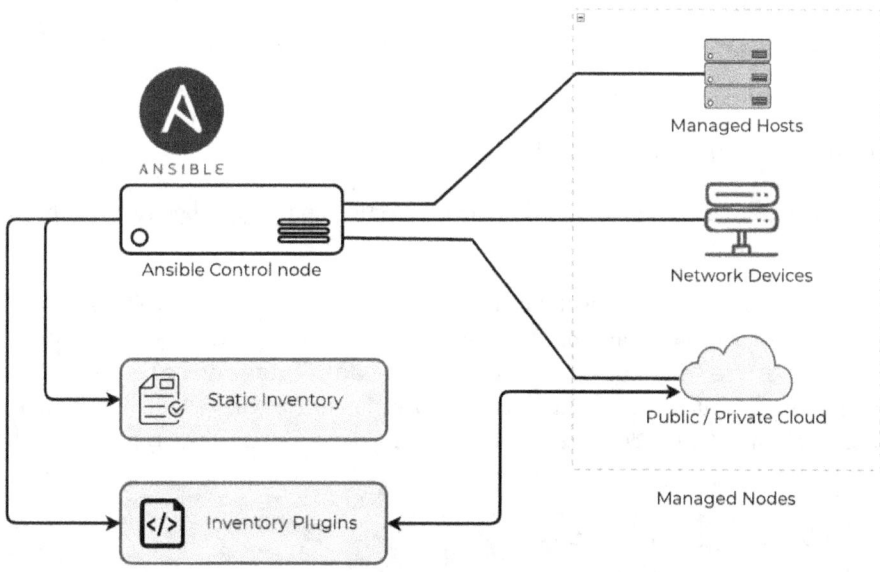

Figure 5.7 – Ansible static and dynamic inventories

You can list all the available dynamic plugins as follows:

```
[ansible@ansible inventories]$ ansible-doc -t inventory -l
```

Please note that you will see additional inventory plugins if you have installed additional Ansible collections, as some of the collections may include inventory plugins as well.

Ansible Inventory Plugins

For more details about Ansible Inventory plugins, you can look them up at `https://docs.ansible.com/ansible/latest/plugins/inventory.html`. Old dynamic inventory scripts are available at `https://docs.ansible.com/ansible/latest/user_guide/intro_dynamic_inventory.html` and `https://github.com/ansible/ansible/tree/stable-2.9/contrib/inventory`.

Using the Ansible dynamic inventory with AWS

In this practice session, you will learn how to install Ansible collections and how to use the Ansible inventory plugin for AWS Cloud. You will use the default AWS inventory plugin available at the official `amazon.aws` collection.

Assumptions

We assume the following:

- You have an AWS account (the Free Tier is enough to proceed with this exercise).

- You have basic knowledge of the AWS platform sufficient to create new users and EC2 instances.

Installing AWS collection and libraries

First of all, we need to install the required Ansible collection and plugins before using the inventory plugin:

1. Configure `ansible.cfg` with a collection path. By default, `ansible-galaxy` will install the Ansible collections (and roles) to a default path, which is under your home directory (for example, `/home/ansible/.ansible/collections/ansible_collections`). In this case, we will tell Ansible to install the collection to a specific path for better management. Configure the COLLECTIONS_PATHS line in your `ansible.cfg` file:

   ```
   [defaults]
   inventory = ./hosts
   remote_user = devops
   ask_pass = false
   COLLECTIONS_PATHS = ./collections
   ```

2. Install the `amazon.aws` collection using the `ansible-galaxy` command:

```
[ansible@ansible Chapter-05]$ ansible-galaxy collection
install amazon.aws
```

3. Verify the installed collection:

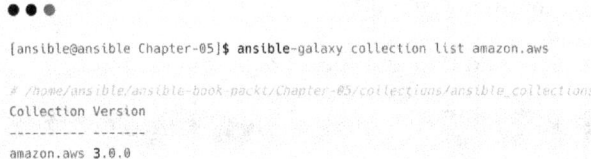

```
[ansible@ansible Chapter-05]$ ansible-galaxy collection list amazon.aws

# /home/ansible/ansible-book-packt/Chapter-05/collections/ansible_collections
Collection Version
---------- -------
amazon.aws 3.0.0
```

Figure 5.8 – Ansible AWS collection installed

You can see that the collection is installed in your `PROJECT_DIRECTORY/collections` path.

4. Verify the installed inventory plugin for AWS:

```
[ansible@ansible Chapter-05]$ ansible-doc -t inventory -l |grep aws
amazon.aws.aws_ec2                          EC2 inventory sourc...
amazon.aws.aws_rds                          rds instance source
```

Figure 5.9 – AWS inventory plugin

5. Install the `python3-boto3` package. AWS modules and plugins require the `boto3` package to be available on the system:

```
[ansible@ansible Chapter-05]$ sudo dnf install python3-
boto3
```

If your Ansible was installed using Python, then install `boto3` using the `pip install boto3` command.

After installing the Ansible AWS collection, you need to create the AWS user with which Ansible will access the AWS platform.

If you are using the automation execution environments with Ansible Automation Platform, then all the dependencies and libraries can be packaged inside the execution environment images. Refer to `https://www.ansible.com/products/execution-environments` for more details.

Creating an AWS user and credential

For the AWS `ec2` inventory plugin to access your AWS account, you need to configure your AWS credentials using standard methods, as shown in *Figure 5.10*:

1. Go to **AWS Console | IAM | Users | Add User** and select **Programmatic access**:

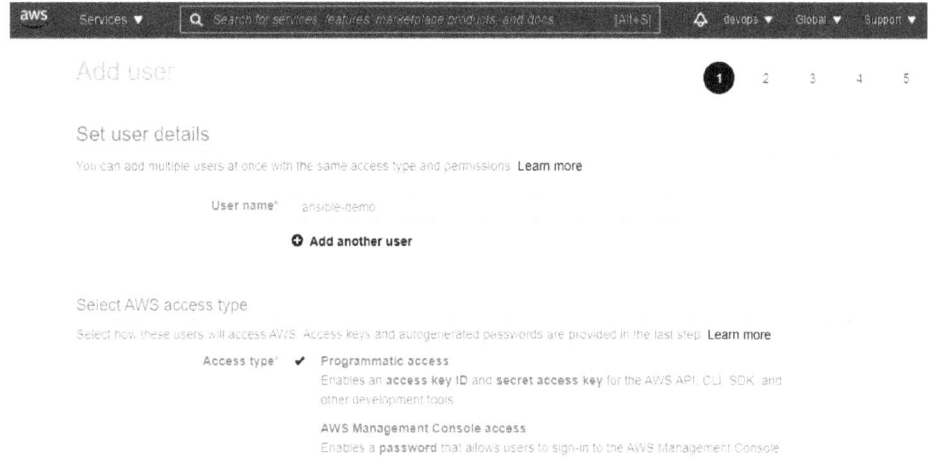

Figure 5.10 – Creating an AWS user with programmatic access

2. Add appropriate permissions for the new user (do not give the user **AdministratorAccess** to your production account), as shown in *Figure 5.11*:

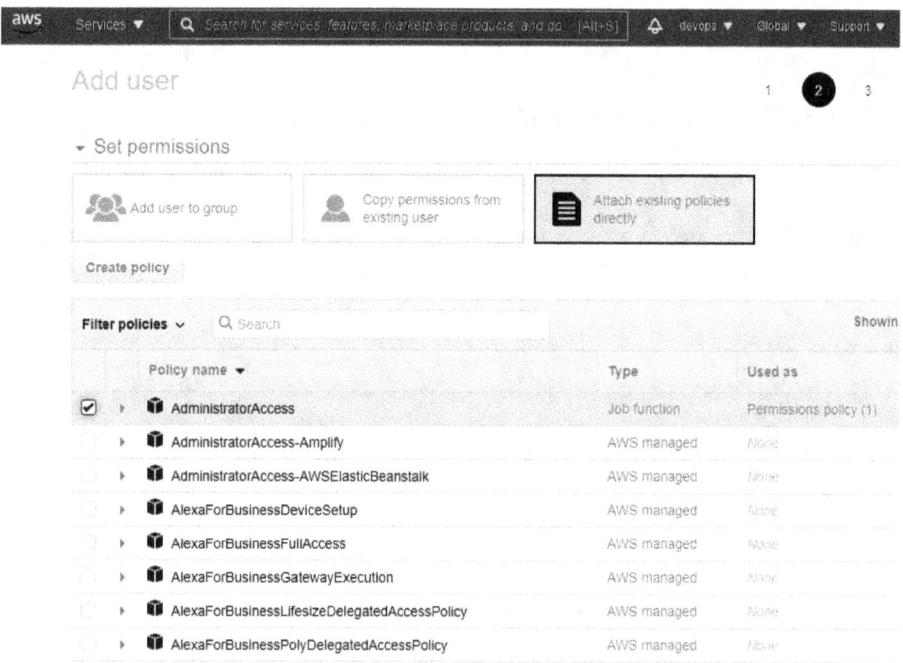

Figure 5.11 – Applying permissions for the new AWS user

3. Add **Tags** if needed and click **Create User**. Please remember to copy the **Access key ID** and **Secret access key** as shown in *Figure 5.12* as we need this information in the next steps.

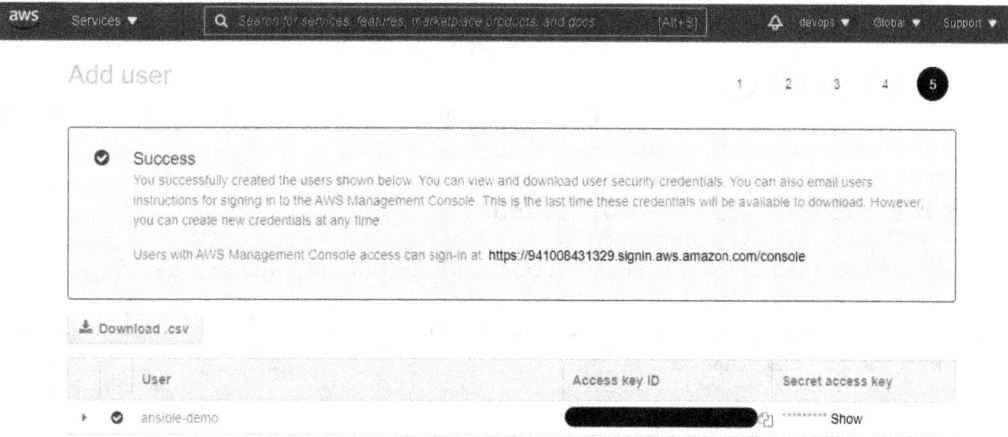

Figure 5.12 – AWS user access key ID and secret access key

Once you create the user account, you need to configure the access key and secret access key on your Ansible control node.

Configuring AWS credentials on the Ansible control node

Now, you need to configure the AWS credential information on the control node:

1. Add new AWS credentials on your Ansible control node machine. If you have multiple credentials, then add this as a new profile as follows (for example, `default` and `ansible` profiles):

```
[ansible@ansible Chapter-05]$ cat ~/.aws/credentials
[default]
aws_access_key_id=EXAMPLEKEY
aws_secret_access_key=EXAMPLEACCESSKEY
[ansible]
aws_access_key_id=EXAMPLEKEY
aws_secret_access_key=EXAMPLEACCESSKEY
```

Figure 5.13 – AWS profile configuration

2. Update the `config` file as shown:

```
[ansible@ansible Chapter-05]$ cat ~/.aws/config
[default]
region=ap-southeast-1 output=json
```

Figure 5.14 – AWS profile config file

Remember to use the correct AWS profile name in your inventory configuration in the next steps (`ansible`, in our case).

Using the AWS aws_ec2 inventory plugin

Once you configure the AWS credentials, you can start testing the dynamic inventory plugin:

1. Create an inventory file with the `ec2` plugin information. We will also include the location to filter for the `ec2` instances:

```
● ● ●

[ansible@ansible Chapter-05]$ mkdir inventories/aws

[ansible@ansible Chapter-05]$ cd inventories/aws/
[ansible@ansible aws]$ cat lab.aws_ec2.yml
# lab.aws_ec2.yml
plugin: amazon.aws.aws_ec2
boto_profile: ansible
regions:
  - ap-southeast-1
```

Figure 5.15 – Inventory file for AWS ec2 instances

2. Verify the dynamic inventory plugin by passing the inventory location (`inventories/aws/`):

```
● ● ●

[ansible@ansible Chapter-05]$ ansible-inventory -i inventories/aws/ --graph
@all:
  |--@aws_ec2:
  |  |--ec2-13-250-108-199.ap-southeast-1.compute.amazonaws.com
  |  |--ec2-13-250-48-91.ap-southeast-1.compute.amazonaws.com
  |  |--ec2-54-179-175-153.ap-southeast-1.compute.amazonaws.com
  |--@ungrouped:
```

Figure 5.16 – Verify AWS dynamic inventory

The inventory plugin will access the AWS platform using the account you have configured in the previous steps and return the result in a format that Ansible can read and use for execution. (We have used the `--graph` option for a summary view, but you can use other options, such as `--list`.)

3. You can add more filters with tags and hostnames as shown here:

```
filters:
    tag:Environment: dev
    tag:Criticality:
        - low
```

4. The plugin will return the result based on the filter you have configured in the inventory YAML file:

```
● ● ●
[ansible@ansible Chapter-05]$ ansible-inventory -i inventories/aws/ --graph
@all:
  |--@aws_ec2:
  |  |--ec2-54-179-175-153.ap-southeast-1.compute.amazonaws.com
  |--@ungrouped:
```

Figure 5.17 – AWS dynamic inventory with additional filters

5. If you have configured the **SSH (Secure Shell)** key access for these ec2 instances (using the AWS ec2 template or other methods), then you can test the access now:

```
● ● ●
[ansible@ansible Chapter-05]$ ansible all -m ping -i inventories/aws/
ec2-13-250-108-199.ap-southeast-1.compute.amazonaws.com | SUCCESS => {
    "ansible_facts": {
        "discovered_interpreter_python": "/usr/bin/python"
    },
    "changed": false,
    "ping": "pong"
}
```

Figure 5.18 – Ansible ping test using the AWS dynamic inventory

> **Secure Shell (SSH)**
>
> SSH is a cryptographic network protocol that helps users to access the target systems in a secure way. Read https://www.techtarget.com/searchsecurity/definition/Secure-Shell to learn more about SSH.

In the preceding exercise, you have learned how to use Ansible dynamic inventory plugins and use the inventory without adding your managed node information inside any static files. The procedure is more or less the same for all other cloud and container platforms, as you can generally find and use suitable inventory plugins to implement the dynamic inventory for them.

> **Ansible Inventory Plugin Options**
>
> Check the official documentation for the inventory plugin's additional options and filters: https://docs.ansible.com/ansible/latest/plugins/inventory.html and https://docs.ansible.com/ansible/latest/collections/amazon/aws/aws_ec2_inventory.html and https://docs.ansible.com/ansible/latest/user_guide/playbooks_variables.html#understanding-variable-precedence.

Summary

In this chapter, you have learned methods to find automation use cases in your workplace and determine the feasibility and usability of these automation use cases. You have also explored the importance of collaboration between teams to implement better automation use cases. Later in the chapter, you learned how to use the Ansible dynamic inventory and practiced using the Ansible inventory plugin with the AWS ec2 inventory plugin.

In the next chapter, we will learn the basics of Windows automation and the remote connection methods available. We will also learn the basics of network automation using Ansible with practice sections.

Further reading

To learn more about the topics covered in this chapter, please visit the following links:

- *Ansible ServiceNow collection*: `https://github.com/ansible-collections/servicenow.itsm`

- *Automating ServiceNow with Red Hat Ansible Automation Platform*: `https://www.ansible.com/blog/certified-collection-servicenow`

- *Ansible Use Cases*: `https://www.ansible.com/use-cases`

- *How to build your inventory*: `https://docs.ansible.com/ansible/latest/user_guide/intro_inventory.html`

- *Inventory tips*: `https://docs.ansible.com/ansible/latest/user_guide/playbooks_best_practices.html#inventory-tips`

- *Working with dynamic inventory*: `https://docs.ansible.com/ansible/latest/user_guide/intro_dynamic_inventory.html`

- *Ansible AWS collection*: `https://docs.ansible.com/ansible/latest/collections/amazon/aws/index.html`

- *Ansible and Windows*: `https://www.ansible.com/for/windows`

- *Ansible Windows Guides*: `https://docs.ansible.com/ansible/latest/user_guide/windows.html`

6

Automating Microsoft Windows and Network Devices

Due to the complexity and wide variety of technologies, there are no one-size-fits-all tools in the information technology space. This is common for automation software as well but fortunately, Ansible can be used for most of your IT automation use cases because of the large community support and contributions from the vendors who provide these services, such as cloud platforms, network appliances, and software platforms.

When we talk about basic system automation, we know how easy it is to automate Linux machines using Ansible. However, we can do the same for Microsoft Windows machines as well. There are community collections and certified Content Collection for managing Microsoft Windows operations, such as user management, firewall, system management, package management, and registry configurations.

Similarly, we have thousands of modules available via different collections for managing network devices such as Cisco, FortiGate, Palo Alto, VyOS, F5, and CheckPoint. To become familiar with network automation, we will discuss the network connection methods and configurations for VyOS and Cisco ASA devices.

In this chapter, we will cover the following topics:

- Ansible remote connection methods
- Automating Microsoft Windows servers using Ansible
- Introduction to network automation
- VyOS information gathering using Ansible
- Creating ACL entries in a Cisco ASA device

First, you will learn about the different connection methods available in Ansible. Then, you will learn how to configure and automate Microsoft Windows tasks using Ansible. Finally, you will learn about how to use Ansible for network automation.

Technical requirements

The following are the technical requirements for this chapter:

- A Linux machine for the Ansible control node

- One or more Linux machines as managed nodes with Red Hat repositories configured (if you are using non-RHEL machines, then make sure you have the appropriate repositories configured to get packages and updates)

- One or more Microsoft Windows machines (we used a Windows 2019 server)

- One or more network devices/virtual appliances (for practicing this chapter's network automation use case)

- Basic administrative knowledge of Microsoft Windows machines, including user creation and package management

- Basic administrative knowledge of network devices, including IP configuration and access configuration

All the Ansible code, playbooks, commands, and snippets for this chapter can be found in this book's GitHub repository at `https://github.com/PacktPublishing/Ansible-for-Real-life-Automation/tree/main/Chapter-06`.

Ansible remote connection methods

By default, Ansible communicates with the remote machine using the SSH protocol (native OpenSSH), as you learned previously in this book. For remote nodes, which do not have SSH server options, it is possible to use other connection methods such as WinRM for Microsoft Windows remote machines or **httpapi** for API-based remote devices (such as Cisco NXAPI and Arista eAPI).

The following diagram shows the different connection methods used by Ansible for automating different devices and platforms:

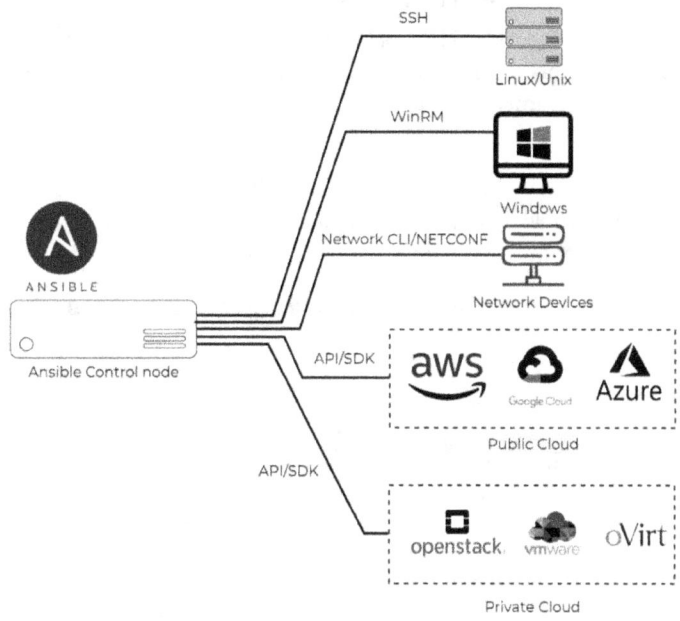

Figure 6.1 – Connection methods used by Ansible

You can find the available Ansible **connection plugins** by using the `ansible-doc` command, as follows:

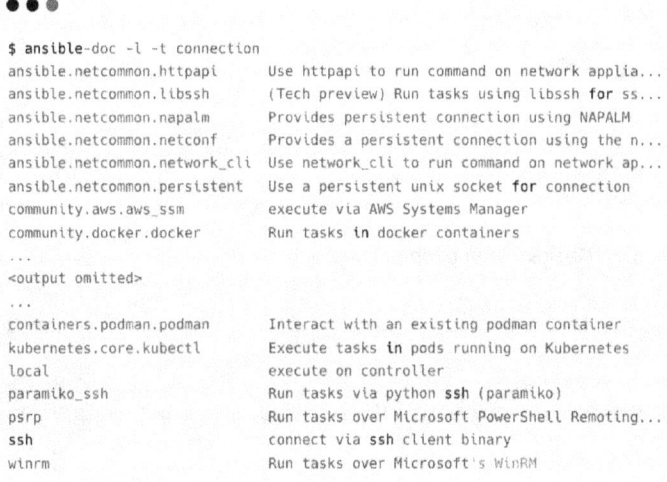

Figure 6.2 – Ansible connection plugins

> **Ansible Inventory and Connection Parameters**
>
> Refer to https://docs.ansible.com/ansible/latest/user_guide/intro_
> inventory.html#connecting-to-hosts-behavioral-inventory-parameters
> for specific connection parameters for connections such as SSH, Docker, and more.

The number of connection plugins on your Ansible control node will depend on the Ansible collections you are using as by default, Ansible only has a few connection options such as ssh, winrm, local, and so on. The remaining connection plugins come with the Ansible collections you have installed.

It is also possible to see details about the **connection plugins** by using the ansible-doc command, as follows:

```
[ansible@ansible Chapter-06]$ ansible-doc -s -t connection community.docker.docker
> COMMUNITY.DOCKER.DOCKER    (/home/ansible/ansible-book-packt/Chapter-06/collections/ansible_collectio>

        Run commands or put/fetch files to an existing docker container. Uses the
        Docker CLI to execute commands in the container. If you prefer to directly
        connect to the Docker daemon, use the community.docker.docker_api
        connection plugin.

OPTIONS (= is mandatory):

- container_timeout
        Controls how long we can wait to access reading output from the container
        once execution started.
        [Default: 10]
        set_via:
          env:
          - name: ANSIBLE_TIMEOUT
          - name: ANSIBLE_DOCKER_TIMEOUT
            version_added_collection: community.docker
          ini:
          - key: timeout
            section: defaults
          - key: timeout
        ;
```

Figure 6.3 – Docker connection plugin details

In the preceding output, we can see details about the community.docker.docker connection plugin, including its usage.

You will learn more about Ansible connection variables and the available options in the next section.

Ansible connection variables

It is possible to control the remote connection details using the Ansible inventory parameters and other variables. Refer to the documentation at `https://docs.ansible.com/ansible/latest/reference_appendices/special_variables.html` to learn more about Ansible special variables. The following screenshot shows the inventory variables section. Here, different remote connection details are mentioned, such as `ansible_connection`, `ansible_port`, and `ansible_user`:

```
---
# inventory variables
ansible_connection: "winrm"
ansible_user: "ansible"
ansible_password: "MySecretWindowsPassword"
ansible_port: "5985"
ansible_winrm_transport: "basic"
ansible_winrm_server_cert_validation: ignore
```

Figure 6.4 – Ansible inventory variable with special variables

It is possible to configure different values and variables, as follows:

- `ansible_connection`: This specifies the connection type to use, such as `ssh`, `local` (for `localhost` nodes), `winrm`, or `docker`:

  ```
  # inventory
  [local]
  localhost ansible_connection=local
  [web]
  node1
  [windows]
  Win2019 ansible_connection: "winrm"
  ```

- `ansible_host`: The actual name or IP address of the remote node if it is different from the inventory name.

- `ansible_user`: The user account to be used for remote node authentication.

- `ansible_password`: The password for `ansible_user` to authenticate. Note that keeping `ansible_password` in plain text is not a best practice; you should consider keeping it encrypted using Ansible Vault (refer to *Chapter 3*, *Automating Your Daily Jobs*, the *Encrypting sensitive data using Ansible Vault* section and *Chapter 13*, *Using Ansible for Secret Management*) or following authentication based on SSH keys (refer to *Chapter 1's*, *Configuring Your Managed Nodes section*).

- `ansible_port`: If the remote connection port is something other than `22` (the default SSH port), then specify the port number to use for the remote connection.

In the next section, we will learn about the SSH connection parameters and how to configure it for managed nodes.

SSH connection parameters

Additionally, there are SSH-specific variables such as `ansible_ssh_private_key_file` and `ansible_ssh_common_args` for assigning different SSH keys for different managed nodes if needed:

```
ansible_ssh_private_key_file=/home/ansible/.ssh/id_rsa
ansible_ssh_common_args='-o StrictHostKeyChecking=no'
```

Figure 6.5 – Ansible SSH-specific variables

With that, you have learned about the Ansible connection methods and connection parameters that are available for controlling the connection. Now, let's learn how to automate Microsoft Windows servers using Ansible.

Automating Microsoft Windows servers using Ansible

As I mentioned earlier, Ansible is only available for Linux/Unix platforms, but that doesn't mean you can't use Ansible to automate Microsoft Windows machines. It is possible to use Ansible on a Linux/Unix machine (the Ansible control node) and automate your Microsoft Windows machines like so:

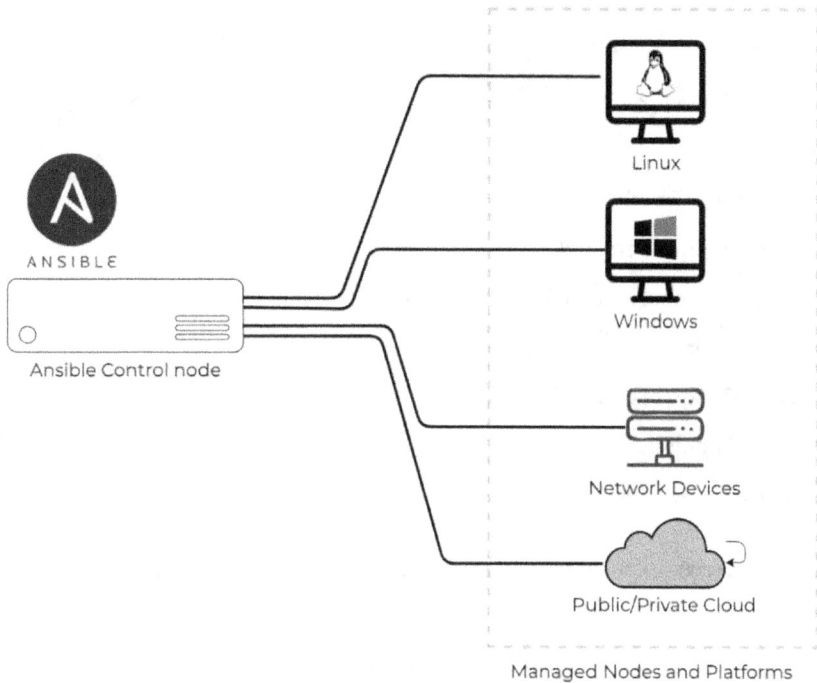

Figure 6.6 – Ansible and managed nodes

Multiple Ansible **Content Collections** can be used for Microsoft Windows automation. Altogether, there are more than 100 Ansible modules available for automating Microsoft Windows tasks:

- Ansible modules for Microsoft Windows from the community (`https://galaxy.ansible.com/community/windows`) contains 84 modules.

- The Ansible Windows module from Red Hat (`https://galaxy.ansible.com/ansible/windows`) contains 40 modules.

In the next few sections, you will learn more about Ansible Windows automation, such as the supported Microsoft Windows versions, prerequisites, credential configuration, and inventory configurations.

> **Ansible Windows Guides**
>
> The official Ansible documentation portal contains details on how to configure and set up Microsoft Windows machines to automate using Ansible. Refer to `https://docs.ansible.com/ansible/latest/user_guide/windows.html` and `https://www.techbeatly.com/ansible-windows` for more details.

Supported Microsoft Windows operating systems

Ansible can manage most general-purpose Microsoft Windows operating system versions, as follows:

- Microsoft Windows 7

- Microsoft Windows 8.1

- Microsoft Windows 10

- Microsoft Windows Server 2008

- Microsoft Windows Server2008 R2

- Microsoft Windows Server 2012

- Microsoft Windows Server 2012 R2

- Microsoft Windows Server 2016

- Microsoft Windows Server 2019

Microsoft Windows automation – Ansible control node prerequisites

There is no special requirement for an Ansible control node other than installing the Python `pywinrm` library, which can be installed as follows:

1. If you are using a Python virtual environment, then remember to activate your virtual environment (skip this step otherwise):

    ```
    $ source ~/python-venv/ansible210/bin/activate
    ```

2. Install the `pywinrm` library:

    ```
    $ python3 -m pip install --user --ignore-installed
    pywinrm
    ```

Make sure that you have installed the `pywinrm` library on the exact Python environment Ansible is using (check `ansible --version` and see which Python version it is using).

Microsoft Windows automation – managed node prerequisites

The Microsoft Windows machine should be installed and configured with the following items:

- PowerShell 3.0 or newer (some of the Ansible modules for Microsoft Windows may require newer versions of PowerShell; refer to the module documentation you use).

- .NET 4.0 or newer.

- A **WinRM** listener should be created and activated – Ansible uses WinRM to connect to the Microsoft Windows machines by default. Microsoft Windows Remote Management is a SOAP-based remote management protocol that communicates over HTTP or HTTPS.

- A Microsoft Windows firewall should be configured to allow traffic on 5985 (HTTP) and/or 5986 (HTTPS). If there is additional firewall or network traffic control between the Ansible control node and Microsoft Windows machines, then make sure the ports are allowed there too.

Configuring the user account and WinRM on a Microsoft Windows machine

In this exercise, you will configure a user and WinRM on the Microsoft Windows machine:

1. Log into the Microsoft Windows machine and create a new user called ansible. Use any method to create this user:

> **Note**
>
> It is possible to use the default **Administrator** account, but it is not a best practice to use the default Administrator account in production or other environments due to security reasons; this is because the Administrator account has full access to the system. Create a standard user account and provide only the required permissions as needed for the automation tasks. In this exercise, a standard user account is being used with Administrator group membership to demonstrate automation without issues. Refer to the documentation at https://docs.microsoft.com/en-us/windows-server-essentials/manage/manage-user-accounts-in-windows-server-essentials to learn more about how to manage user accounts in Windows Server.

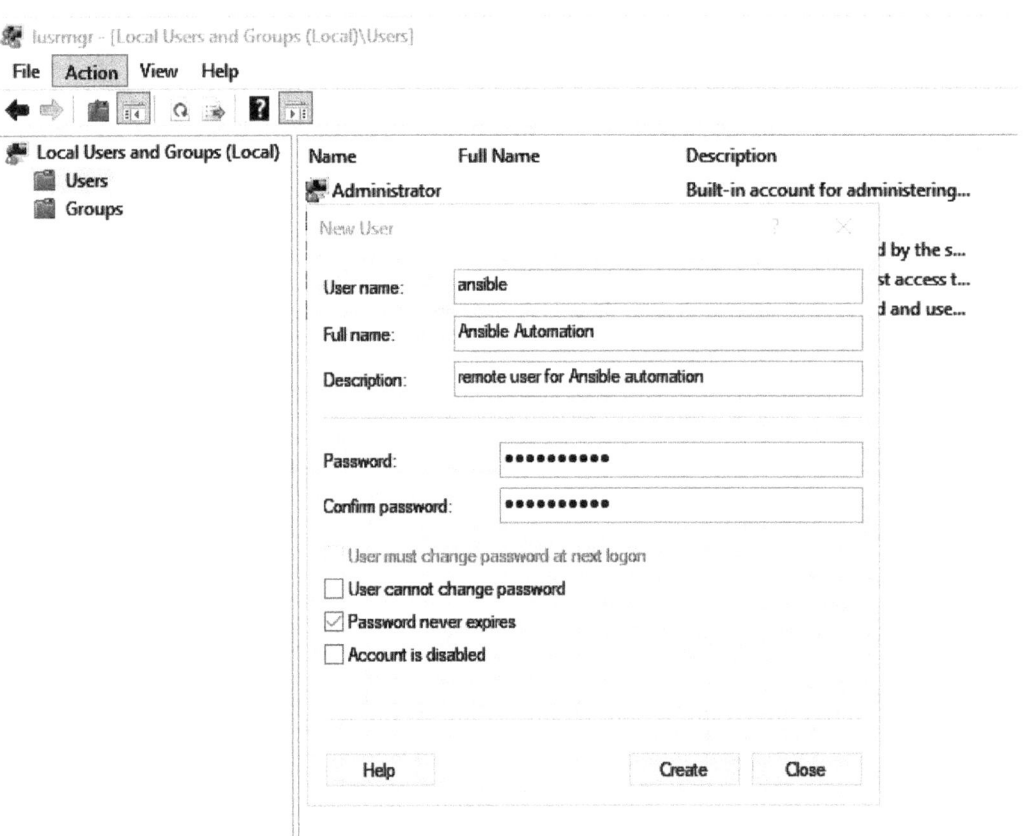

Figure 6.7 – Creating a new user account on the Microsoft Windows server

2. Like `sudo` access in Linux, you need administrator rights for the Microsoft Windows user that you are using to connect from Ansible to the Microsoft Windows machine. Add the new `ansible` user to the **Administrators** group, as shown here:

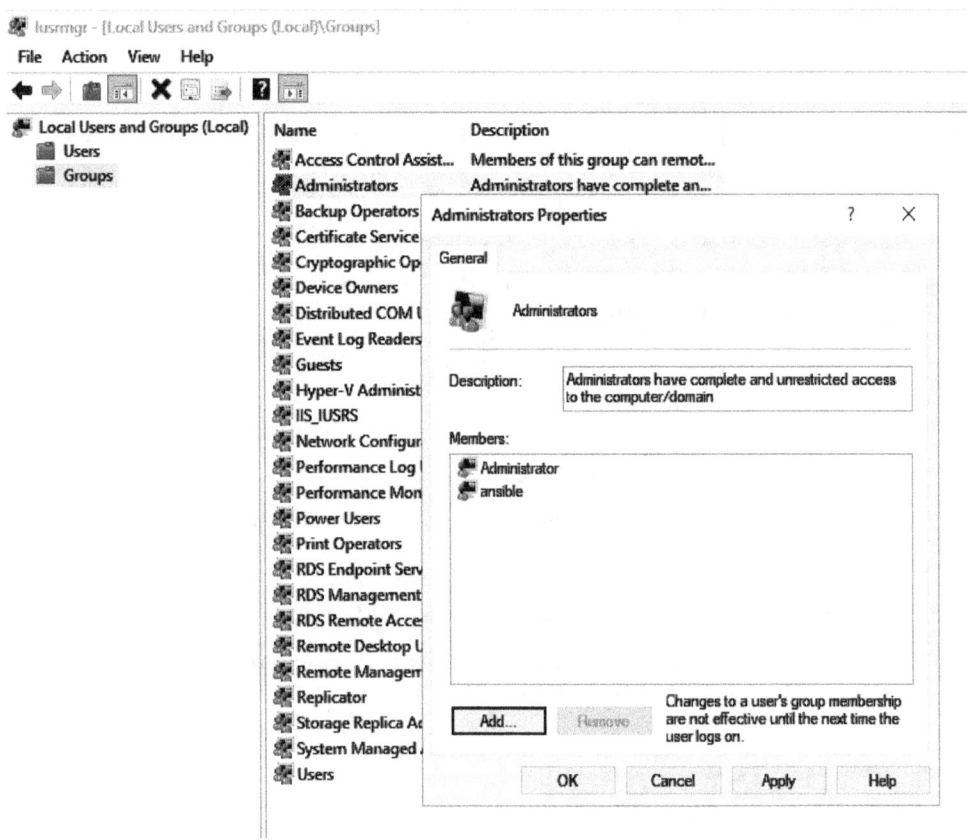

Figure 6.8 – Adding the Ansible user to the Administrators group

3. Verify the PowerShell version using the (Get-Host).Version command:

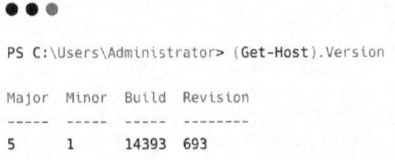

```
PS C:\Users\Administrator> (Get-Host).Version

Major  Minor  Build  Revision
-----  -----  -----  --------
5      1      14393  693
```

Figure 6.9 – Checking the PowerShell version

4. The next step is to configure the WinRM listener and enable the WinRM service on the Microsoft Windows managed node. Instead of executing multiple commands on PowerShell, use the ready-to-use script available in the Ansible ConfigureRemotingForAnsible.ps1 repository (https://github.com/ansible/ansible/blob/devel/examples/scripts/ConfigureRemotingForAnsible.ps1).

Download the script and execute it on the Microsoft Windows machine, as follows:

● ● ●

```
PS C:\Users\Administrator> $url =
"https://raw.githubusercontent.com/ansible/ansible/devel/examples/scripts/ConfigureRemotingForAnsible.ps1"
PS C:\Users\Administrator> $file = "$env:temp\ConfigureRemotingForAnsible.ps1"
PS C:\Users\Administrator> (New-Object -TypeName System.Net.WebClient).DownloadFile($url, $file)
PS C:\Users\Administrator> powershell.exe -ExecutionPolicy ByPass -File $file
Self-signed SSL certificate generated; thumbprint: DD2BFCC45E7503BC9C05BA9174326B593614C733

wxf                 : http://schemas.xmlsoap.org/ws/2004/09/transfer
a                   : http://schemas.xmlsoap.org/ws/2004/08/addressing
w                   : http://schemas.dmtf.org/wbem/wsman/1/wsman.xsd
lang                : en-US
Address             : http://schemas.xmlsoap.org/ws/2004/08/addressing/role/anonymous
ReferenceParameters : ReferenceParameters

Ok.
```

Figure 6.10 – Configuring WinRM on the Microsoft Windows machine using a script

If your Microsoft Windows machine is in a disconnected environment, then download the script from some other machine and transfer it to the Microsoft Windows machine.

5. Verify the WinRM configuration using the `winrm e winrm/config/listener` command once the script has been executed successfully:

● ● ●

```
C:\Users\Administrator>winrm e winrm/config/listener
Listener
    Address = *
    Transport = HTTP
    Port = 5985
    Hostname
    Enabled = true
    URLPrefix = wsman
    CertificateThumbprint
    ListeningOn = 10.0.2.15, 127.0.0.1, 192.168.99.103, ::1, fe80::5efe:10.0.2.15%3, fe80::5efe:192.168.99.103%13,
fe80::785d:9659:c4d4:9b0f%16

Listener
    Address = *
    Transport = HTTPS
    Port = 5986
    Hostname = WIN-CCUQI8Q4RMH
    Enabled = true
    URLPrefix = wsman
    CertificateThumbprint = 64E69568BD75F3068BDCBF7ED819E4EA9ED1FDA3
    ListeningOn = 10.0.2.15, 127.0.0.1, 192.168.99.103, ::1, fe80::5efe:10.0.2.15%3, fe80::5efe:192.168.99.103%13,
fe80::785d:9659:c4d4:9b0f%16
```

Figure 6.11 – Verifying the WinRM configuration

6. Then verify the WinRM configuration using the `winrm get winrm/config` command.

7. Verify port access from the Ansible control node:

● ● ●

```
[ansible@ansible Chapter-06]$ nc -vz 192.168.56.22 5985
Connection to 192.168.56.22 5985 port [tcp/wsman] succeeded!

[ansible@ansible Chapter-06]$ nc -vz 192.168.56.22 5986
Connection to 192.168.56.22 5986 port [tcp/wsmans] succeeded!
```

Figure 6.12 – Verifying the Ansible to Microsoft Windows connection using WinRM

> **Configure Your Microsoft Windows Host to be Managed by Ansible**
>
> Refer to `https://www.techbeatly.com/configure-your-windows-host-to-manage-by-ansible/` or `https://docs.ansible.com/ansible/latest/user_guide/windows_winrm.html` for additional reading.

Now, the Microsoft Windows managed node is ready to connect using WinRM. Now, you need to configure these details on the Ansible side. In the next section, you will learn how to configure your Ansible inventory to connect to a Microsoft Windows managed node.

Configuring Ansible to access the Microsoft Windows machine

In this exercise, you will configure the Ansible control node with Microsoft Windows user and other access details:

1. Create a host entry for Microsoft Windows in the Ansible inventory (replace the IP address as needed for your machine):

    ```
    [windows]
    win2019 ansible_host=192.168.56.22
    ```

2. Create a directory for Ansible group variables:

    ```
    [ansible@ansible Chapter-06]$ mkdir group_vars
    ```

3. Add the `group_vars/windows` group variable file to configure `ansible_connection`, `ansible_port`, and user credentials. These are special variables; see the `ansible_` prefix for all of them (refer to the `https://docs.ansible.com/ansible/latest/reference_appendices/special_variables.html` documentation to learn more about special variables):

```
● ● ●
[ansible@ansible Chapter-06]$ cat group_vars/windows
---
ansible_user: "ansible"
ansible_password: "MySecretWindowsPassword"
ansible_port: "5985"
ansible_connection: "winrm"
ansible_winrm_transport: "basic"
ansible_winrm_server_cert_validation: ignore
```

Figure 6.13 – Ansible group variable for Windows

Note that it is not a best practice to use basic authentication using a username and password in critical environments. For production and critical environments, you are encouraged to use a password that's been encrypted using Ansible Vault or keep credentials in Ansible Automation Controller (or Ansible Tower). Also, you may need to create a different user rather than an Administrator. You should also consider using SSL certificates and other secure methods to connect Microsoft Windows machines from the Ansible control node.

> **Ansible Windows Management Using HTTPS and SSL**
>
> Consider using SSL certificates and other secure methods to connect Microsoft Windows machines from the Ansible control node (or Automation Controller). Refer to `https://www.techbeatly.com/ansible-windows-management-using-https-and-ssl` to learn more.

4. Verify the Ansible to Microsoft Windows machine connection:

```
● ● ●
[ansible@ansible Chapter-06]$ ansible windows -m win_ping
win2019 | SUCCESS => {
    "changed": false,
    "ping": "pong"
}
```

Figure 6.14 – Ansible to Microsoft Windows connection test using the win_ping module

Like the `ping` module (for Linux machines), the `win_ping` module will establish a connection to the target machine and display the `pong` message if the connection is successful.

Microsoft Windows automation – using Ansible to create a Windows user

In this exercise, you will create a new user in the Microsoft Windows machine using Ansible:

1. Create a new Ansible playbook called `Chapter-06/windows-create-user.yaml` and add the following details:

```
---
- name: "Create New user on Windows Machine"
  hosts: "{{ NODES }}"
  vars:
    windows_username: "john"
    windows_password: "MyP4ssw0rd"
  tasks:
    - name: Create a new User
      win_user:
        name: "{{ windows_username }}"
        password: "{{ windows_password }}"
        state: present
        groups:
          - Users
      when: ansible_os_family == 'Windows'
```

Figure 6.15 – Ansible playbook to create a user in Microsoft Windows

2. Execute the Ansible playbook:

```
[ansible@ansible Chapter-06]$ ansible-playbook windows-create-user.yaml -e "NODES=windows"
```

Figure 6.16 – Executing the Ansible playbook to create a user in Microsoft Windows

> **Inventory Nodes as Extra Variables**
>
> In the preceding playbook, we did not hardcode the `hosts` information. Instead, we passed the `windows` host group while executing the playbook.

3.   Verify the Microsoft Windows machine to see if the user has been created or not:

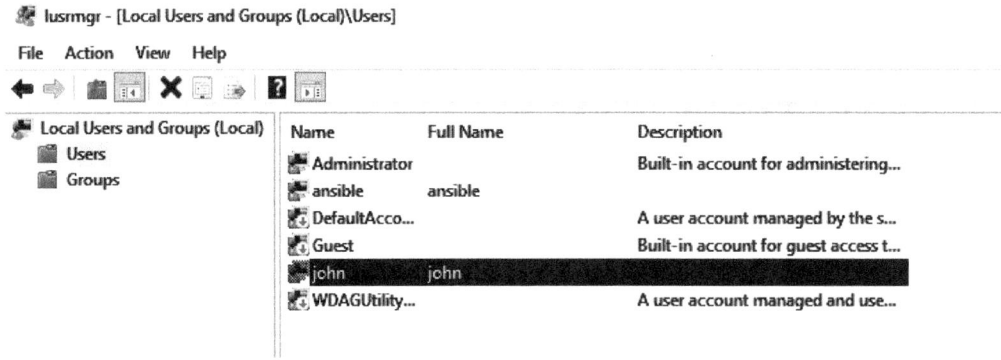

Figure 6.17 – New Windows user account created using Ansible

Find more Microsoft Windows automation use cases from your workplace, such as package deployment, group policy update, Active Directory operations, firewall management, service management, or even executing PowerShell scripts.

> **Ansible for Windows Automation**
>
> Refer to https://www.ansible.com/for/windows to learn more about Ansible Windows automation. Visit https://aap2.demoredhat.com/exercises/ansible_windows to find workshops and practice sessions for Microsoft Windows automation using Ansible.

In the next section, you will learn the basics of network automation using Ansible.

Introduction to network automation

Network automation using Ansible is based on different connection methods. There are some differences between Ansible network automation compared to Linux/Unix and Microsoft Windows automation. Also, note that Ansible can be used to automate the existing network automation tools such as Cisco ACI using the available Cisco ACI modules (https://docs.ansible.com/ansible/latest/scenario_guides/guide_aci.html).

Task execution on an Ansible control node

Previously, you learned that Ansible is built on top of Python, so a remote node must be installed with Python to execute the automation tasks (Microsoft Windows modules are written in PowerShell and a winrm connection must be set to use PowerShell modules). Unlike Linux/Microsoft Windows nodes, many network devices do not have Python and cannot run Python scripts. Hence, the network automation modules are processed and executed in the Ansible control node; all actual commands will be executed on the target network devices.

Different connection methods

Network task execution can support multiple communication methods, depending on the operating system of the network device and its version:

Value of ansible_connection	Protocol	Requires	Persistent?
ansible.netcommon.network_cli	CLI over SSH	network_os setting	yes
ansible.netcommon.netconf	XML over SSH	network_os setting	yes
ansible.netcommon.httpapi	API over HTTP/HTTPS	network_os setting	yes
local	depends on provider	provider setting	no

Figure 6.18 – Ansible network communication protocols (source: https://docs.ansible.com)

You need to specify the appropriate method to use for the network device connection, the privilege escalation method (become_method such as enable), and the operating system of the network device, as shown in the following screenshot:

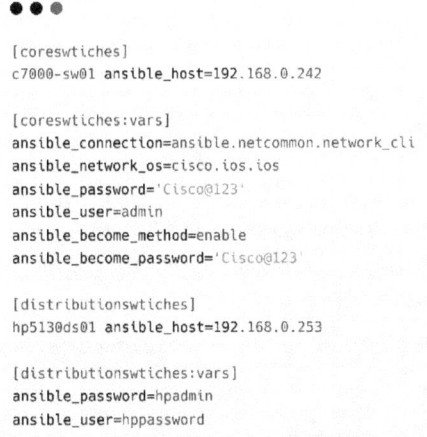

```
[coreswtiches]
c7000-sw01 ansible_host=192.168.0.242

[coreswtiches:vars]
ansible_connection=ansible.netcommon.network_cli
ansible_network_os=cisco.ios.ios
ansible_password='Cisco@123'
ansible_user=admin
ansible_become_method=enable
ansible_become_password='Cisco@123'

[distributionswtiches]
hp5130ds01 ansible_host=192.168.0.253

[distributionswtiches:vars]
ansible_password=hpadmin
ansible_user=hppassword
```

Figure 6.19 – Ansible network device inventory

In the preceding screenshot, we used the following:

- `ansible_connection=ansible.netcommon.network_cli` to specify the connection plugin to use
- `ansible_become_method=enable` to use `enable` as the privilege escalation method
- `ansible_network_os=cisco.ios.ios` to indicate the network device operating system

> **Network Communication Protocols**
>
> Refer to `https://docs.ansible.com/ansible/latest/network/getting_started/network_differences.html#multiple-communication-protocols` for different communication protocols available for network automation using Ansible.

In the next section, you will learn how to create an Ansible playbook to gather information from a VyOS network device.

VyOS information gathering using Ansible

This is an optional exercise for you to become familiar with network automation using Ansible. We assume that you have the basic knowledge to install and configure the **VyOS** appliance inside a virtual machine with your choice of virtualization platform.

Download the VyOS image from `https://support.vyos.io/en/downloads` and install it as a virtual appliance (refer to the VyOS documentation at `https://support.vyos.io/en/kb` more for details).

> **VyOS Network Operating System**
>
> VyOs is an open source network operating system based on Debian Linux. VyOS provides most networking functionalities, such as routing, **Virtual Private Networks** (**VPNs**), firewalls, **Network Address Translation** (**NAT**), and so on. Refer to `https://vyos.io` for more details.

In the following exercise, you will create a simple Ansible playbook to collect the operating system information from a VyOS device (or virtual appliance):

1. Add the VyOS virtual machine details to the Ansible inventory:

    ```
    [ansible@ansible Chapter-06]$ mkdir inventories/
    [ansible@ansible Chapter-06]$ cd inventories/
    ```

2. Update the inventory as follows:

```
[ansible@ansible inventories]$ cat network
...output omitted...

[vyos]
vyos-01 ansible_host=192.168.56.201

[vyos:vars]
ansible_connection=ansible.netcommon.network_cli
ansible_user=vyosuser
ansible_password=vyispassword
ansible_network_os=vyos.vyos.vyos
```

Figure 6.20 – Updating the Ansible inventory with VyOS device information

3. Create a playbook called Chapter-06/vyos-facts.yaml to gather the VyOS facts:

```
---
## Chapter-06/vyos-facts.yaml

- name: Collecting VyOS facts
  connection: ansible.netcommon.network_cli
  gather_facts: false
  hosts: vyos
  tasks:
    - name: Fetching VyOS details
      vyos.vyos.vyos_facts:
        gather_subset: all

    - name: Display fact output
      debug:
        msg: "VyOS version: {{ ansible_net_version }}"
```

Figure 6.21 – Ansible playbook for collecting details from the VyOS device

4.   Execute the playbook, as follows:

```
[ansible@ansible Chapter-06]$ ansible-playbook -i inventories/ vyos-facts.yaml

PLAY [Collecting VyOS facts] ********************************************************

TASK [Fetching VyOS details] ********************************************************
ok: [vyos-01]

TASK [Display fact output] **********************************************************
ok: [vyos-01] => {
    "msg": "VyOS version: VyOS 1.4-rolling-202202130317"
}

PLAY RECAP *************************************************************************
vyos-01                    : ok=2    changed=0    unreachable=0    failed=0    skipped=0    rescued=0    ignored=0
```

Figure 6.22 – VyOS fact-gathering playbook output

5.   Expand the playbook by collecting more facts and generating reports using the methods you learned about in *Chapter 3*, *Automating Your Daily Jobs*.

In the next section, you will learn about some more advanced network automation use cases by creating **access control list** (**ACL**) entries in a Cisco ASA device.

Creating ACL entries in a Cisco ASA device

Cisco ASA is a security device with the capabilities of firewall, antivirus, intrusion prevention, and VPN. Refer to `https://www.cisco.com/c/en/us/products/security/adaptive-security-appliance-asa-software/index.html` to learn more about Cisco ASA.

The Cisco ASA collection (`https://galaxy.ansible.com/cisco/asa`) provides modules and plugins to automate Cisco ASA operations. In this section, you will learn how to use Cisco ASA modules to create ACL entries in a Cisco ASA device.

The first task is to install the Cisco ASA collection using the `ansible-galaxy` command, as follows:

```
$ ansible-galaxy collection install cisco.asa
```

Like you have configured the VyOS connection variables, you need to configure the Cisco ASA device connection variables, as follows:

```
[asa]
ciscoasa ansible_host=192.168.57.121

[asa:vars]
ansible_user=adminasa
ansible_ssh_pass=password
ansible_become=true
ansible_become_method=ansible.netcommon.enable
ansible_become_pass=password
ansible_connection=ansible.netcommon.network_cli
ansible_network_os=cisco.asa.asa
```

Figure 6.23 – Cisco ASA inventory variables

As usual, remember to encrypt the password using Ansible Vault (or a credential in Ansible Automation Controller) instead of keeping it as a plain text value inside the file. Also, notice the connection variables we have used for the Cisco ASA device, such as `ansible_network_os=cisco.asa.asa` and `ansible_connection=ansible.netcommon.network_cli`.

A typical ACL entry creation task includes multiple steps, as shown in the following diagram:

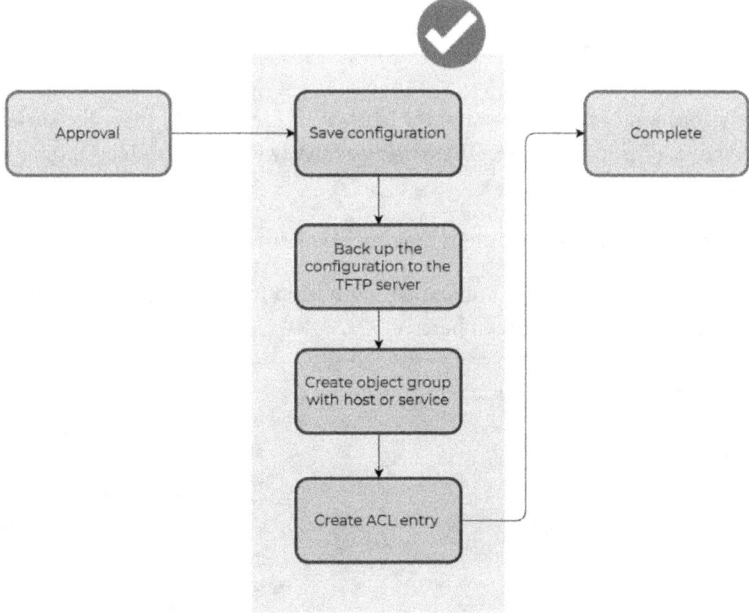

Figure 6.24 – ACL entry creation steps for Cisco ASA

Follow the standard pre-tasks and post-tasks as per the organization's policies and procedures. It is a best practice to create Ansible roles based on specific operations, such as creating an object group, backup configuration, or creating an ACL entry. To make this easier to understand, I have used a single playbook for this demonstration, as follows:

```
---
- name: Cisco ASA Create ACL Entry
  hosts: "{{ nw_devices }}"
  gather_facts: no

  vars:
    take_backup: "Yes"
    tftp_server: 192.168.57.106
    tftp_server_port: 69

    acl_identifier: Demo-ACL
    acl_type: extended
    acl_action: permit #or deny
    acl_entry_source_ip: 10.1.20.11
    acl_entry_source_mask: 255.255.255.255

    asa_object_group_name: DEMO-NETWORK-TEAM-NEW
    asa_object_group_type: network #or service, security etc
    asa_object_group_host: 192.0.50.4
```

Figure 6.25 – Cisco ACL playbook with variables

In the actual environment, take those variables (shown in *Figure 6.25*) from the variable file in the repository (as a source of truth) or collect them using a Survey form in Ansible Automation Controller (or Ansible Tower).

Add the tasks to the playbook to implement the steps explained in *Figure 6.24*.

First, you must take the backup of the current configuration to the TFTP server using the `cisco.asa.asa_command` module, as shown here:

```
tasks:
  - name: Set backup filename
    ansible.builtin.set_fact:
      backup_filename: "{{ inventory_hostname }}_{{ lookup('pipe', 'date +%Y%m%d-%H%M%S') }}_backup.cfg"

  - name: Save configuration and take device Backup to tftp
    cisco.asa.asa_command:
      commands:
        - write memory
        - copy /noconfirm running-config tftp://{{ tftp_server }}/{{ backup_filename }}
      when: take_backup == "Yes"
```

Figure 6.26 – Playbook task to take the Cisco ASA backup

Add a task to create the object group using the `cisco.asa.asa_ogs` module, as shown here:

```
- name: Merge module attributes of given object-group
  cisco.asa.asa_ogs:
    config:
    - object_type: network
      object_groups:
        - name: "{{ asa_object_group_name }}"
          network_object:
            host:
              - "{{ asa_object_group_host }}"
    state: merged
```

Figure 6.27 – Playbook task to create the object group

Finally, add the task to create the ACL entry using the `cisco.asa.asa_acls` module, as shown here:

```
- name: Add new ACL Entry and Merge configuration with device configuration
  cisco.asa.asa_acls:
    config:
      acls:
        - name: "{{ acl_identifier }}"
          acl_type: "{{ acl_type }}"
          aces:
          - grant: "{{ acl_action }}"
            protocol_options:
              tcp: true
            source:
              address: "{{ acl_entry_source_ip }}"
              netmask: "{{ acl_entry_source_mask }}"
            destination:
              object_group: "{{ asa_object_group_name }}"
    state: merged
```

Figure 6.28 – Playbook task to create the ACL entry

Execute the playbook and verify that the task has been completed:

```
[ansible@ansible Chapter-06]$ ansible-playbook cisco-asa-acl-create.yaml -e "nw devices=asa"

PLAY [Cisco ASA Create ACL Entry] ************************************************************

TASK [Set backup filename] *******************************************************************
ok: [ciscoasa]

TASK [Save configuration and take device Backup to tftp] ************************************
ok: [ciscoasa]

TASK [Merge module attributes of given object-group] ***************************************
changed: [ciscoasa]

TASK [Add new ACL Entry and Merge configuration with device configuration] *****************
changed: [ciscoasa]

PLAY RECAP *********************************************************************************
ciscoasa                   : ok=4    changed=2    unreachable=0    failed=0    skipped=0    rescued=0    ignored=0
```

Figure 6.29 – Cisco ASA ACL playbook execution

Once the playbook has been executed successfully, verify its details from the Cisco ASA device by logging in. The following screenshot shows the sample output:

```
$ ssh adminasa@192.168.57.121
adminasa@192.168.57.121's password:
User adminasa logged in to ciscoasa
Logins over the last 1 days: 2.  Last login: 03:49:07 UTC May 29 2022 from 192.168.57.1
Failed logins since the last login: 0.
Type help or '?' for a list of available commands.
ciscoasa> en
Password: ********
ciscoasa#
ciscoasa#
ciscoasa# show running-config object-group | include DEMO-NETWORK-TEAM-NEW
object-group network DEMO-NETWORK-TEAM-NEW
ciscoasa#
ciscoasa# show running-config access-list | include Demo-ACL
access-list Demo-ACL extended permit tcp host 10.1.20.11 object-group DEMO-NETWORK-TEAM-NEW
```

Figure 6.30 – Verifying the ACL entry's creation from the Cisco ASA device console

Then, log in to the TFTP server and verify that the backups have been created from the playbook tasks:

● ● ●

```
[operator@tftp-prod tftpboot]$ ls -lrt
total 48
-rw-r--r--. 1 nobody nobody 8368 May 29 03:54 ciscoasa_20220529-115451_backup.cfg
-rw-r--r--. 1 nobody nobody 8368 May 29 03:55 ciscoasa_20220529-115531_backup.cfg
-rw-r--r--. 1 nobody nobody 8400 May 29 03:55 ciscoasa_20220529-115553_backup.cfg
-rw-r--r--. 1 nobody nobody 8432 May 29 03:56 ciscoasa_20220529-115610_backup.cfg
```

Figure 6.31 – Verifying the configuration backups in the TFTP server

This use case can be expanded by adding multiple rules, object groups, and other backup methods. Refer to the Cisco ASA module documentation at https://docs.ansible.com/ansible/latest/collections/cisco/asa/index.html to learn more.

Expand your knowledge with other network devices you have such as Cisco, Arista, Juniper Network, or HPE devices, but remember to use the appropriate connection methods and parameters.

Ansible for Network Automation

Refer to the documentation at https://docs.ansible.com/ansible/latest/network/index.html to learn more about network automation basics. Also, check out *Network Getting Started* (https://docs.ansible.com/ansible/latest/network/getting_started/index.html) to start configuring network automation.

Visit https://www.ansible.com/use-cases/network-automation?hsLang=en-us to learn more about Red Hat Ansible Network Automation. Also, check out the Ansible Network Automation Workshop (https://aap2.demoredhat.com/exercises/ansible_network) for practice guides.

Summary

In this chapter, you learned about different remote connection methods and connection variables available in Ansible. After that, you explored Microsoft Windows automation using Ansible. You learned how to connect to a Microsoft Windows machine from Ansible and create a new user account using an Ansible playbook.

You also learned the difference in network automation between Linux and Windows. You explored simple network automation using a VyOS appliance and collected system information using a fact-gathering playbook.

Finally, you learned how to use a Cisco ASA collection and implemented a use case for creating an ACL entry in a Cisco ASA device.

In the next chapter, you will learn how to use Ansible to automate your virtualization and cloud platforms, such as VMware, AWS, and Google Cloud Platform.

Further reading

To learn more about the topics that were covered in this chapter, take a look at the following resources:

- *Cisco ACI Guide*: `https://docs.ansible.com/ansible/latest/scenario_guides/guide_aci.html`

- *Self-paced interactive hands-on labs with Ansible Automation Platform*: `https://www.redhat.com/en/engage/redhat-ansible-automation-202108061218`

- *Network Automation with Ansible*: `https://www.ansible.com/integrations/networks`

- *Ansible Network Automation Workshop*: `https://aap2.demoredhat.com/exercises/ansible_network/`

7
Managing Your Virtualization and Cloud Platforms

Since the introduction of virtualization and cloud computing, organizations can handle their IT infrastructure using programmatic methods since most of the IT components are software-defined, such as **software-defined data centers (SDDC)**, **software-defined storage (SDS)**, **software-defined networking (SDN)**, and others. But this additional layer of technologies also made infrastructure management more complex as engineers need to handle both the underlying infrastructure and the overcloud virtual components.

Ansible can help you automate both the underlying cloud infrastructure as well as the overcloud virtual components such as the automated cluster configurations of virtualization platforms (VMware, OpenStack, Red Hat Virtualization, and others). It can also help you provision virtual components such as virtual machines, virtual networks, and virtual storage.

In this chapter, we will cover the following topics:

- Introduction to Infrastructure as Code
- Managing cloud platforms using Ansible
- Automating VMware vSphere resources using Ansible
- Using Ansible as an IaC tool for AWS
- Creating resources in GCP using Ansible

First, you will learn how to configure the necessary authentication and provision resources in virtualization platforms such as VMware vCenter and public cloud platforms such as GCP and AWS.

Technical requirements

The following are the technical requirements for this chapter:

- A Linux machine for the Ansible control node (with internet access)

- Knowledge of managing cloud platforms (VMware, GCP, and AWS)

- Access to the VMware vCenter console and API (for the VMware use case)

- Access to the AWS console and API

- Access to the GCP console and API

All the Ansible code, playbooks, commands, and snippets for this chapter can be found in this book's GitHub repository at `https://github.com/PacktPublishing/Ansible-for-Real-life-Automation/tree/main/Chapter-07`.

Introduction to Infrastructure as Code

Infrastructure as Code (**IaC**) is a method that's used to provision and manage infrastructure details and configurations as software code and make changes inside the code instead of changing the infrastructure whenever required. There are many dedicated tools and software for IaC, including Ansible. Instead of manually deploying the infrastructure components, such as virtual machines, storage, network, policies, and so on, it is possible to develop IaC and use tools to deploy the infrastructure automatically. The following diagram shows some typical IaC components regarding the following:

- Infrastructure component details will be stored as code in a specific format (for example, YAML playbooks).

- IaC tools (for example, Ansible) will create and manage the infrastructure component in the private or public cloud based on the infrastructure code:

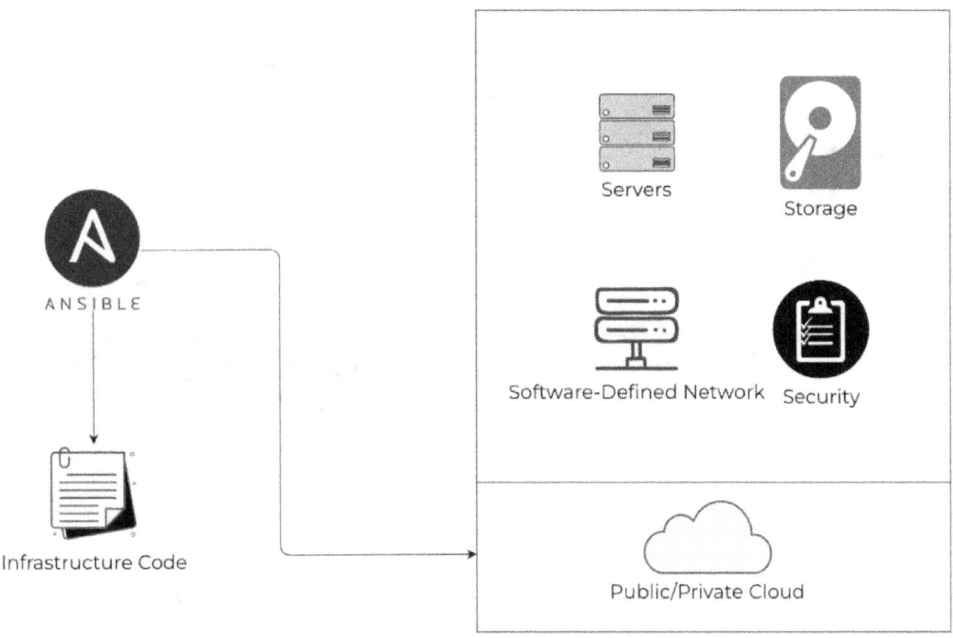

Figure 7.1 – Typical Infrastructure as Code components

In the next sections, you will learn how to use Ansible as an IaC tool for deploying and managing infrastructure in private and public cloud platforms.

Managing cloud platforms using Ansible

As you learned in the previous chapters, Ansible can manage both Linux, Windows, and network devices. But virtualization platforms work differently and you cannot use SSH-based connections and operations to automate such platforms. Most of these platforms offer **application programming interface** (**API**) and **software development kit** (**SDK**)-based access to help us access and control such platforms over HTTP (or HTTPS). Since Ansible can use SDK (Python libraries) and communicate over HTTP/HTTPS, it is possible to automate any platforms that offer such access.

The following diagram shows the different connection methods used by Ansible to communicate with the managed devices and platforms:

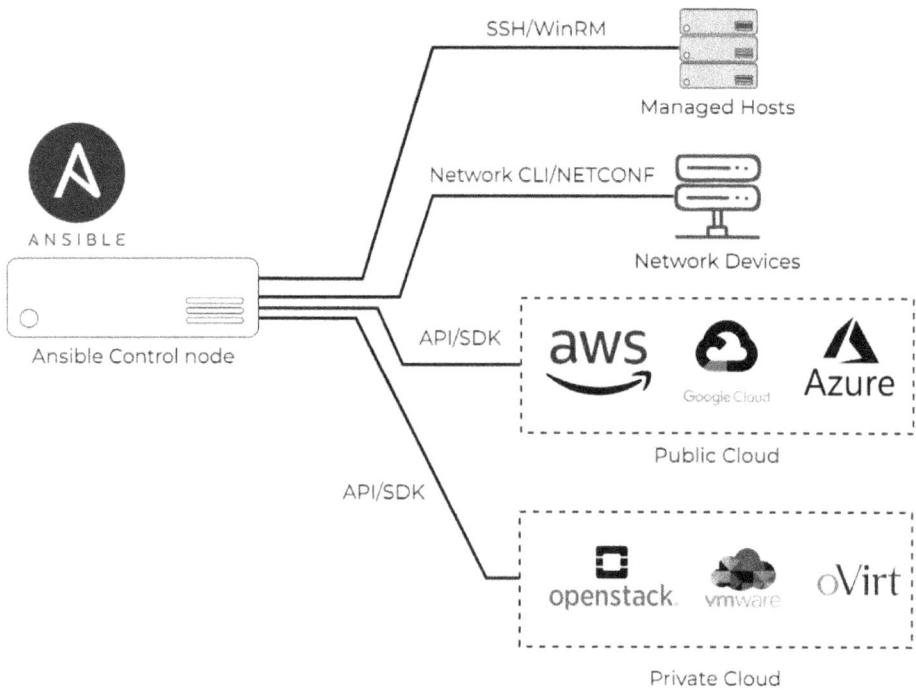

Figure 7.2 – Ansible connection methods

Application Programming Interface

An API is a connection or protocol that allows one system to communicate with another using a dedicated set of instructions and results. Unlike command-line utilities, which are meant for human interaction with the system, API-based access can help automate and control such systems programmatically and in a controlled way. Read more about APIs at `https://en.wikipedia.org/wiki/API`.

Ansible has hundreds of modules (refer to *Figure 7.3*) and plugins that support cloud automation and they are available as collections for specific cloud or virtualization platforms. With these modules, it is possible to create and manage cloud resources such as virtual machines, virtual private networks, virtual disks, access policies, serverless components, containers, and more:

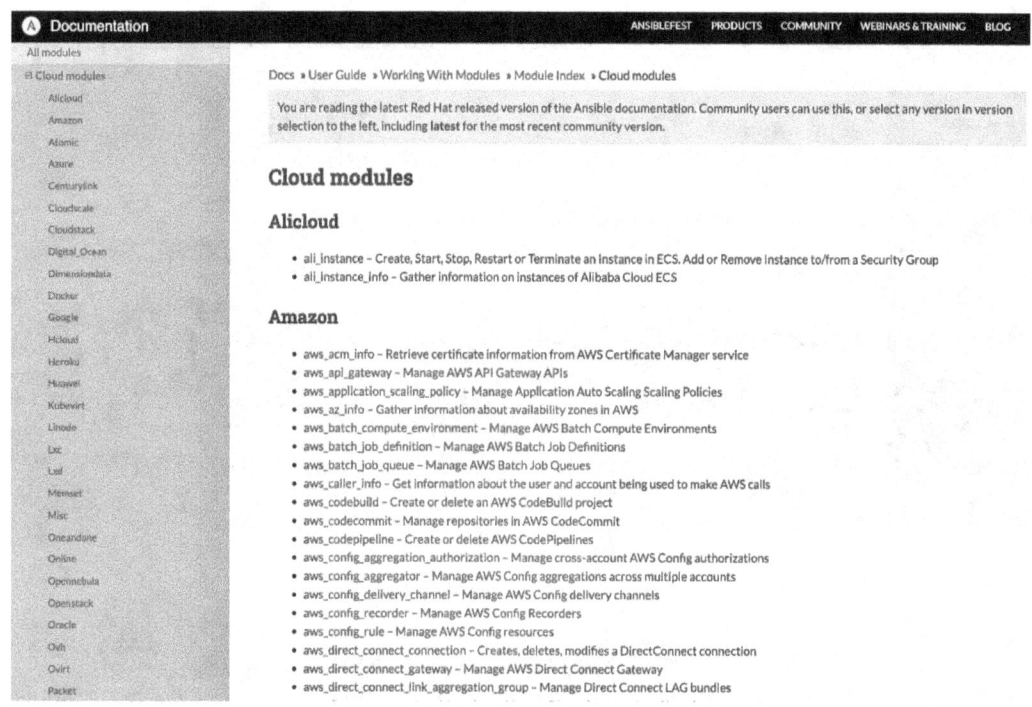

Figure 7.3 – Ansible cloud modules

Find the full list of cloud modules at `https://docs.ansible.com/ansible/2.9/modules/list_of_cloud_modules.html` (for Ansible 2.9). Since the introduction of Ansible collections, these modules have been migrated to the respective cloud collections. Check out the collection pages in Ansible Galaxy (`https://galaxy.ansible.com`) to see all the plugins, roles, and playbooks that are part of the Ansible content collection. For example, the following screenshot shows the Ansible collection page of the VMware collection by the community (`https://galaxy.ansible.com/community/vmware`):

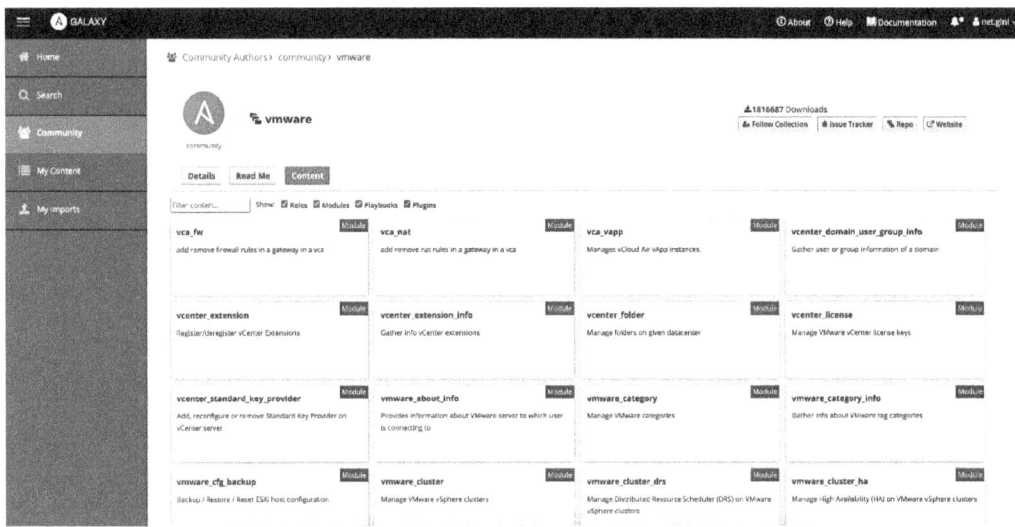

Figure 7.4 – Ansible VMware collection

Search for the collection in Ansible Galaxy and use the appropriate collection either from the community or from the vendors themselves. The following screenshot shows the search result for the aws collection in the Ansible Galaxy portal:

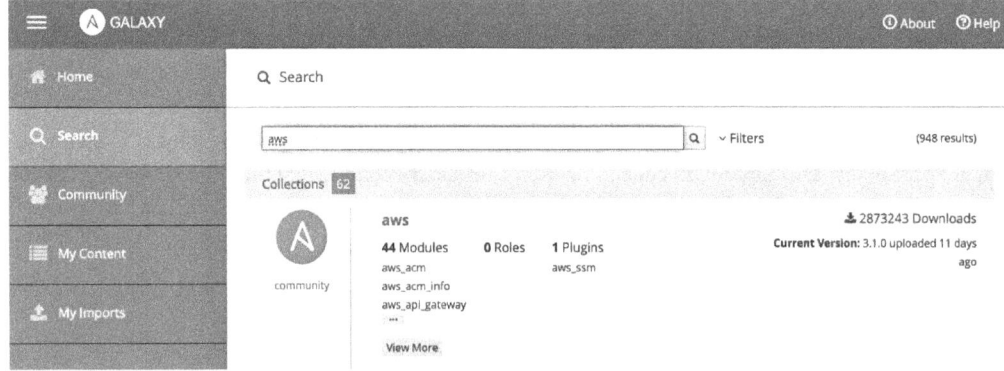

Figure 7.5 – AWS collection result in Ansible Galaxy

By exploring the Ansible Galaxy portal, you will find thousands of contributions from the community and vendors. By using Ansible collections, it is possible to save playbook development time by using existing roles, modules, and playbooks from existing collections.

In the next section, you will learn how to manage VMware using Ansible by using connection methods, modules, and more.

Automating VMware vSphere resources using Ansible

We will start with some simple automation use cases for VMware, such as provisioning of virtual machines, managing **high availability** (**HA**), network creation, and managing snapshots. The Ansible VMware collection (`community.vmware`) contains around 150 modules and other plugins:

Figure 7.6 – Ansible VMware collection by the community

The `community.vmware` collection relies on the `pyvmomi` and **vSphere Automation SDK for Python** libraries. Hence, to use the `community.vmware` collection, you need to install appropriate packages for Ansible to use it.

VMware has already introduced the **vSphere REST API** for vSphere 6.0 and later. A new Ansible collection was introduced (`vmware.vmware_rest`) to manage the operations using a REST API instead of Python libraries and SDKs. `vmware.vmware_rest` contains around 130 modules and other plugins:

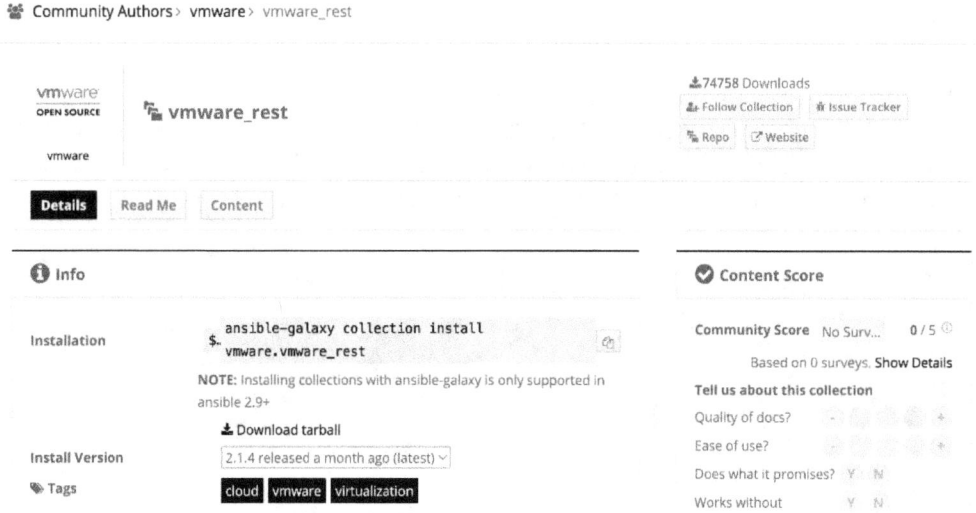

Figure 7.7 – Ansible VMware REST API collection by VMware

Use any of these Ansible collections, depending on your vSphere version and module requirements.

> **VMware Ansible Collections**
>
> The Ansible VMware community collection based on the Python library is available at `https://galaxy.ansible.com/community/vmware`. The collection based on the REST API is available at `https://galaxy.ansible.com/vmware/vmware_rest`.

Ansible VMware automation prerequisites

To use the VMware collection, you need to meet the following requirements:

- Install the Ansible VMware collection.

- Install the required libraries.

> **VMware vCenter Lab**
>
> If you do not have access to a VMware lab to test and practice, then get an evaluation copy of VMware vSphere (`https://customerconnect.vmware.com/en/web/vmware/evalcenter?p=vsphere-eval-7`) and install it on your home server. Refer to *How to Install VMware vSphere Hypervisor ESXi* (`https://www.techbeatly.com/how-to-install-vmware-vsphere-hypervisor-esxi`) and *How to Install VMware vCenter Server Appliance* (`https://www.techbeatly.com/how-to-install-vmware-vcenter-server-appliance`) to set up a VMware home lab.

Installing an Ansible VMware collection

Use the `ansible-galaxy` command to install a collection. By default, `ansible-galaxy` will store the collection's content in the `HOME_DIRECTORY/.ansible/collections` path. It is a best practice to store the collection in a project directory instead of a home directory to avoid any dependency issues. Configure the collection path in `ansible.cfg`, as follows:

```
[ansible@ansible Chapter-07]$ cat ansible.cfg
[defaults]
inventory = ./hosts
remote_user = devops
ask_pass = false

COLLECTIONS_PATHS = ./collections
```

Figure 7.8 – ansible.cfg with collection_paths

When you execute the `ansible-galaxy` command, the collection will be downloaded and stored in the project directory as per the configuration, as shown in the following screenshot (refer to the **VMware Ansible collections** page at `galaxy.ansible.com` to see the installation command; refer to the previous information box to find the link):

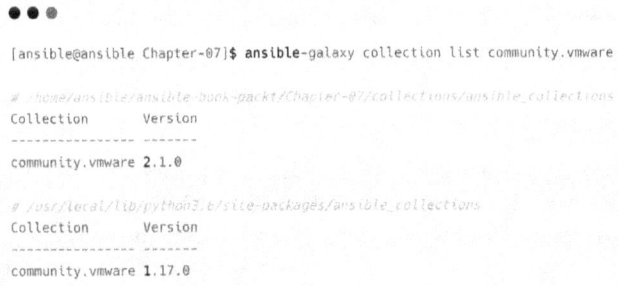

```
[ansible@ansible Chapter-07]$ ansible-galaxy collection install community.vmware
Starting galaxy collection install process
Process install dependency map
Starting collection install process
Downloading https://galaxy.ansible.com/download/community-vmware-2.1.0.tar.gz to
/home/ansible/.ansible/tmp/ansible-local-2922ya8yiroz/tmpgwmpkp4a/community-vmware-2.1.0-hk7iqmqh
Installing 'community.vmware:2.1.0' to '/home/ansible/ansible-book-packt/Chapter-
07/collections/ansible_collections/community/vmware'
community.vmware:2.1.0 was installed successfully
```

Figure 7.9 – Installing the VMware collection

Verify the collection's details using the `ansible-galaxy collection list` command, as follows:

```
[ansible@ansible Chapter-07]$ ansible-galaxy collection list community.vmware

# /home/ansible/ansible-book-packt/Chapter-07/collections/ansible_collections
Collection        Version
----------------- -------
community.vmware  2.1.0

# /usr/local/lib/python3.6/site-packages/ansible_collections
Collection        Version
----------------- -------
community.vmware  1.17.0
```

Figure 7.10 – Listing the installed collections

If you have installed the Ansible community package, then the default collection will be there as part of Python's `site-packages`. The latest collection that's been installed inside your custom collection directory can be seen in the preceding screenshot.

Installing the required Python libraries

The list of required Python libraries and SDKs will be stored inside the `requirements.txt` file in the collection folder, as shown here:

```
[ansible@ansible Chapter-07]$ cat collections/ansible_collections/community/vmware/requirements.txt
pyVmomi>=6.7
git+https://github.com/vmware/vsphere-automation-sdk-python.git ; python_version >= '2.7'
```

Figure 7.11 – Collection dependencies in requirements.txt

Use the `pip install` command to install all the dependencies for the Ansible collection, as follows:

```
[ansible@ansible Chapter-07]$ pip install -r collections/ansible_collections/community/vmware/requirements.txt
```

Figure 7.12 – Installing the requirements for the collection

If you are using Python virtual environments for Ansible, then remember to activate the appropriate virtual environment and install the necessary libraries inside the Python virtual environment.

> **Automation Execution Environments**
>
> Red Hat Ansible Tower used Python virtual environments to handle dependencies such as Python libraries, collections, and so on. In **Ansible Automation Platform 2.x** (**AAP**) the automation **execution environment** (**EE**) is the standard way to package and deploy the automation environment. The execution environment can contain all the necessary dependencies, including Python packages, libraries, `ansible-core`, and the required collections. Refer to `https://www.ansible.com/products/execution-environments` for more details. Learn how to build new execution environments (container images) using the `ansible-builder` utility (`https://docs.ansible.com/automation-controller/latest/html/userguide/execution_environments.html`).

Provisioning VMware virtual machines in Ansible

In this exercise, you will create Ansible content to provision virtual machines in a VMware cluster.

You must have the following configurations and details in place before proceeding with this exercise:

- A virtual machine template in VMware vCenter (Linux or Windows)

- A default user account inside the virtual machine template to access new virtual machines once provisioned

- A VMware vCenter username and password with adequate access to create and manage resources in the cluster

- Details of the data center, cluster, datastore, and folder to save the VM in

Ensure the appropriate permissions are in place by logging into VMware vCenter and creating some test machines using the VMware VM template. Once you have confirmed the access and permissions, start developing Ansible artifacts for VMware VM provisioning.

Creating Ansible automation artifacts for VM provisioning

You need to declare and pass a few variables for the vCenter access and VM creation. Keep these variables inside the playbook or in a separate file. Follow these steps:

1. Store the vCenter credentials inside your Ansible Vault file – that is, `vars/vmware-credential.yaml`:

```
[ansible@ansible Chapter-07]$ mkdir vars
[ansible@ansible Chapter-07]$ cd vars/
[ansible@ansible vars]$ ansible-vault create vmware-credential.yaml
New Vault password:
Confirm New Vault password:
```

Figure 7.13 – Creating an Ansible Vault file for your VMware credentials

2. Add the VMware vCenter username and password inside the file and save the vault file:

```
vcenter_username: yourvmwareadmin
vcenter_password: yoursecretpassword
```

Remember the vault password as you will need this while executing the playbook.

3. Create another variable file called `vars/common-vars.yml` for storing details about the new virtual machine (refer to `Chapter-07` in this book's GitHub repository for more details):

```
# details about the cluster
vcenter_hostname: vcenter.lab.local
vmware_datacenter: DC1
vmware_cluster_name: 'AZ1'
vmware_datastore: 'datastore1'
vm_folder: '/'

# details for the new VM
vm_name: 'DC1AZ1POC101'
vm_template_name: 'RHEL7-New'
# disk details
vm_disk_size_gb: '40'
vm_disk_type: 'thin'
vm_disk_datastore_name: 'datastore1'
# capacity and hardware
vm_memory_size_mb: '8192'

...<output omitted for brevity>...
```

Figure 7.14 – Variables for the VMware cluster and VM

It is possible to add multiple VM details inside and loop the task. However, we only covered using a single VM here.

4. Create a role for virtual machine provisioning by using the `ansible-galaxy role init` command, as follows:

```
[ansible@ansible Chapter-07]$ mkdir roles

[ansible@ansible Chapter-07]$ cd roles
[ansible@ansible roles]$ ansible-galaxy role init vmware-provision-vm-from-template
- Role vmware-provision-vm-from-template was created successfully
```

Figure 7.15 – Initializing a new role using the ansible-galaxy role init command

5. Add the necessary tasks to `roles/vmware-provision-vm-from-template/tasks/main.yml`:

```
• • •

---
# tasks file for vmware-provision-vm-from-template
- name: Check VM exist or not
  include_tasks: vmware-provisioning-pre-check.yaml

- name: Provision VM
  include_tasks: vmware-provisioning-task.yaml
  when: vm_check.failed
```

Figure 7.16 – main.yaml with a subtasks file

tasks/main.yml contains two tasks for calling the subtasks file. The first task does a pre-check to ensure a VM with the same name does not exist before you proceed with VM creation. If one doesn't exist, then the next task will be executed based on the when: vm_check.failed condition. This is a best practice as you need to add all possible validations and error handling inside the playbook.

6. Create the necessary content for roles/vmware-provision-vm-from-template/ tasks/vmware-provisioning-pre-check.yaml. Also, display a message if the virtual machine already exists:

```
• • •

---
# vmware-provisioning-pre-check.yaml
- name: Check if VM exist with same name
  no_log: true
  community.vmware.vmware_guest_find:
    hostname: "{{ vcenter_hostname }}"
    username: "{{ vcenter_username }}"
    password: "{{ vcenter_password }}"
    validate_certs: no
    name: "{{ vm_name }}"
  delegate_to: localhost
  register: vm_check
  ignore_errors: yes

- name: If VM with same name already exist
  debug:
    msg: "The virtual machine {{ vm_name }} already exist. Skipping tasks..."
  when: not vm_check.failed
```

Figure 7.17 – vmware-provisioning-pre-check.yaml

7. Create roles/vmware-provision-vm-from-template/tasks/vmware- provisioning-task.yaml and add a task to create a new VMware VM (refer to this book's GitHub repository for the full code), as shown in the following screenshot:

```
...
# vmware-provisioning-task.yml
- name: "Provisioning New VM using template {{ vm_template_name }}"
  vmware_guest:
    hostname: "{{ vcenter_hostname }}"
    username: "{{ vcenter_username }}"
    password: "{{ vcenter_password }}"
    validate_certs: no
    datacenter: "{{ vmware_datacenter }}"
    cluster: "{{ vmware_cluster_name }}"
    folder: "{{ vm_folder }}"
    #guest_id: "{{ os_guest_id }}"
    name: "{{ vm_name }}"
    template: "{{ vm_template_name }}"
    state: poweredon
    wait_for_ip_address: "{{ vm_wait_for_ip_connection }}"
    wait_for_customization: "{{ vm_wait_for_customization }}"
    #customization_spec: "{{ vm_base_profile }}"
    datastore: "{{ vmware_datastore }}"
    ...<remved code for brevity>...
```

Figure 7.18 – VMware VM creation task file

8. It is a best practice to keep the default values of variables in `roles/vmware-provision-vm-from-template/defaults/main.yml` and pass the actual values while executing the playbook:

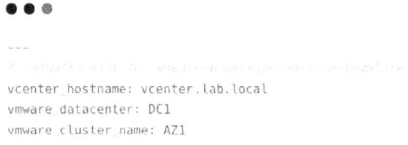

```
---
# roles/vmware-provision-vm-from-template/defaults/main.yml
vcenter_hostname: vcenter.lab.local
vmware_datacenter: DC1
vmware_cluster_name: AZ1
```

Figure 7.19 – Default variable for the VMware role

9. Create the main playbook, `Chapter-07/vmware-provision-vm-from-template.yml`. Since the execution happens on a local machine for API and HTTP-based platforms, make sure `hosts: localhost` has been configured, as follows:

```
## vmware-provision-vm-from-template.yml
---
- name: "Provision VM from Template"
  hosts: localhost
  gather_facts: no
  become: no
  connection: local
  vars_files:
    - vars/common-vars.yml          # other common variables
    - vars/vmware-credential.yaml    # vcenter credential
  tasks:
    - name: Deploy new VM in vCenter
      ansible.builtin.include_role:
        name: vmware-provision-vm-from-template
      tags: provisionvmfromtemplate

    - name: Waits for SSH (VM UP and Running)
      ansible.builtin.wait_for:
        host: "{{ vm_net1_ip_address }}"
        port: 22
        delay: "{{ vm_wait_for_ssh_time }}"
        timeout: 300
        state: started
      when: vm_wait_for_ip_connection == "yes"
```

Figure 7.20 – The main playbook to create the VMware infrastructure

It is possible to keep the variables under `group_vars` or `host_vars` as needed. However, here we have used a variable file approach to demonstrate this use case easily.

In the preceding snippet, the `ansible.builtin.wait_for` module is being used to wait for the newly created virtual machine to come online with SSH. We used the IP address of the virtual machine, port 22 (SSH), and instructed Ansible to wait for some period (`vm_wait_for_ssh_time`). Refer to `https://docs.ansible.com/ansible/latest/collections/ansible/builtin/wait_for_module.html` to learn more about the `wait_for` module.

10. Once the VM is up and running, add the new VM to a dynamic host group, `vmwarenewvms`, for post-provisioning tasks, as shown here:

```
●  ●  ●

## vmware_provision_vm_and_post_configure.yml

  .

  .

  - name: Add newly created VMs to a host group
    no_log: true
    ansible.builtin.add_host:
      name: "{{ vm_net1_ip_address }}"
      groups: "vmwarenewvms"
      ansible_ssh_extra_args: ' -o StrictHostKeyChecking=no '
      ansible_user: "{{ vm_ansible_user_name }}"
      ansible_password: "{{ vm_ansible_user_password }}"
      var_vm_os_family: "{{ vm_os_family }}"
      var_vm_user_name_list: "{{ vm_user_name_list }}"
      var_vm_user_password: "{{ vm_user_password }}"
    when: vm_os_family == "RHEL"
```

Figure 7.21 – Adding the newly created VM to the Ansible inventory

Instead of adding hosts, it is possible to use the dynamic inventory plugin for VMware to detect the newly created virtual machines and use them in the next set of tasks. Refer to `https://docs.ansible.com/ansible/latest/collections/community/vmware/vmware_vm_inventory_inventory.html` for more details. Also, refer to *Chapter 5, Expanding Your Automation Landscape*, the *Ansible dynamic inventory* section, to learn how to use an Ansible dynamic inventory with AWS.

11. Create another play in the same Ansible playbook to execute the post-provisioning tasks such as creating new users, configuring system files, and installing new packages, as shown here:

```
●●●
## vmware-provision-vm-from-template.yml
.
.
## 2nd play for post-configurations
- name: RHEL VM Post-Provisioning Configurations
  hosts: vmwarenewvms
  gather_facts: no
  become: yes
  tasks:
    - name: Waiting for SSH
      wait_for:
        host: "{{ inventory_hostname }}"
        port: 22
        delay: 1
        timeout: 300
        state: started
      when: var_vm_os_family is defined
      become: no
      vars:
        ansible_connection: local

    - name: New VM post-provisioning configurations
      debug:
        msg: "You can include additonal tasks to execute inside the new VM as post provisioning configurations"
      when: var_vm_os_family is defined
```

Figure 7.22 – Second play in the playbook for post-provisioning tasks

When you use Ansible Automation Controller (part of Ansible Automation Platform) or Ansible AWX, it is possible to split the playbook into multiple job templates and create job workflows, as shown in the following diagram:

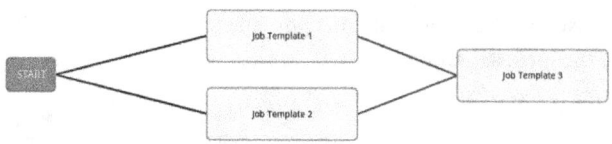

Figure 7.23 – Workflow templates (source: docs.ansible.com)

You will learn about workflow job templates and Ansible Automation Controller in *Chapter 12, Integrating Ansible with Your Tools*. Refer to https://docs.ansible.com/automation-controller/latest/html/userguide/workflows.html to learn more about workflow templates.

12. Execute the `ansible-playbook vmware-provision-vm-from-template.yml` playbook with `--ask-vault-password` and provide the vault password, as follows:

```
[ansible@ansible Chapter-07]$ ansible-playbook vmware-provision-vm-from-template.yml  --ask-vault-password
Vault password:
```

Figure 7.24 – Executing the playbook to create the VMware VM

Add more tasks to the post-provisioning section to automate the VM configuration workflows and make it a single workflow for end-to-end VM provisioning. In the next section, you will learn about cloud management for **Amazon Web Services** (**AWS**) and how to use Ansible as an IaC tool.

> **Create a Template in the vSphere Client**
>
> Refer to `https://docs.vmware.com/en/VMware-vSphere/6.0/com.vmware.vsphere.hostclient.doc/GUID-40BC4243-E4FA-4A46-8C8B-F50D92C186ED.html` for VMware template creation procedures.

Using Ansible as an IaC tool for AWS

In this section, you will create Ansible content to provision and manage AWS resources using Ansible. Let's assume that whenever you need to create a new EC2 instance, you need to follow multiple manual procedures such as creating a new **Virtual Private Cloud** (**VPC**), a new security group, network access policies, and many other items. You also need to do post-provisioning steps such as creating new user accounts, installing packages, configuring applications, and more.

With the help of the Ansible AWS collection, it is possible to automate all of these tasks and manage the entire life cycle of the infrastructure.

AWS Free Tier

To practice AWS and Ansible use cases, it is possible to use AWS Free Tier, which provides more than 100 AWS resources free of charge. Visit `https://aws.amazon.com/free` (*Figure 7.25*) and sign up for a free AWS Free Tier account to find them:

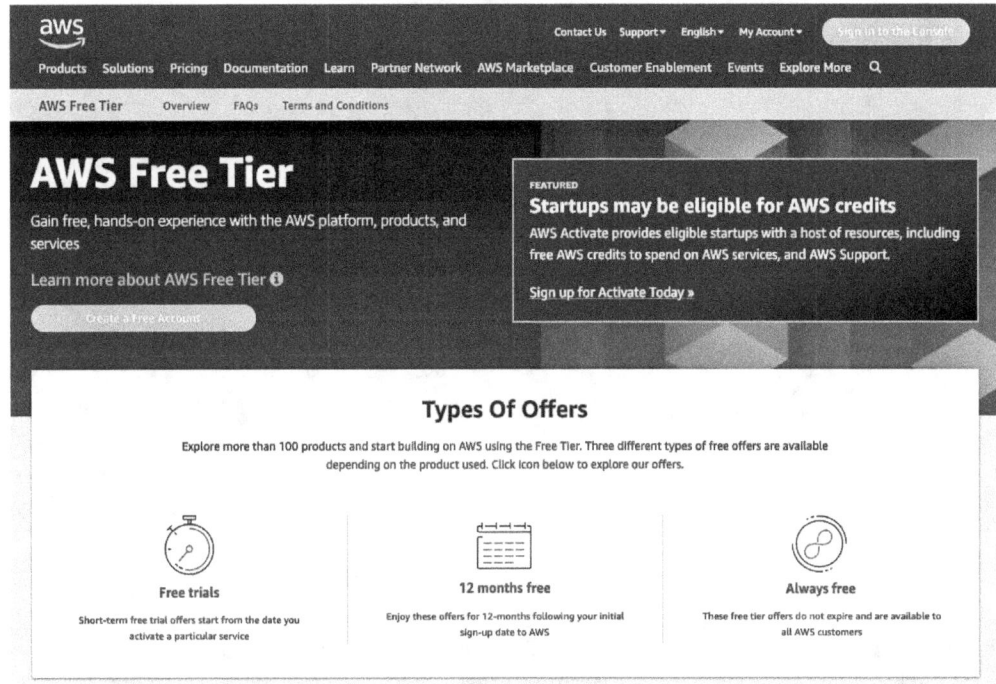

Figure 7.25 – AWS Free Tier access with more than 100 services

It is important to ensure that your AWS resource usage is within the AWS Free Tier limit to avoid the credit or debit card connected with your account being billed. Refer to `https://aws.amazon.com/aws-cost-management/aws-budgets/` to learn how to configure budget alerts and other billing details for your AWS account.

I have ensured that the following demonstration only uses the AWS Free Tier-based resources so that no additional costs will be incurred to your AWS Free Tier account. But it is also important to delete the resource from the AWS account once you have finished testing.

Installing the Ansible AWS collection

As you learned in the previous section, it is possible to use the `ansible-galaxy` command to install the Ansible AWS collection. If you need multiple collections for the project, then create a `requirements.yaml` file and mention the required collections and roles inside it:

```
[ansible@ansible Chapter-07]$ cat requirements.yaml
---
collections:
  # Install a collection from Ansible Galaxy.
  - name: amazon.aws
    version: 3.0.0
    source: https://galaxy.ansible.com
  - name: community.vmware
    version: 2.1.0
    source: https://galaxy.ansible.com
  - name: google.cloud
    version: 1.0.2
    source: https://galaxy.ansible.com
  - name: community.general
    version: 4.0.1
    source: https://galaxy.ansible.com
  - name: ansible.posix
    version: 1.3.0
    source: https://galaxy.ansible.com
```

Figure 7.26 – The requirements.yaml file with collection details

Once `requirements.yaml` has been updated, install the collection using the `ansible-galaxy` command by calling the `requirements.yaml` file:

```
[ansible@ansible Chapter-07]$ [ansible@ansible Chapter-07]$ ansible-galaxy install -r requirements.yaml
```

Figure 7.27 – Installing the collection using requirements.yaml

The collection will be installed in the `Chapter-07/collections` path that you configured in `ansible.cfg`. Verify the collection, as follows:

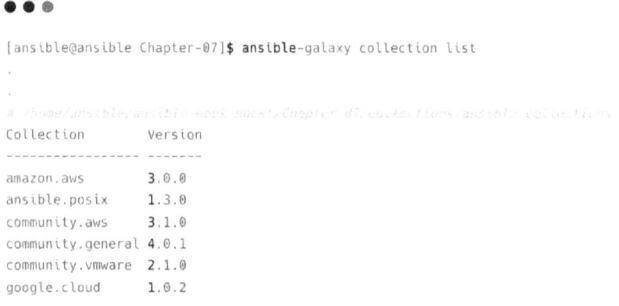

```
[ansible@ansible Chapter-07]$ ansible-galaxy collection list
.
.
# /home/ansible/ansible-book-marx/Chapter-07/collections/ansible_collections
Collection        Version
----------------- -------
amazon.aws        3.0.0
ansible.posix     1.3.0
community.aws      3.1.0
community.general 4.0.1
community.vmware  2.1.0
google.cloud      1.0.2
```

Figure 7.28 – Ansible collection list

The Ansible collection for AWS automation is now ready to use. Now, start developing your Ansible playbook.

Creating Ansible IaC content for the AWS infrastructure

In this scenario, two web servers will be installed with website content and configured to serve behind an **elastic load balancer** (**ELB**), as shown in the following diagram:

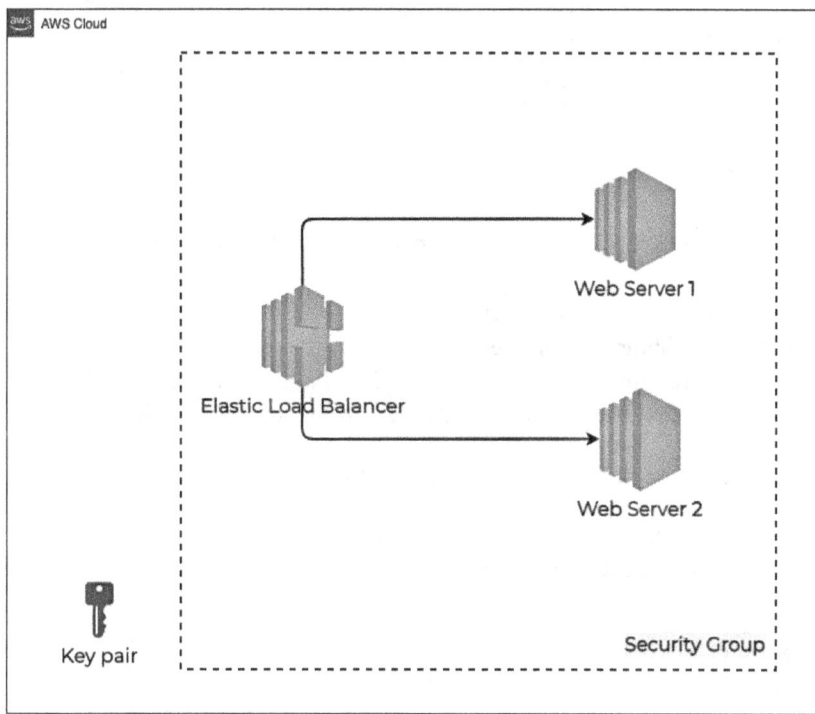

Figure 7.29 – Build a web server infrastructure and application

Multiple roles can be created to handle multiple resources in this Ansible automation content, as shown in the following screenshot:

```
[ansible@ansible Chapter-07]$ ls -l roles/
total 0
drwxrwxr-x. 4 ansible ansible  35 Jul 30 08:00 aws-create-ec2
drwxrwxr-x. 5 ansible ansible  61 Jul 30 08:00 aws-create-elb
drwxrwxr-x. 3 ansible ansible  19 Jul 30 08:00 aws-create-keypair
drwxrwxr-x. 3 ansible ansible  19 Jul 30 08:00 aws-create-sg
drwxrwxr-x. 3 ansible ansible  19 Jul 30 08:00 aws-create-targetgrp
drwxrwxr-x. 3 ansible ansible  19 Jul 30 08:00 aws-delete-ec2
drwxrwxr-x. 3 ansible ansible  19 Jul 30 08:00 aws-delete-elb
drwxrwxr-x. 3 ansible ansible  19 Jul 30 08:00 aws-delete-keypair
drwxrwxr-x. 3 ansible ansible  19 Jul 30 08:00 aws-delete-sg
drwxrwxr-x. 3 ansible ansible  19 Jul 30 08:00 aws-delete-targetgrp
drwxrwxr-x. 3 ansible ansible  19 Jul 30 08:00 aws-get-vpc-details
drwxrwxr-x. 3 ansible ansible  19 Jul 30 08:00 aws-remove-web
drwxrwxr-x. 3 ansible ansible  19 Jul 30 08:00 deploy-web-server
drwxrwxr-x. 8 ansible ansible 124 Jul 30 08:00 vmware-provision-vm-from-template
```

Figure 7.30 – Ansible roles for managing cloud platforms

Check out the amazon.aws collection document (https://galaxy.ansible.com/amazon/aws), to see the Python library requirements for installing the boto and boto3 Python libraries. You learned how to do this in *Chapter 5, Expanding Your Automation Landscape*. Please refer to this chapter to learn how to configure the AWS credential for this automation job.

Creating a variable for the AWS environment

Details about the AWS region, VPC, and subnets are stored in the Chapter-07/vars/aws-common-vars.yml file (refer to the course repository on GitHub):

```
## aws-common-vars.yml
---
# variables for aws environment
vpc_id: ""
vpc_subnet_list: []

region: ap-southeast-1
aws_region: ap-southeast-1
elbgroupname: webtarget

inventory_webgroup: ec2webservers
existing_ec2_list: []
new_ec2_list: []
existing_ec2_public_ips: []
...<removed for brevity>...
```

Figure 7.31 – AWS-related variables in vars/aws-common-vars.yml

Once the details about the AWS cloud have been stored in a variable, you must create a variable to store the EC2 instance details.

Default VPC in AWS

A default VPC will be present in each AWS region when you start using AWS VPC, which comes with the following resources:

- Subnets (public) in each **Availability Zone (AZ)**

- An internet gateway

- DNS resolution

In this demonstration, we will use the **default VPC** instead of creating a new VPC. It is possible to create a new VPC (and manage its subnets) using the amazon.aws.ec2_vpc_net Ansible module. Refer to the official documentation (https://docs.aws.amazon.com/vpc/latest/userguide/default-vpc.html) to learn more.

Listing the EC2 instances

Details about the new EC2 instances are stored in the vars/aws-ec2-new.yml file:

```
## vars/aws-ec2-new.yml
---
# list of ec2 instances
ec2_new_list:
  aws_web_101:
    name: AWS_WEB_101
    key_name: "{{ aws_demo_key }}"
    group: SG-Ansible-Demo
    instance_type: t2.micro
  aws_web_102:
    name: AWS_WEB_102
    key_name: "{{ aws_demo_key }}"
    group: SG-Ansible-Demo
    instance_type: t2.micro
```

Figure 7.32 – New EC2 details in the variable file

The number of EC2 instances can be controlled in this variable as you will be using a loop to provision them. Now, you must configure the AWS security group variables.

Creating an AWS security group

Use the amazon.aws.ec2_group module to create a new security group with the required rules in roles/aws-create-sg/tasks/main.yml, as shown here:

```
# tasks file for aws-create-sg
- name: Create Security group
  amazon.aws.ec2_group:
    profile: "{{ aws_boto_profile }}"
    name: "{{ aws_security_group }}"
    description: 'Security Group with SSH and HTTP rules'
    vpc_id: "{{ aws_vpc_id }}"
    region: "{{ aws_region }}"
    rules:
      - proto: tcp
        ports:
        - 80
        cidr_ip: 0.0.0.0/0
        rule_desc: allow all on port 80
      - proto: tcp
        ports:
        - 22
        cidr_ip: 0.0.0.0/0
        rule_desc: allow all on port 22
```

Figure 7.33 – Security group task

If your application is serving over different ports, add the ports accordingly and adjust the playbook as needed. Once you've done this, you must create the SSH key pair task.

Creating the SSH key pair

A new key can be created using the `amazon.aws.ec2_key` module. A local key pair (`~/.ssh/id_rsa.pub`) will be used for this in `roles/aws-create-keypair/tasks/main.yml`:

```
# tasks file for aws-create-keypair
- name: Create key pair
  amazon.aws.ec2_key:
    name: "{{ aws_demo_key }}"
    key_material: "{{ lookup('file', '~/.ssh/id_rsa.pub') }}"
    profile: "{{ aws_boto_profile }}"
    region: "{{ aws_region }}"
```

Figure 7.34 – Creating the SSH key resource

This key pair is important as you will need it later to access the EC2 instances and configure web servers and other details. In the preceding code block, we used an existing SSH key pair on the workstation (`~/.ssh/id_rsa.pub`) and its public key to create this new key pair resource in AWS so that we do not need to download the private key again. Now, you must add tasks for creating an **Elastic Load Balancer (ELB)**.

Creating an Elastic Load Balancer

A new ELB can be created by using the `amazon.aws.ec2_elb_lb` module in `roles/aws-create-elb/tasks/main.yml`, as shown here:

```
●  ●  ●

---
# tasks file for aws-create-elb
- name: Create Amazon ELB
  amazon.aws.ec2_elb_lb:
    profile: "{{ aws_boto_profile }}"
    name: "{{ aws_elb_app_lb }}"
    region: "{{ aws_region }}"
    zones:
      - "{{ ap_zone1 }}"
      - "{{ ap_zone2 }}"
    listeners:
      - protocol: http
        load_balancer_port: 80
        instance_port: 80
        proxy_protocol: True
    state: present
  register: elbcreated

- name: Collect ELB Public DNS
  ansible.builtin.set_fact:
    elb_public_dns: "{{ elbcreated.elb.dns_name }}"
```

Figure 7.35 – Creating an Amazon ELB

If you are using a different version of the AWS Ansible collection, then check out the appropriate Ansible module for creating load balancers and change the playbook accordingly. Now, you are ready to create the EC2 instance tasks.

Creating EC2 instances

Finally, start creating the EC2 instances. However, you must ensure that no duplicate VMs are created. It is a best practice to add validations inside the playbook to avoid any mistakes happening automatically as AWS will not complain about creating multiple instances with the same name.

Create the necessary tasks in `roles/aws-create-ec2/tasks/main.yml` to collect the existing instances. Only proceed with EC2 creation if the instances don't exist:

```
● ● ●
---
# tasks file for aws-create-vm

- name: Fetch Instances by tag, subnet and type
  amazon.aws.ec2_instance_info:
    profile: "{{ aws_boto_profile }}"
    region: "{{ aws_region }}"
    filters:
      "tag:Name": "{{ item.value.name }}"
      #network.interface.subnet-id: "{{ item.value.vpc_subnet_id }}"
      instance-type: "{{ item.value.instance_type }}"
      instance-state-name: ["running", "stopped", "stopping", "starting", "pending"]
  loop: "{{ lookup('dict', ec2_new_list, wantlist=True) }}"
  register: ec2_collected

- name: Collect ec2 in a list
  set_fact:
    existing_ec2_list: "{{ existing_ec2_list }} + {{ item }}"
  loop: "{{ ec2_collected | json_query('results[*].instances[*].tags.Name') }}"
  #loop: "{{ ec2_collected | json_query('results[*].instances[*].aws_instance_id') }}"
```

Figure 7.36 – Fetching EC2 instances

If the EC2 instance already exists, then you should not create a new EC2 instance again. So, based on `existing_ec2_list`, create new EC2 instances, as shown here:

```
● ● ●
---
# tasks file for aws-create-vm
.
.
- name: Launching EC2 instances
  amazon.aws.ec2_instance:
    profile: "{{ aws_boto_profile }}"
    key_name: "{{ aws_demo_key }}"
    security_group: "{{ aws_security_group }}"
    instance_type: "{{ item.value.instance_type }}"
    image_id: "{{ aws_ami_id }}"
    state: running
    wait: true
    #wait_timeout: 500
    #count: 1
    region: "{{ aws_region }}"
    tags:
      Name: "{{ item.value.name }}"
    detailed_monitoring: no
    vpc_subnet_id: "{{ vpc_subnet_list | random }}"
    network:
      assign_public_ip: yes
  loop: "{{ lookup('dict', ec2_new_list, wantlist=True) }}"
  when: "not item.value.name in existing_ec2_list"
  register: created_ec2
```

Figure 7.37 – Creating an EC2 instance if one does not exist

After creating the EC2 instances, you need to collect the newly created EC2 instances' details and wait for them to boot up and respond as being online, as shown here:

```
● ● ●

---
# tasks file for aws-create-vm
.
.
- name: Collect newly created ec2 in a list
  ansible.builtin.set_fact:
    new_ec2_list: "{{ new_ec2_list }} + [ '{{ item.instances[0].public_ip }}' ]"
  when: item.instances[0].public_ip is defined
  loop: "{{ created_ec2.results }}"

- name: Status
  ansible.builtin.debug:
    msg: "{{ item }} : Waiting for instances online..."
  with_items: "{{ new_ec2_list }}"

- name: Wait for SSH
  ansible.builtin.wait_for:
    host: "{{ item }}"
    port: 22
    delay: 3
    connect_timeout: 180
    sleep: 5
    state: started
  with_items: "{{ new_ec2_list }}"
```

Figure 7.38 – Collecting information about the newly created EC2 instances

Once the EC2 instances are up and running, collect their details so that you can update them to ELB:

```
● ● ●

---
# tasks file for aws-create-vm
.
.
- name: Fetch Instances by tag, subnet and type
  amazon.aws.ec2_instance_info:
    profile: "{{ aws_boto_profile }}"
    region: "{{ aws_region }}"
    filters:
      "tag:Name": "{{ item.value.name }}"
      #network-interface.subnet-id: "{{ item.value.vpc_subnet_id }}"
      instance-type: "{{ item.value.instance_type }}"
      instance-state-name: ["running", "stopped", "stopping", "starting", "pending"]
  loop: "{{ lookup('dict', ec2_new_list, wantlist=True) }}"
  register: ec2_existing_collected
```

Figure 7.39 – Collecting the new EC2 instances' details

Using the collected details (`ec2_existing_collected`), update the ELB with instance items in the backend, as shown here:

```yaml
---
# tasks file for aws/role=aws

- name: Update Amazon ELB and add instance ids
  amazon.aws.ec2_elb_lb:
    profile: "{{ aws_boto_profile }}"
    name: "{{ aws_elb_app_lb }}"
    region: "{{ aws_region }}"
    zones:
      - "{{ ap_zone1 }}"
      - "{{ ap_zone2 }}"
    listeners:
      - protocol: http
        load_balancer_port: 80
        instance_port: 80
        proxy_protocol: True
    instance_ids:
      - "{{ item.instances[0].instance_id }}"
    state: present
  register: elbcreated
  loop: "{{ ec2_existing_collected.results }}"
```

Figure 7.40 – Updating the ELB with instance details in the backend

The ELB and backend instances are ready and connected, but nothing is running inside the EC2 instances. Now, you need to install the web server and its content inside it. For that, collect the newly created EC2 instances' details in a dynamic host group for **post-provisioning** tasks, as shown here:

```yaml
---
# tasks file for aws/role=aws

- name: Collect ec2 Public IP in a list
  ansible.builtin.set_fact:
    existing_ec2_public_ips: "{{ existing_ec2_public_ips }} + [ '{{ item.instances[0].public_ip_address }}' ] "
  loop: "{{ ec2_existing_collected.results }}"

- name: Add ec2 instances to a host group
  ansible.builtin.add_host:
    name: "{{ item }}"
    groups: "{{ inventory_webgroup }}"
    ansible_ssh_extra_args: ' -o StrictHostKeyChecking=no '
  loop: "{{ existing_ec2_public_ips }}"
```

Figure 7.41 – Updating the inventory with new EC2 instance details

Instead of collecting the host details, as shown in the preceding screenshot, it is possible to utilize an Ansible dynamic inventory. Refer to *Chapter 5*, *Expanding Your Automation Landscape*, the *Ansible dynamic inventory* section to learn how to use an Ansible dynamic inventory with AWS.

Once you have prepared all of the tasks in the roles, use these roles in the appropriate order inside the main playbook.

The main playbook for integrating the provisioning workflow for AWS

The IaC provisioning workflow is written inside the main playbook – that is, Chapter-07/ aws-infra-provisioning.yaml (refer to this book's GitHub repository for the full code). The variables can be stored under group_vars or host_vars as a best practice. However, we have used the variable file approach to demonstrate this use case easily. Use the previously created roles in the appropriate order, as follows:

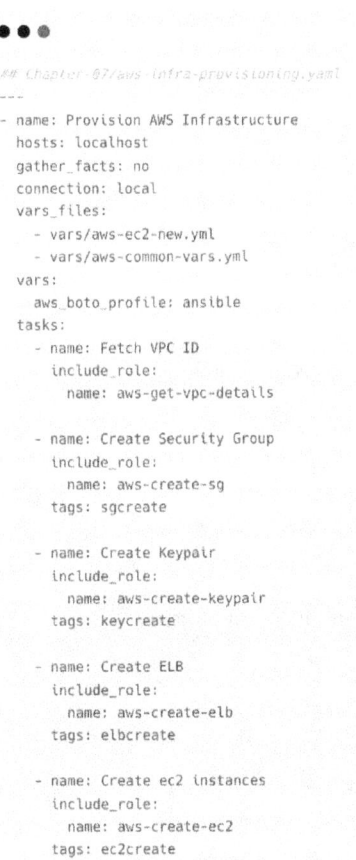

```
## Chapter-07/aws-infra-provisioning.yaml
---
- name: Provision AWS Infrastructure
  hosts: localhost
  gather_facts: no
  connection: local
  vars_files:
    - vars/aws-ec2-new.yml
    - vars/aws-common-vars.yml
  vars:
    aws_boto_profile: ansible
  tasks:
    - name: Fetch VPC ID
      include_role:
        name: aws-get-vpc-details

    - name: Create Security Group
      include_role:
        name: aws-create-sg
      tags: sgcreate

    - name: Create Keypair
      include_role:
        name: aws-create-keypair
      tags: keycreate

    - name: Create ELB
      include_role:
        name: aws-create-elb
      tags: elbcreate

    - name: Create ec2 instances
      include_role:
        name: aws-create-ec2
      tags: ec2create
```

Figure 7.42 – AWS infrastructure provisioning playbook

In the same playbook, the post-provisioning tasks have been added under different plays called `Deploy Webserver to EC2 instances` and `hosts: ec2webservers` (the dynamic host group created in `roles/aws-create-ec2/tasks/main.yml`).

The `deploy-web-server` role will install the `httpd` web server and install `firewalld` and configure the web server with the default website content. The following screenshot shows the post-provisioning play and an optional play to show the ELB address so that you do not need to log into the AWS console and check for the URL to access the website:

```
## Chapter 02 aws infra provisioning....st
.
.
## 2nd play to deploy webserver on new ec2 instance
- name: Deploy Webserver to EC2 instances
  hosts: ec2webservers
  remote_user: ec2-user
  become: true
  tasks:
    - name: Deploy Web service
      include_role:
        name: deploy-web-server

## 3rd play to display ELB details
- name: IaC Summary
  hosts: localhost
  tasks:
    - debug:
        msg: "Website is accessible on Appication ELB: {{ elb_public_dns }} (It may take some time to get the
backend instance to come InService)"
```

Figure 7.43 – Post-provisioning tasks in the second play

When you execute the playbook, the workflow will be triggered for provisioning the infrastructure and application:

```
[ansible@ansible Chapter-07]$ ansible-playbook aws-infra-
provisioning.yaml
```

Finally, the following message will be displayed after the post-provisioning tasks:

```
[ansible@ansible Chapter-07]$ ansible-playbook aws-infra-provisioning.yaml

...<output omitted for brevity>...

TASK [debug] ****************************************************************
ok: [localhost] => {
    "msg": "Website is accessible on Appication ELB: ansible-iac-demo-elb-app-lb-893112002.ap-southeast-
1.elb.amazonaws.com (It may take some time to get the backend instance to come InService)"
}
```

Figure 7.44 – The Ansible playbook displaying the ELB details

Add more resources to the provisioning workflow, such as additional EBS volumes to the server, Network Access Control rules, and images from custom AMIs or snapshots. Post-provisioning can be expanded with more configurations as necessary.

Now, you need to create the playbook's content to delete the resources to complete the full IaC life cycle.

Completing the IaC life cycle by using a destroy playbook

The Chapter-07/aws-infra-destroy.yaml playbook will take care of destroying the resources and housekeeping jobs (refer to this book's GitHub repository for the full code):

```
## Chapter-07/aws-infra-destroy.yaml
---
- name: Destroy AWS Infrastructure
  hosts: localhost
  gather_facts: no
  connection: local
  vars_files:
    - vars/aws-ec2-new.yml
    - vars/aws-common-vars.yml
  vars:
    aws_boto_profile: ansible
  tasks:
    - name: Fetch VPC ID
      include_role:
        name: aws-get-vpc-details

    - name: Delete ec2 instances
      include_role:
        name: aws-delete-ec2
      tags: ec2delete

    - name: Delete App ELB
      include_role:
        name: aws-delete-elb
      tags: elbdelete

...<omitted for brevity>...
```

Figure 7.45 – AWS infrastructure destroy playbook

If the resources are not required, then destroy the entire infrastructure using the destroy playbook:

```
[ansible@ansible Chapter-07]$ ansible-playbook aws-infra-
destroy.yaml
```

In this section, you learned how to create AWS resources and manage the end-to-end life cycle for implementing IaC practices. In the next section, you will explore similar automation for **Google Cloud Platform (GCP)**.

> **Ansible AWS Integration**
>
> Read more about Ansible AWS integration at `https://www.ansible.com/integrations/cloud/amazon-web-services`.

Creating resources in GCP using Ansible

Like VMware and AWS, it is possible to create and manage GCP resources with the help of the Ansible GCP content collection.

Prerequisite for Ansible GCP automation

Before you start, you need to ensure the prerequisites have been configured for Ansible GCP automation.

As you learned for VMware, AWS, and other platforms, you need to install the relevant Ansible content collection. In this case, you must install the `google.cloud` collection if you haven't done so yet:

```
[ansible@ansible Chapter-07]$ ansible-galaxy collection install
google.cloud
```

The `google.cloud` collection contains around 170 modules, roles, and other plugins to automate the GCP infrastructure and its resources.

The modules in the `google.cloud` collection require the following Python libraries to be installed on the system:

```
$ pip install requests google-auth
```

Make sure you are installing the libraries into the correct path if you are using a Python virtual environment for Ansible.

> **Ansible for Google Cloud**
>
> Learn more about Ansible GCP automation by reading the *Google Cloud Platform Guide*: `https://docs.ansible.com/ansible/latest/scenario_guides/guide_gce.html`. Also, check out the Google Cloud Ansible collection at `https://galaxy.ansible.com/google/cloud` to explore Ansible modules for GCP automation.

GCP free trial

To practice GCP and Ansible use cases, it is possible to get a GCP free trial with more than 20 GCP services to use free of cost. Visit `https://cloud.google.com/free` and sign up for a GCP Free Tier account. I have ensured that the following demonstration only uses the GCP Free Tier-based resources so that no additional costs will be incurred on your account. It is also important to delete the resource from your GCP account once you have finished testing.

Configuring GCP credentials

Use your GCP service accounts or machine accounts to configure the platform access for Ansible automation (refer to the following information box for more information). Once you created the service account and downloaded the JSON file, configure Ansible to use the JSON file as the credential for GCP access.

It is possible to configure the credential as module parameters or as environment variables, as follows:

```
GCP_AUTH_KIND
GCP_SERVICE_ACCOUNT_EMAIL
GCP_SERVICE_ACCOUNT_FILE
GCP_SCOPES
```

Depending on the environment and practices, follow an appropriate method, similar to what we did for VMware and AWS.

> **Creating a GCP Service Account**
>
> Refer to `https://developers.google.com/identity/protocols/oauth2/service-account#creatinganaccount` to learn how to create a GCP service account and create and download JSON key files.

Creating a GCP instance using Ansible

In this exercise, you will create Ansible content to create a simple compute instance in GCP:

1. Create a variable file called `vars/gcp-details.yaml` so that you can store the GCP credential details, as follows:

```
[ansible@ansible Chapter-07]$ cat vars/gcp-details.yaml
gcp_auth_kind: serviceaccount
gcp_service_account_email: ansible-demo@ansible-automation-demo.iam.gserviceaccount.com
gcp_service_account_file: ~/.config/ansible-automation-demo-bce5e5cf69d0.json
gcp_project: ansible-automation-demo
gcp_scopes:
  - https://www.googleapis.com/auth/compute
```

Figure 7.46 – GCP configuration variables for Ansible

2. Create the `Chapter-07/gcp-create-instance.yml` playbook. Create a new VPC network using `google.cloud.gcp_compute_network` if required and configure the new GCP instance so that it uses the new VPC network. In this exercise, you will be using the `default` VPC network and adding a task to fetch the `default` VPC network details:

```
## Chapter-07/gcp-create-instance.yml
---
- name: "Provision new GCP instance"
  hosts: localhost
  gather_facts: no
  become: no
  connection: local
  vars_files:
    - vars/gcp-details.yaml         # GCP credentials and details
  tasks:
    - name: Get info about default VPC network
      gcp_compute_network_info:
        project: "{{ gcp_project }}"
        auth_kind: "{{ gcp_auth_kind }}"
        service_account_file: "{{ gcp_service_account_file }}"
        filters:
        - name = default
      register: default_network_details
```

Figure 7.47 – GCP provisioning playbook

3. Add a task to the same playbook to create the GCP instance boot disk with the Debian 9 operating system (use other public images or your own private images as needed):

```
## Chapter-07/gcp-create-instance.yml

  - name: Create a disk with OS
    google.cloud.gcp_compute_disk:
      project: "{{ gcp_project }}"
      auth_kind: "{{ gcp_auth_kind }}"
      service_account_file: "{{ gcp_service_account_file }}"
      name: demo-disk
      size_gb: 10
      source_image: projects/debian-cloud/global/images/family/debian-9
      zone: us-central1-a
      state: present
    register: instance_source_disk
```

Figure 7.48 – Creating a GCP disk for the new VM

The `instance_source_disk` variable will contain details about the disk. We will pass this information to the instance creation task in the next step.

4. Add a task to create the GCP instance and pass the network and disk details, as follows:

```
## Chapter-07/gcp-create-instance.yml

  - name: Create GCP instance
    google.cloud.gcp_compute_instance:
      project: "{{ gcp_project }}"
      auth_kind: "{{ gcp_auth_kind }}"
      service_account_file: "{{ gcp_service_account_file }}"
      zone: us-central1-a
      state: present
      name: demo-instance
      machine_type: n1-standard-1
      disks:
      - auto_delete: 'true'
        boot: 'true'
        source: "{{ instance_source_disk }}"
      labels:
        environment: production
      network_interfaces:
      - network: "{{ default_network_details.resources[0] }}"
```

Figure 7.49 – Creating a GCP instance with network and disk details

5. Execute the Ansible `gcp-create-instance.yml` playbook:

    ```
    [ansible@ansible Chapter-07]$ ansible-playbook
    gcp-create-instance.yml
    ```

6. Verify the instance details on the GCP console, as shown here:

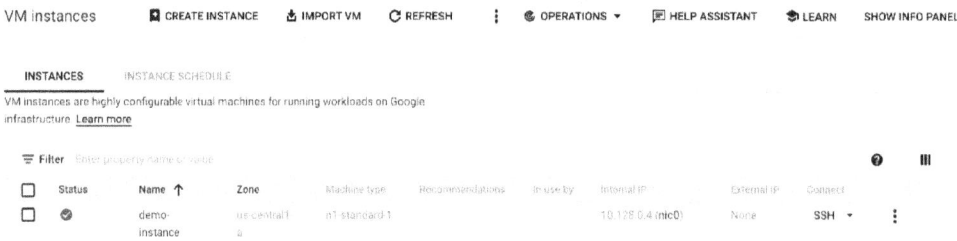

Figure 7.50 – Verifying the instance on the GCP dashboard

7. Also, verify the disk details, as shown here:

Figure 7.51 – Verifying the disk details on the GCP dashboard

With around 170 modules available, it is possible to manage almost all cloud resources in GCP, including computing, disk, networks, load balancers, firewalls, routing, cloud builds, autoscaling, DNS, databases, Spanner, and more.

> **Ansible GCP Modules**
>
> Refer to the official documentation to see the available Ansible GCP modules: `https://docs.ansible.com/ansible/latest/collections/google/cloud/index.html`.

Summary

In this chapter, you learned about IaC concepts and how to use Ansible as an IaC tool. You also learned about how Ansible can manage virtualization and cloud platforms such as VMware, AWS, and GCP. Then, you learned about the different methods and credential configurations for these platforms so that Ansible can access and execute automated operations.

Next, you explored the Ansible modules and collections that are available for VMware, AWS, and GCP. By developing the basic playbooks for creating new virtual machines (EC2 instances or GCP instances), you have started your journey in infrastructure automation and management. Expand the playbook's content to build use cases suitable for your cloud and virtualization environment.

In the next chapter, you will learn how to help non-platform teams use Ansible for their automation use cases, such as building and managing databases using Ansible.

Further reading

To learn more about the topics that were covered in this chapter, take a look at the following resources:

- *What is Infrastructure as Code (IaC)?*: https://www.redhat.com/en/topics/automation/what-is-infrastructure-as-code-iac

- *Ansible VMware Guide*: https://docs.ansible.com/ansible/2.5/scenario_guides/guide_vmware.html

- *VMware API and SDK Documentation*: https://www.vmware.com/support/pubs/sdk_pubs.html

- *Introducing the VMware REST Ansible Content Collection*: https://www.ansible.com/blog/introducing-the-vmware-rest-ansible-content-collection

- *Ansible Amazon Web Services Guide*: https://docs.ansible.com/ansible/latest/collections/amazon/aws/docsite/guide_aws.html

Helping the Database Team with Automation

Stateful applications, by definition, must save data persistently. So, when we talk about stateful applications, data will come into the picture, and hence, database servers. Choose any supported database software, depending on the type of data you want to store. This includes the number of transactions, the performance that's required for your application, high availability and failover support, and many other factors. However, there are more important concerns, such as preparing the datastore, installing the necessary dependencies, packages or libraries. In terms of maintenance, this will be a continuous process as we need to take care of backups, data dumps, snapshots, and restoration in case of failure.

Ansible can help you in such situations. There are hundreds of Ansible database modules available that can help you implement your database automation tasks, including database installation, deployment, managing tables, managing users, and many other tasks.

In this chapter, we will cover the following topics:

- Ansible for database operations
- Installing database servers
- Creating and managing databases using Ansible
- Automating PostgreSQL operations
- Automating a password reset using ITSM and Ansible

We will learn how to install the PostgreSQL database servers, create databases, configure database tables, user authentication, and more. You will also learn about the integration opportunities for zero-touch automation while using Ansible and **IT Service Management** (**ITSM**) tools.

Technical requirements

The following are the technical requirements to proceed with this chapter.

- A Linux machine for the Ansible control node (with internet access)
- A Linux machine for installing and configuring the PostgreSQL server
- Basic knowledge about databases (PostgreSQL) and servers

All the Ansible code, playbooks, commands, and snippets for this chapter can be found in this book's GitHub repository at `https://github.com/PacktPublishing/Ansible-for-Real-life-Automation/tree/main/Chapter-08`.

Ansible for database operations

Database operations not only involve deploying database servers but also counting the day-to-day operations, such as managing databases, tables, database users, permissions or access. Ansible can talk to most of the well-known database services using the appropriate Ansible modules, as shown in the following diagram:

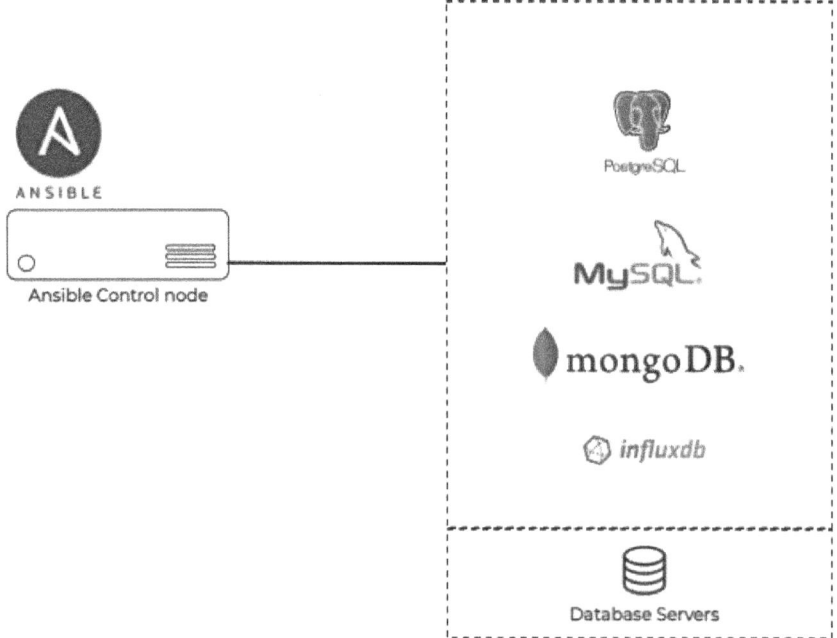

Figure 8.1 – Ansible database automation

Some of the most common database automation use cases are as follows:

- Deploying standalone database servers
- Configure **high availability** (**HA**) database clusters
- Creating databases and tables
- Managing user accounts
- Managing permissions
- Managing database and server access
- Backup and restore operations
- Implementing data replication and mirroring
- Automated database failovers

With the help of Ansible database collections and modules, we can automate most of these operations. In the next section, you will learn how to install a PostgreSQL database server using Ansible.

Please refer to the *Further reading* section at the end of this chapter for more resources.

Installing database servers

If you are a database administrator or if you know how database servers work, then you know the pain and struggle of managing and maintaining the services and data as per the application's requirements. Since the introduction of virtualized and cloud-based platforms, provisioning virtual machines, disks, and other resources has become less of a headache. However, we still need automated options to provision database servers and database instances. There are single-click deployment solutions from public **cloud service providers** (**CSPs**) known as **managed database solutions** but in most cases, we do not have much control and transparency over such services if we have more strict requirements. Hence, organizations are forced to use self-hosted database servers and follow manual deployment and management processes.

In *Chapter 7, Managing Your Virtualization and Cloud Platforms*, you learned how to automate infrastructure provisioning, including virtual machines and disks. In this chapter, we will explore how to automate database tasks, such as installing database servers, or provisioning databases or tables, as shown in the following diagram:

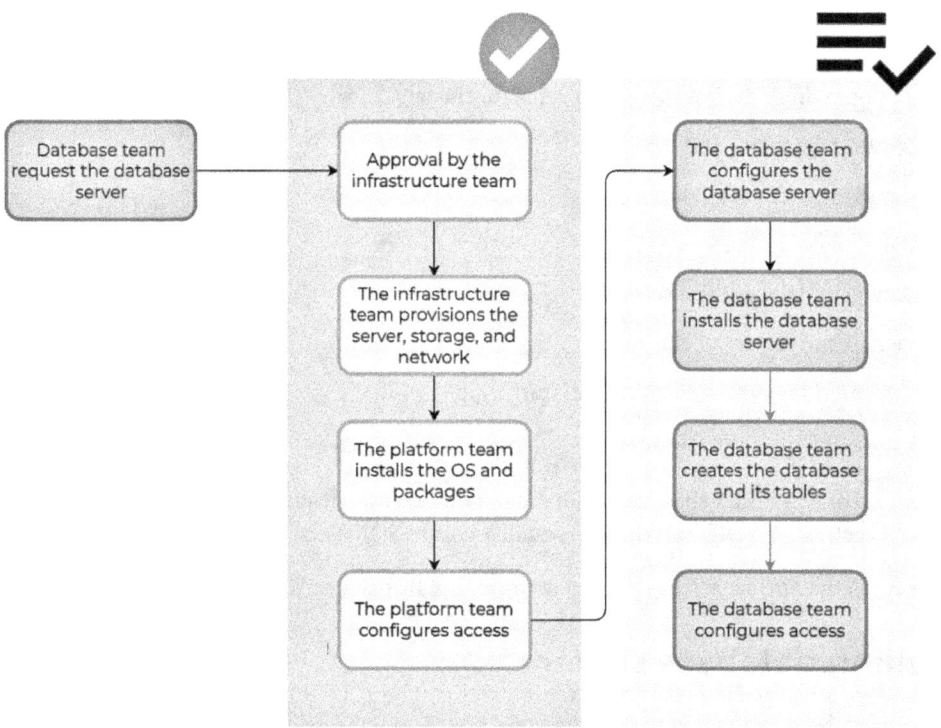

Figure 8.2 – Basic database operations

Fortunately, Ansible has a good collection of modules and plugins for deploying and managing database servers such as Microsoft SQL, MySQL, PostgreSQL, InfluxDB, MongoDB, ProxySQL or Vertica. You will learn about the basics of Ansible-based database deployment in the following sections.

Installing PostgreSQL using Ansible

Installing PostgreSQL is simple if you refer to the official documentation. However, you need to install all the dependencies and libraries that are required for PostgreSQL. You also need to configure the database server details. Fortunately, there are well-written Ansible roles available in Ansible Galaxy that we can download and use to install and configure PostgreSQL servers (and other database servers). The following screenshot shows us searching for the `postgresql` role in Ansible Galaxy:

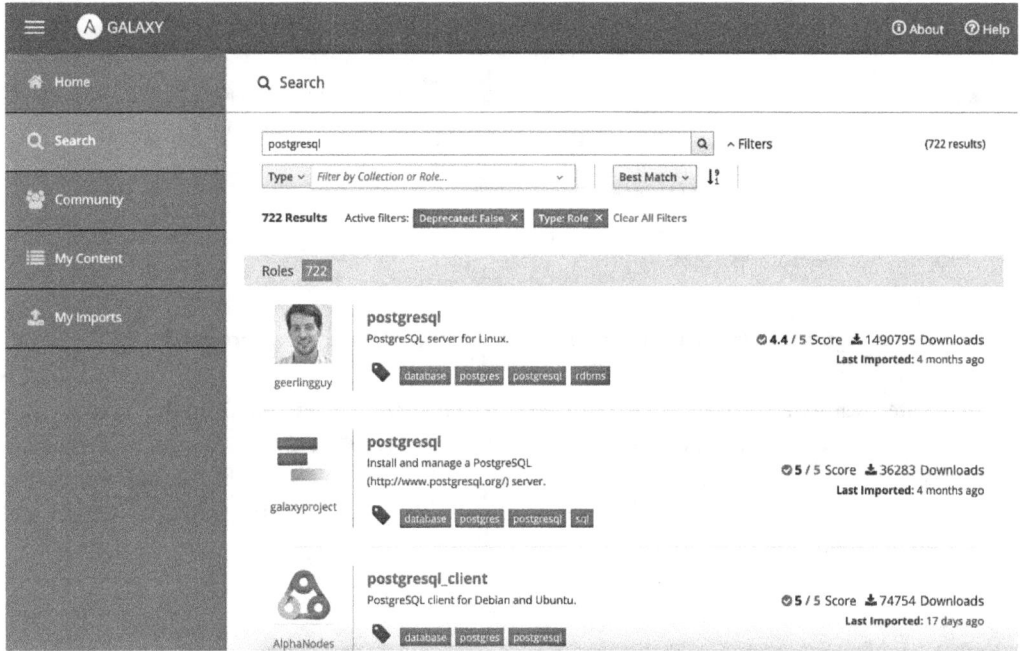

Figure 8.3 – The postgresql role search in Ansible Galaxy

In this exercise, you will install a simple standalone PostgreSQL server using the `geerlingguy.postgresql` Ansible role, which was contributed by community member *Jeff Geerling* (https://galaxy.ansible.com/geerlingguy/postgresql). Follow these steps:

1. Make sure that your `ansible.cfg` file has been configured with project-specific roles and a collection page:

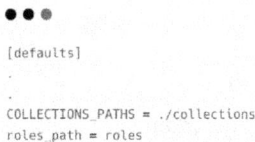

```
[defaults]
.
.
COLLECTIONS_PATHS = ./collections
roles_path = roles
```

Figure 8.4 - Configure ansible.cfg

2. Install the role using the `ansible-galaxy` command, as follows:

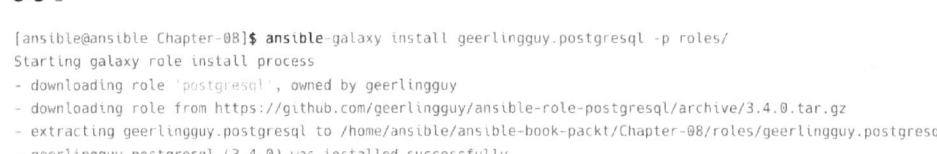

```
[ansible@ansible Chapter-08]$ ansible-galaxy install geerlingguy.postgresql -p roles/
Starting galaxy role install process
- downloading role 'postgresql', owned by geerlingguy
- downloading role from https://github.com/geerlingguy/ansible-role-postgresql/archive/3.4.0.tar.gz
- extracting geerlingguy.postgresql to /home/ansible/ansible-book-packt/Chapter-08/roles/geerlingguy.postgresql
- geerlingguy.postgresql (3.4.0) was installed successfully
```

Figure 8.5 – Installing an Ansible role using the ansible-galaxy command

3. Create a variable file called `Chapter-08/vars/postgres.yaml` so that you can pass some user details, the database to create, and `hba` entries to update the `geerlingguy.postgresql` role. The role will create the resources automatically based on the variables you are passing to the playbook. Skip this step if you do not wish to create such entries and configurations automatically:

```
# vars/postgres.yaml
postgresql_databases:
  - name: database_demo
postgresql_users:
  - name: demouser
    password: password
postgresql_hba_entries:
  - { type: local, database: all, user: all, auth_method: peer }
  - { type: host, database: all, user: all, address: '0.0.0.0/0', auth_method: md5 }
```

Figure 8.6 – Variables for PostgreSQL database

> **Important Note**
>
> In the preceding code snippet, the database username and passwords for `postgresql_users` have been specified in plain text for demonstration purposes. You should consider using encrypted passwords using Ansible Vault or other appropriate methods.

4. Create the `Chapter-08/postgres-deploy.yaml` playbook, as follows:

```
• • •

---
## Chapter-08/postgres-deploy.yaml
- name: Deploying PostgreSQL Database Server
  hosts: "{{ NODES }}"
  become: true
  vars_files:
    - vars/postgres.yaml
  tasks:
    - name: Install and configure PostgreSQL
      include_role:
        name: geerlingguy.postgresql
      .
      .
```

Figure 8.7 – Playbook to deploy PostgreSQL server

5. Add a task that allows remote connections for PostgreSQL and restart the PostgreSQL service. Finally, allow database port 5432 in the firewall, as shown in the following code snippet. Use other firewall service modules such as community.general.ufw if you are using a different firewall:

```
• • •

---
## Chapter-08/postgres-deploy.yaml
  .
  .
    - name: Allow remote connection for PostgreSQL
      ansible.builtin.lineinfile:
        path: /var/lib/pgsql/data/postgresql.conf
        regexp: '^listen_addresses'
        line: "listen_addresses = '*'"
        insertbefore: '^#port = 5432'

    - name: restart postgresql
      service:
        name: postgresql.service
        state: restarted
        sleep: 5

    - name: Allow 5432 port for PostgreSQL on firewall
      ansible.posix.firewalld:
        port: 5432/tcp
        zone: public
        permanent: yes
        state: enabled
        immediate: yes
```

Figure 8.8 – Tasks for opening the port and database service

6. Execute the playbook and deploy the PostgreSQL server:

● ● ●

```
[ansible@ansible Chapter-08]$ ansible-playbook postgres-deploy.yaml -e "NODES=node1"
```

Figure 8.9 – Execute Ansible playbook for PostgreSQL deployment

7. Log in to the database server (node1) and verify the database server's details by switching to the postgres user, as follows:

● ● ●

```
[devops@node-1 ~]$ sudo su - postgres
Last login: Tue Mar 15 09:59:35 UTC 2022 on pts/1

[postgres@node-1 ~]$ postgres -V
postgres (PostgreSQL) 10.17
```

Figure 8.10 – Logging in to the PostgreSQL database server

8. Open the psql client as the postgres user, as follows:

● ● ●

```
[postgres@node-1 ~]$ psql
psql (10.17)
Type "help" for help.

postgres=
```

Figure 8.11 – Open psql client on the database server

9. List the existing databases with the \l command and verify that database_demo was created (as per the variable configuration in the vars/postgres.yaml file):

● ● ●

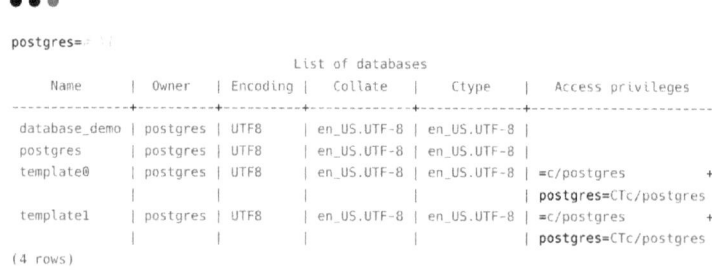

```
postgres=#
                                   List of databases
     Name      |  Owner   | Encoding |  Collate   |   Ctype    |   Access privileges
---------------+----------+----------+------------+------------+-----------------------
 database_demo | postgres | UTF8     | en_US.UTF-8 | en_US.UTF-8 |
 postgres      | postgres | UTF8     | en_US.UTF-8 | en_US.UTF-8 |
 template0     | postgres | UTF8     | en_US.UTF-8 | en_US.UTF-8 | =c/postgres          +
               |          |          |            |            | postgres=CTc/postgres
 template1     | postgres | UTF8     | en_US.UTF-8 | en_US.UTF-8 | =c/postgres          +
               |          |          |            |            | postgres=CTc/postgres
(4 rows)
```

Figure 8.12 – Listing the existing databases in the psql command line

10. Verify the users list by using the \du command as follows:

● ● ●

```
postgres=# \du
                            List of roles
    Role name |                    Attributes                    | Member of
 -----------+--------------------------------------------------+------------
    demouser  |                                                  | {}
    postgres  | Superuser, Create role, Create DB, Replication, Bypass RLS | {}
```

Figure 8.13 – Verifying users in the psql command line

11. Exit the psql console by using the *Ctrl + D* keyboard shortcut or the \q command:

● ● ●

```
postgres=# \q
[postgres@node-1 ~]$
```

Figure 8.14 – Exiting the psql console

12. Also, verify the pg_hba.conf file; the geerlingguy.postgresql role will configure this file based on the content of your variable. Check out the /var/lib/pgsql/data/pg_hba.conf file, as follows:

● ● ●

```
[postgres@node-1 ~]$ cat /var/lib/pgsql/data/pg_hba.conf
#
# Ansible managed
#
# PostgreSQL Client Authentication Configuration File
# ===================================================
#
# See: https://www.postgresql.org/docs/current/static/auth-pg-hba-conf.html

local all all     peer
host all all 0.0.0.0/0    md5
```

Figure 8.15 – Verifying the /var/lib/pgsql/data/pg_hba.conf file

If you want to access PostgreSQL from remote nodes, you need to make sure that the pg_hba.conf file entries have been configured appropriately.

It is a best practice to use existing Ansible roles so that you can save a lot of time and effort while developing your automation content. Also, you need to make sure that the Ansible role and its methods are suitable for your environment instead of blindly using them.

In the next section, you will learn how to update the password for the default `postgres` user.

Configuring a password for a default postgres user

The default user, `postgres`, is configured with no password and the default authentication method is **ident** (`https://www.postgresql.org/docs/11/auth-ident.html`). However, to perform automated operations using Ansible, we need to pass the password for the `postgres` user (or the passwords for other admin users). Set or update the password for the `postgres` user as follows:

1. Switch to the `postgres` user:

```
[devops@node-1 ~]$ sudo su - postgres
Last login: Tue Mar 15 08:59:39 UTC 2022 on pts/1
[postgres@node-1 ~]$

# open psql command line
[postgres@node-1 ~]$ psql
psql (10.17)
Type "help" for help.
```

Figure 8.16 – Switch to postgres user and open psql cli

2. Change the password, exit the `psql` console and then exit the `postgres` user as follows:

```
postgres=# ALTER USER postgres WITH ENCRYPTED PASSWORD 'pass123';
ALTER ROLE

postgres=# \q

[postgres@node-1 ~]$ exit
logout
[devops@node-1 ~]$
```

Figure 8.17 – Change the password and exit from the postgres account

These steps can also be automated in your database installation playbook if required, though this depends on your organization's requirements.

It is also possible to automate other database server installations using Ansible, such as Microsoft SQL Server. We will look at this in the next section.

Installing Microsoft SQL Server on Linux

In 2016, Microsoft announced via their blog (refer to the following information box) that Microsoft SQL Server would run on Linux platforms. So, without a Windows server, you can install and use MSSQL databases, and the installation will support the most common Linux platforms, such as **Red Hat Enterprise Linux (RHEL)**, **SUSE Enterprise Linux Server (SLES)**, Ubuntu, and so on.

Installing MSSQL is pretty straightforward for Linux, but still, there are several steps involved in terms of configuration and services. When several database servers must be installed as part of the deployment, instead of configuring each manually, it is possible to use Ansible roles and playbooks to achieve this. Follow the steps as per documentation (refer to the following information box) or use any existing contributions from the community; for example, the role available at `galaxy.ansible.com/microsoft/sql` can be used to install MSSQL on Linux:

```
[ansible@ansible Chapter-08]$ ansible-galaxy collection install microsoft.sql
```

Figure 8.18 – Installing microsoft.sql collection

The `community.general.mssql_db` community module can be used to add or remove MSSQL databases, as follows:

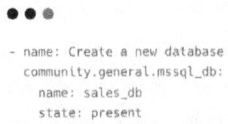

```
- name: Create a new database
  community.general.mssql_db:
    name: sales_db
    state: present
```

Figure 8.19 – Creating Microsoft SQL database

Explore more automation content for MSSQL in Ansible Galaxy and use it based on your environment's requirements.

> **Announcing SQL Server on Linux**
>
> Microsoft's announcement about SQL Server availability for Linux platforms can be found at `https://blogs.microsoft.com/blog/2016/03/07/announcing-sql-server-on-linux/`.
>
> Installation guidance for SQL Server on Linux can be found at `https://docs.microsoft.com/en-us/sql/linux/sql-server-linux-setup`.

Once the database server has been installed and configured, automate additional tasks, such as creating a new database, creating new tables, adding users or permissions. Use the Ansible collections and modules that are available, such as `community.postgresql`, `community.mysql`, `community.cockroachdb`, `community.cassandra`, and more. We will learn about PostgreSQL database automation in the next section while using the `community.postgresql` Ansible collection.

Creating and managing databases using Ansible

The community collection for PostgreSQL comes with more than 20 modules and a few plugins. It is possible to use these modules and plugins to automate PostgreSQL database operations, including creating, dropping, and updating databases, tables, users, and other resources in the database server.

Ansible community.postgresql prerequisites

If you are accessing PostgreSQL from a remote node (for example, an Ansible control node), then you need to install the `psycopg2` Python library on this machine to use these PostgreSQL modules:

```
$ pip install psycopg2
```

In the next section, we will execute tasks from the database node itself (`node1`) using Ansible. This library is not required as the database server has already been configured with the required dependencies.

In the next section, you will learn how to manage database operations using Ansible and the `community.postgresql` collection.

Managing the database life cycle

In this section, you will learn how to create a database, create tables inside the new database, and then configure users and permissions using Ansible. Follow these steps:

1. Create the `postgres-manage-database.yaml` playbook and add the variables that provide details about the database to be created, tables to be configured, users to be added, and so on. Remember to use Ansible Vault to encrypt sensitive items such as the username and password as follows:

```
# Chapter-08/postgres-manage-database.yaml
---
- name: Deploying PostgreSQL Database Server
  hosts: "{{ NODES }}"
  vars:
    ansible_become_user: postgres
    postgres_user: postgres
    postgres_password: 'PassWord'
    postgres_host: localhost
    postgres_database: db_sales
    postgres_table: demo_table
    postgres_new_user_name: devteam
    postgres_new_user_password: 'DevPassword'
```

Figure 8.20 – Playbook to manage the database operations

2. Add a task to create a new PostgreSQL database, as follows:

```
# Chapter-08/postgres-manage-database.yaml
  .
  .
  tasks:
    - name: Create a new database
      community.postgresql.postgresql_db:
        login_user: "{{ postgres_user }}"
        login_password: "{{ postgres_password | default(omit) }}"
        login_host: "{{ postgres_host | default('localhost') }}"
        name: "{{ postgres_database }}"
```

Figure 8.21 – Task to create database

3. Now, add another task to create the table with columns:

```
# Chapter-08/postgres-manage-database.yaml
  .
  .
    - name: Create table with few columns
      community.postgresql.postgresql_table:
        login_user: "{{ postgres_user }}"
        login_password: "{{ postgres_password }}"
        login_host: "{{ postgres_host }}"
        db: "{{ postgres_database }}"
        name: "{{ postgres_table }}"
        columns:
          - id bigserial primary key
          - num bigint
          - stories text
        ssl_mode: disable
```

Figure 8.22 – Task to create table inside the database

4. Add one more task for creating users and grant access to the newly created database:

●●●

```
# Chapter-08/postgres-manage-database.yaml
.
.
  - name: Create user and grant access to database
    community.postgresql.postgresql_user:
      login_user: "{{ postgres_user }}"
      login_password: "{{ postgres_password }}"
      login_host: "{{ postgres_host }}"
      db: "{{ postgres_database }}"
      name: "{{ postgres_new_user_name }}"
      password: "{{ postgres_new_user_password }}"
      encrypted: yes
      priv: "CONNECT/{{ postgres_table }}:ALL"
      expires: "Dec 31 2022"
      comment: "Developer user access"
      state: present
```

Figure 8.23 – Task to create user and grant access to the database

5. Execute the playbook to create the database and other resources:

●●●

```
[ansible@ansible Chapter-08]$ ansible-playbook postgres-manage-database.yaml -e "NODES=node1"
```

Figure 8.24 – Execute playbook to create database, table and user

6. Once the playbook has been created, verify the database and resources on the database server (for example, node1).

7. Log in to node1, switch to the postgres user and open the psql console. List the databases and you will see the db_sales database and the access privilege for devteam:

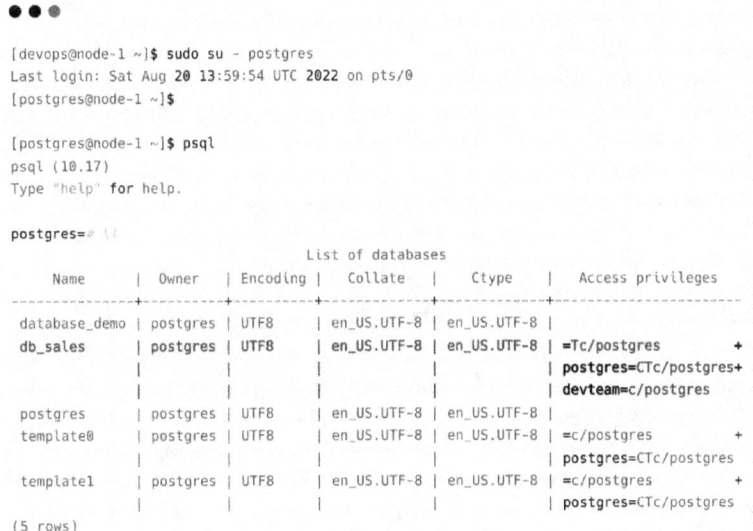

```
[devops@node-1 ~]$ sudo su - postgres
Last login: Sat Aug 20 13:59:54 UTC 2022 on pts/0
[postgres@node-1 ~]$

[postgres@node-1 ~]$ psql
psql (10.17)
Type "help" for help.

postgres=#
                          List of databases
       Name      |  Owner   | Encoding |  Collate    |   Ctype     |   Access privileges
---------------+----------+----------+-------------+-------------+------------------------
 database_demo | postgres | UTF8     | en_US.UTF-8 | en_US.UTF-8 |
 db_sales      | postgres | UTF8     | en_US.UTF-8 | en_US.UTF-8 | =Tc/postgres         +
               |          |          |             |             | postgres=CTc/postgres+
               |          |          |             |             | devteam=c/postgres
 postgres      | postgres | UTF8     | en_US.UTF-8 | en_US.UTF-8 |
 template0     | postgres | UTF8     | en_US.UTF-8 | en_US.UTF-8 | =c/postgres          +
               |          |          |             |             | postgres=CTc/postgres
 template1     | postgres | UTF8     | en_US.UTF-8 | en_US.UTF-8 | =c/postgres          +
               |          |          |             |             | postgres=CTc/postgres
(5 rows)
```

Figure 8.25 – Log in to the database server and verify details

8. Verify the user details in psql, as follows:

```
postgres=# \du
                               List of roles
 Role name |                         Attributes                         | Member of
-----------+------------------------------------------------------------+-----------
 demouser  |                                                            | {}
 devteam   | Password valid until 2022-12-31 00:00:00+00                | {}
 postgres  | Superuser, Create role, Create DB, Replication, Bypass RLS | {}
```

Figure 8.26 – Listing and verifying the newly created user

9. Verify that the tables have been created as per the playbook.

First, connect to the newly created database from the psql console using the \c command and then list the tables inside the database using the \dt command, as follows:

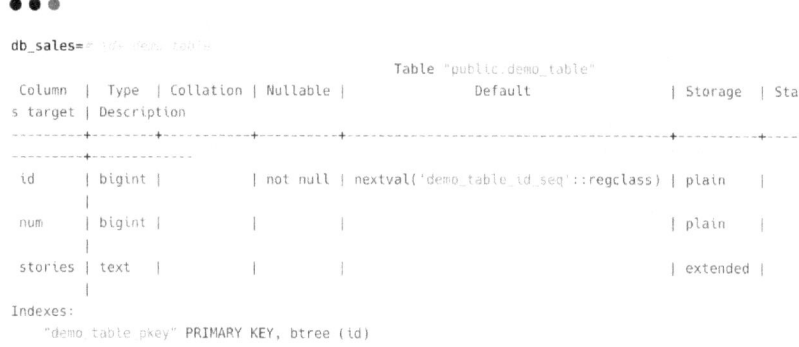

```
postgres=# \c db_sales
You are now connected to database "db_sales" as user "postgres".
db_sales=#
db_sales=# \dt
           List of relations
 Schema |    Name    | Type  |  Owner
--------+------------+-------+----------
 public | demo_table | table | postgres
(1 row)
```

Figure 8.27 – Connecting to the newly created database and list tables

10. Ensure that the columns that you have created using the Ansible playbook are in the table, as shown here:

```
db_sales=# \d+ demo_table
                                        Table "public.demo_table"
 Column |  Type  | Collation | Nullable |                 Default                 | Storage  | Stat
s target | Description
--------+--------+-----------+----------+-----------------------------------------+----------+-----
---------+-------------
 id     | bigint |           | not null | nextval('demo_table_id_seq'::regclass) | plain    |
        |
 num    | bigint |           |          |                                         | plain    |
        |
 stories | text  |           |          |                                         | extended |
        |
Indexes:
    "demo_table_pkey" PRIMARY KEY, btree (id)
```

Figure 8.28 – Database table details

11. Also verify access for the new user by using the psql console with the relevant username and password as follows:

```
[postgres@node-1 ~]$ psql -U devteam -h localhost -d db_sales
Password for user devteam:
psql (10.17)
Type "help" for help.

db_sales=> \dt
           List of relations
 Schema |    Name    | Type  |  Owner
--------+------------+-------+----------
 public | demo_table | table | postgres
(1 row)
```

Figure 8.29 – Verifying new user access and the list tables

Expand the playbook with more details, such as the columns that are required for the tables, more users, permissions, and so on. Refer to the module's documentation at `https://docs.ansible. com/ansible/latest/collections/community/postgresql/` for more details.

In the next section, you will learn more about database operations, such as how to manage remote access by automating `pg_hba` configurations, taking database dumps, and so on.

Automating PostgreSQL operations

With the help of the modules in the `community.postgresql` collection, it is possible to automate more database maintenance and operations. Let's take a closer look.

Managing PostgreSQL remote access

Database servers are accessed by applications on remote nodes and this access needs to be configured appropriately and securely. For a test environment, allow wildcard entries (for example, `0.0.0.0/0`), but this is not a recommended practice for production servers. You need to configure the correct IP address or hostname to allow or restrict access to the database. This operation can be automated using the `community.postgresql.postgresql_pg_hba` module, as follows:

```
- name: Grant users access to databases
  community.postgresql.postgresql_pg_hba:
    dest: /var/lib/postgres/data/pg_hba.conf
    contype: host
    users: johnt
    source: 192.168.0.100/24
    databases: db_sales
    method: peer
    create: true
```

Figure 8.30 – Grant user access to database

Managing the `pg_hba` entries using Ansible will allow you to handle the entire life cycle of the database and its access.

Next, we will learn how to take automated database backups.

Database backup and restore

Taking database backups is critical for sensitive and important data. Use Ansible to automate this database dump and schedule daily, weekly, or monthly database backups:

● ● ●

```
- name: Dump existing database to a file
  community.postgresql.postgresql_db:
    login_user: "{{ postgres_user }}"
    login_password: "{{ postgres_password }}"
    login_host: "{{ postgres_host }}"
    name: "{{ postgres_database }}"
    state: dump
    target: /data/db_dumps/daily_prod_db_sales.sql
```

Figure 8.31 – Database backup using Ansible

The backup will be saved on the managed node. Customize the destination or automatically copy the backups to remote locations such as NFS volumes or cloud storage.

Similarly, we can automate the database restore operation as follows:

● ● ●

```
- name: Restore backup from file to database
  community.postgresql.postgresql_db:
    login_user: "{{ postgres_user }}"
    login_password: "{{ postgres_password }}"
    login_host: "{{ postgres_host }}"
    name: "{{ postgres_database }}"
    state: restore
    target: /tmp/test.sql
```

Figure 8.32 – Restore database from backup file

Notice state: restore in the preceding example. This instructs Ansible to perform a restore operation using the file or archive mentioned in the target parameter. The following screenshot shows the full playbook, which can perform backup or restore operations based on the db_action value:

```
# Chapter-08/postgres-backup-restore.yaml
---
- name: Deploying PostgreSQL Database Server
  hosts: "{{ NODES }}"
  vars:
    ansible_become_user: postgres
    postgres_user: postgres
    postgres_password: 'PassWord'
    postgres_host: localhost
    postgres_database: db_sales

    db_action: 'restore' # 'backup'
  tasks:
    - name: Dump existing database to a file
      community.postgresql.postgresql_db:
        login_user: "{{ postgres_user }}"
        login_password: "{{ postgres_password }}"
        login_host: "{{ postgres_host }}"
        name: "{{ postgres_database }}"
        state: dump
        target: /tmp/test.sql
      when: db_action == 'backup'

    - name: Restore backup from file to database
      community.postgresql.postgresql_db:
        login_user: "{{ postgres_user }}"
        login_password: "{{ postgres_password }}"
        login_host: "{{ postgres_host }}"
        name: "{{ postgres_database }}"
        state: restore
        target: /tmp/test.sql
      when: db_action == 'restore'
```

Figure 8.33 – PostgreSQL database backup and restore playbook

Please refer to the `https://docs.ansible.com/ansible/latest/collections/community/postgresql/postgresql_db_module.html` documentation to learn more about the `community.postgresql.postgresql_db` module.

Ansible PostgreSQL and MySQL Collection

Please refer to `https://galaxy.ansible.com/community/postgresql` for the PostgreSQL community collection and `https://docs.ansible.com/ansible/latest/collections/community/postgresql/` for the documentation. Also, check out `https://galaxy.ansible.com/community/mysql` for the Ansible MySQL collection and modules.

In the next section, you will learn about automated database password reset request handling, which you can do using Ansible and your ITSM tool.

Automating a password reset using ITSM and Ansible

With the help of ITSM tools and Red Hat Ansible Automation Platform (or community Ansible AWX), it is possible to implement zero-touch automation use cases such as database user password resets, database provisioning, and so on. Users will interact with the ITSM tool, and the tool will interact with Ansible Automation Platform to implement the task implementation, as shown in the following figure:

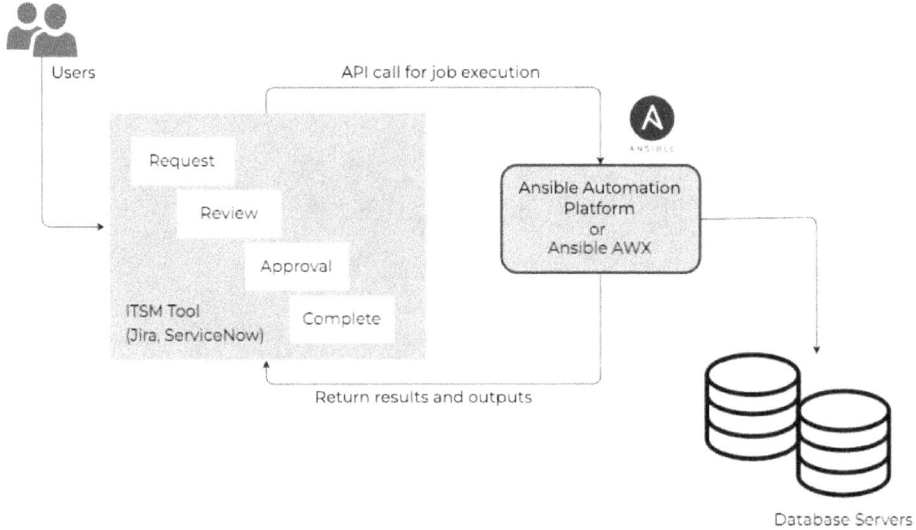

Figure 8.34 – ITSM and Ansible Automation Platform integration for database operations

This **programmatic automation** is one of the best features of Ansible and helps organizations scale their automation landscape by integrating with existing tools and software.

Use customized forms or ticketing systems in the ITSM tool, as shown in the following screenshot:

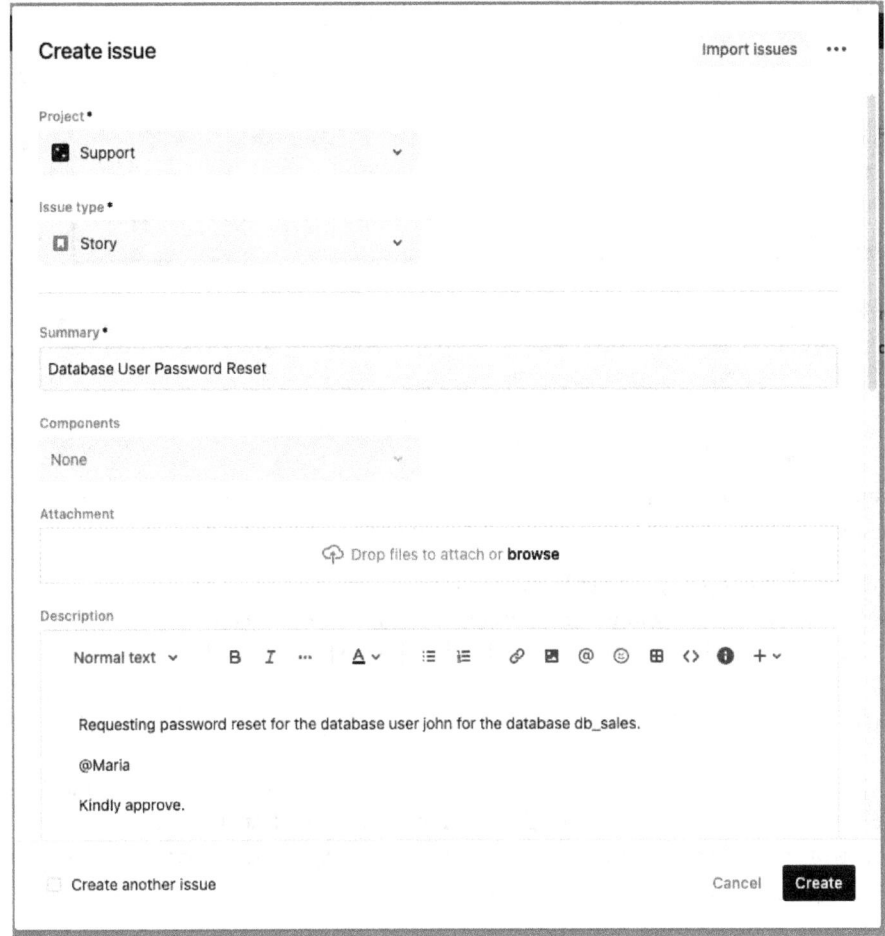

Figure 8.35 – Jira ticket and its details

The ITSM tool, such as Jira or ServiceNow, can also be configured with custom fields to collect information, such as the database server's name, database name, username, and more, as shown in the following screenshot:

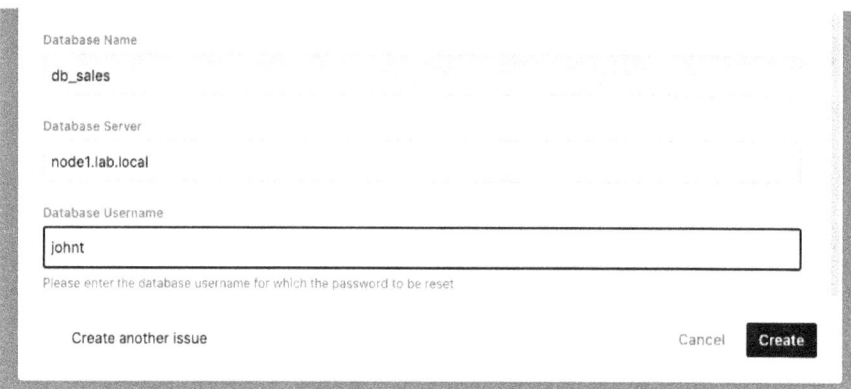

Figure 8.36 – Jira ticket with custom fields

Automation will not skip your ITSM processes or procedures; the review and approvals will be in place before the task is executed, as shown in *Figure 8.34*.

Once the approval happens, the ITSM tool will send a notification containing data (such as the database hostname, database name, and username) to Ansible Automation Platform and trigger the automation job. We will look at this in more detail in the next section.

Ansible playbook for resetting passwords

The Ansible playbook will be triggered from Ansible Automation Platform based on the Ansible job template and the input data from Jira. Collect the input data from Jira with the custom fields shown in the following screenshot (refer to `Chapter-08/postgres-password-reset.yaml` for more details):

```
# Chapter-08/postgres-password-reset.yaml
---
- name: Deploying PostgreSQL Database Server
  ## Collect the database server name from Jira
  hosts: "{{ DATABASE_NODE }}"
  vars:
    ansible_become_user: postgres
    postgres_user: postgres
    postgres_password: 'PassWord'
    postgres_host: localhost

    ## Collect the database name from Jira
    postgres_database: "{{ DATABASE_NAME }}"
    ## Collect the database user name from Jira
    db_user_name: "{{ DATABASE_USER_NAME }}"
    ## Generate random password
    db_user_password: "{{ lookup('password', '/dev/null chars=ascii_lowercase,digits length=8') }}"
```

Figure 8.37 – Collecting details from Jira in the playbook

The following task will set the new password for the user:

```
# Chapter-08/postgres-password-reset.yaml

  - name: Set user's password with no expire date
    community.postgresql.postgresql_user:
      login_user: "{{ postgres_user }}"
      login_password: "{{ postgres_password | default(omit) }}"
      login_host: "{{ postgres_host | default('localhost') }}"
      db: "{{ postgres_database }}"
      name: "{{ db_user_name }}"
      password: "{{ db_user_password }}"
      priv: "CONNECT/products:ALL"
      expires: infinity
```

Figure 8.38 – A task for setting a new password

Once the password reset operation is successful, the following task will update the Jira ticket with the output of the password reset operation:

```
# Chapter-08/postgres-password-reset.yaml

  - name: Comment on Jira issue
    community.general.jira:
      uri: '{{ jira_server }}'
      username: '{{ jira_user }}'
      password: '{{ jira_pass }}'
      issue: '{{ issue.meta.key }}'
      operation: comment
      comment: 'Password has been reset for the user {{ db_user_name }}, for the database {{ postgres_database
}}'
```

Figure 8.39 – Updating the Jira ticket using the community.general.jira module

Expand the playbook by adding tasks that send the new password to the user via email using the community.general.mail module. Read more about Ansible Jira module from documentation (https://docs.ansible.com/ansible/latest/collections/community/general/jira_module.html).

Summary

In this chapter, you learned about how Ansible can help you install database servers and manage database operations such as creating databases, creating tables, assigning user permissions, taking database backups, and configuring pg_hba. You also learned about the integration opportunities that are provided by the ITSM tools for implementing zero-touch automation with Ansible Automation Platform.

In the next chapter, you will learn how to integrate Ansible with your DevOps practices for deployment, rolling updates, IaC provisioning, and more.

Further reading

To learn more about the topics that were covered in this chapter, please visit the following links:

- *How to send emails using Ansible and Gmail*: `https://www.techbeatly.com/ansible-gmail`

- *Using Ansible to deploy Microsoft SQL Server 2019 on Red Hat Enterprise Linux 8*: `https://www.redhat.com/sysadmin/mssql-linux-easy`

- *Community.Postgresql collection*: `https://docs.ansible.com/ansible/latest/collections/community/postgresql/index.html`

- *Ansible Database modules*: `https://docs.ansible.com/ansible/2.9/modules/list_of_database_modules.html`

- *Automating IT Service Management with ServiceNow and Red Hat Ansible Automation Platform*: `https://www.ansible.com/integrations/it-service-management/servicenow`

- *MongoDB Collection for Ansible*: `https://galaxy.ansible.com/community/mongodb`

9
Implementing Automation in a DevOps Workflow

DevOps is a combination of practices, tools, and philosophies that can help increase the speed, efficiency, and security of software development, application delivery, and infrastructure management processes. DevOps practices and methods are common in organizations now due to several advantages, such as faster and frequent deployments, improvement in quality, fewer errors, and high transparency via automation. By combining automation, collaboration, and integration, it is possible to develop and implement efficient DevOps practices, ensuring much higher quality output from your IT operations team.

Due to the numerous integrations, supported plugins, and modules, Ansible is a great tool for automating the tasks in your DevOps workflows. Ansible can help you automate different stages in the **software development life cycle** (**SDLC**), such as building applications, scanning the source code, storing artifacts in repositories, deploying the application, configuring application services, and more. Automating such application life cycle processes is known as **continuous integration** and **continuous delivery** (**CI/CD**). There are several choices for CI/CD tools and frameworks, such as Jenkins, CircleCI, GitLab, GitHub Actions, Bamboo, and others.

This chapter will focus on using Ansible inside the CI/CD and DevOps workflow to deploy and manage applications rather than using Ansible as a CI/CD tool.

In this chapter, we will cover the following topics:

- A quick introduction to DevOps
- Serving applications using a load balancer
- Rolling updates using Ansible
- Using Ansible as a provisioning tool in Terraform

First, you will learn how to use Ansible to deploy applications to servers, including the load balancer configuration. You will also learn how to implement rolling updates using Ansible to deploy the application without downtime and interruption. Finally, you will learn how to use Ansible as a provisioner and configuration management tool with the infrastructure management tool Terraform.

Technical requirements

The following are the technical requirements for this chapter:

- A Linux machine for the Ansible control node (with internet access)
- Three Linux machines for installing and configuring applications
- Basic knowledge of DevOps methodologies, CI/CD tools (Jenkins), and the Git workflow
- Basic knowledge of Terraform

All the Ansible code, playbooks, commands, and snippets for this chapter can be found in this book's GitHub repository at `https://github.com/PacktPublishing/Ansible-for-Real-life-Automation/tree/main/Chapter-09`.

A quick introduction to DevOps

In simple words, **DevOps** is the combination of **development** (**Dev**) and **operations** (**Ops**), but in reality, DevOps is a combination of ideas, tools, and practices that help increase the speed and efficiency of software development, delivery, and infrastructure management processes. There are several known best practices we can follow and include in the DevOps workflow, as follows:

- Team collaboration and transparent communication
- CI/CD
- **Infrastructure as code** (**IaC**) and automated infrastructure management
- Containerization and microservices
- Logging, monitoring, and feedback loops

One of the key concepts in DevOps practices is to reduce the time and effort required for application life cycle management, such as integration, build, test, release, and deployment. Using DevOps methodologies and tools, it is possible to automate this process. This is known as CI/CD.

> **Learning about DevOps**
>
> Refer to the following guides to understand and learn more about DevOps and CI/CD processes:
>
> - What is DevOps?: `https://aws.amazon.com/devops/what-is-devops/`
> - DevOps explained: `https://about.gitlab.com/topics/devops/`
> - Understanding DevOps: `https://www.redhat.com/en/topics/devops`

In a typical CI/CD workflow, the developer will push code to the central code repository (a Git server, for example) and whenever there is a change in the repository's content, a trigger will be sent to the CI/CD tool (such as Jenkins, CircleCI, GitHub Actions, and so on). The following diagram shows a typical CI/CD environment:

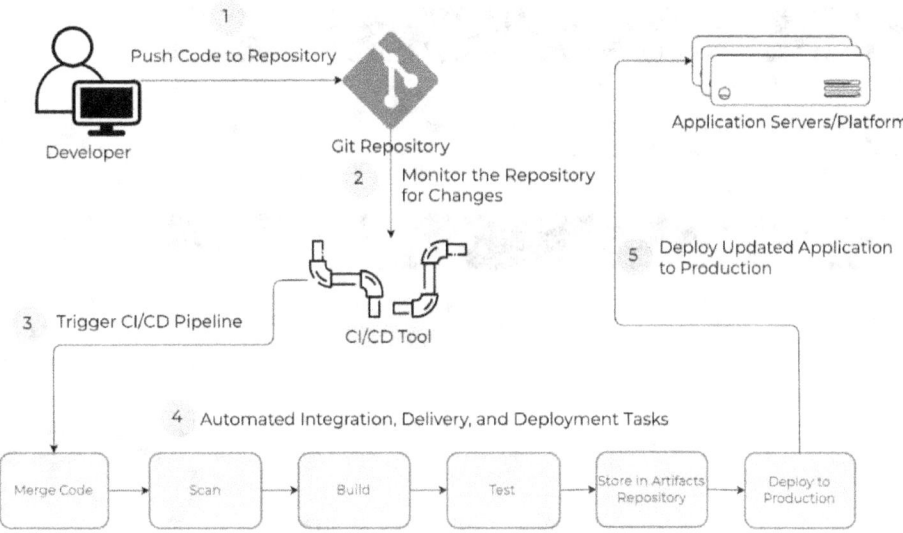

Figure 9.1 – A typical workflow in a CI/CD environment

Several tasks are involved in the build, test, delivery, and deployment processes, depending on the application type, application platform, and other environmental factors.

The following diagram shows the typical manual and automated tasks for CI, CD, and continuous deployment processes:

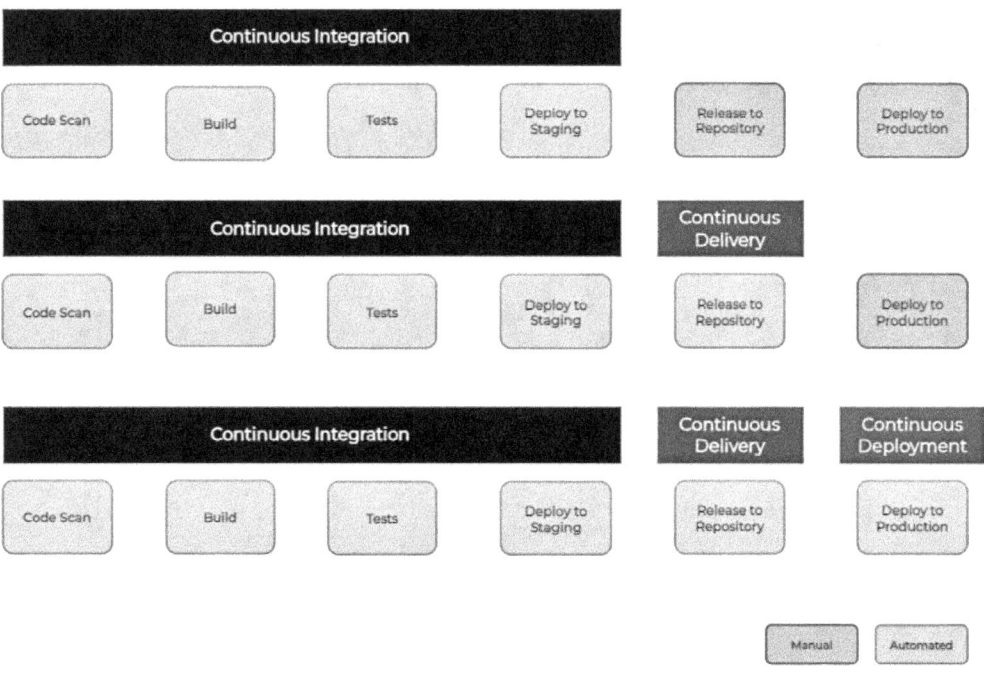

Figure 9.2 – Typical tasks in the CI/CD process

The application can be a simple JAR file, a compressed image, a container image, or in any other format (we will learn about container management using Ansible in *Chapter 10, Managing Containers Using Ansible*).

CI helps developers merge the software code changes regularly and complete the testing and scanning processes automatically and quickly. The CI process also helps detect the defects, bugs, and security issues in the code quicker and more effectively.

The **CD** process involves automated software life cycle operations such as testing the application, scanning, and preparing the application so that it's ready for the production environment (release).

Once the application has been built, scanned, and made available in the application repository (application artifacts), it needs to be deployed to production (or the development environment) as per the process. This is the next **CD** process or **continuous deployment** task and depending on the environment, this can be implemented as an automated or semi-automated deployment.

Continuous delivery versus continuous deployment

CD helps in the application life cycle by deploying the application to production so that the latest change in the application will reach the end users automatically as part of the CI/CD process without any manual intervention. The same CI/CD tool or a dedicated tool can be used for the continuous deployment process, depending on your application's nature, environment, and dependencies.

The deployment can be part of the CI/CD pipelines or a separate trigger for the deployment tool, such as **Ansible Automation Platform** (**AAP**) (you will learn how to integrate Ansible inside the CI/CD pipeline using Jenkins in *Chapter 12, Integrating Ansible with Your Tools*).

Ansible inside CI/CD tasks

Ansible can be used as the tool for most of the tasks in the CI/CD workflows, as follows:

- Scanning the application
- Building application artifacts
- Running unit and integration tests
- Promoting and testing the application in the staging environment
- Storing application artifacts in the artifacts repository
- Deploying the application to production

In the next section, you will learn how to use Ansible to deploy applications to production servers as a continuous deployment tool.

> **AAP as a CI/CD Tool**
>
> It is possible to use AAP as a CI/CD tool and manage the full life cycle of an application using Ansible playbooks and job templates. You will learn more about this in *Chapter 12, Integrating Ansible with Your Tools*.

It is possible to reduce the software deployment time by using Ansible. Since Ansible can manage the application deployment effectively and efficiently, inside the CI/CD pipeline, Ansible can be used as the primary tool to deploy applications. The following diagram shows how Ansible is used in Jenkins pipeline jobs for deployment purposes (continuous deployment):

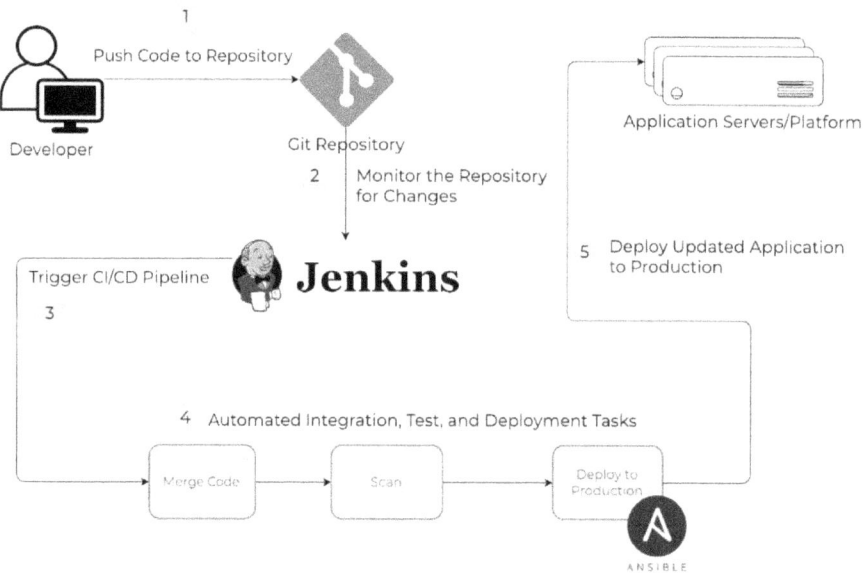

Figure 9.3 – Ansible inside a Jenkins pipeline job

The software build and CI/CD workflow can be triggered by several native Jenkins methods and also by using additional plugins. For example, to activate the build trigger based on Git repository changes and to execute Ansible playbooks from Jenkins, you must perform a few mandatory steps, as follows:

1. First, you must install and configure the Ansible plugin for Jenkins to use Ansible inside the Jenkins pipeline job. Refer to the documentation at `https://www.jenkins.io/doc/pipeline/steps/ansible/` to learn more about the Ansible plugin for Jenkins.

> **Ansible in Jenkins**
>
> Refer to `https://plugins.jenkins.io/ansible` to learn more about the Ansible plugin for Jenkins.

2. Then, you must install and configure Ansible and its required packages on the Jenkins server (or the Jenkins agent machine) as the Ansible playbook will be executed from the Jenkins machine. (Later, in *Chapter 12, Integrating Ansible with Your Tools*, you will learn how to use Jenkins to call automation jobs in AAP.)

3. To trigger the Jenkins pipeline job, the build trigger must be configured on the Jenkins job, as shown in the following screenshot. Copy the URL (`JENKINS_URL/job/ansible-demo/build?token=TOKEN_NAME`) and the `TOKEN` value and use them in the webhook configuration in the Git server:

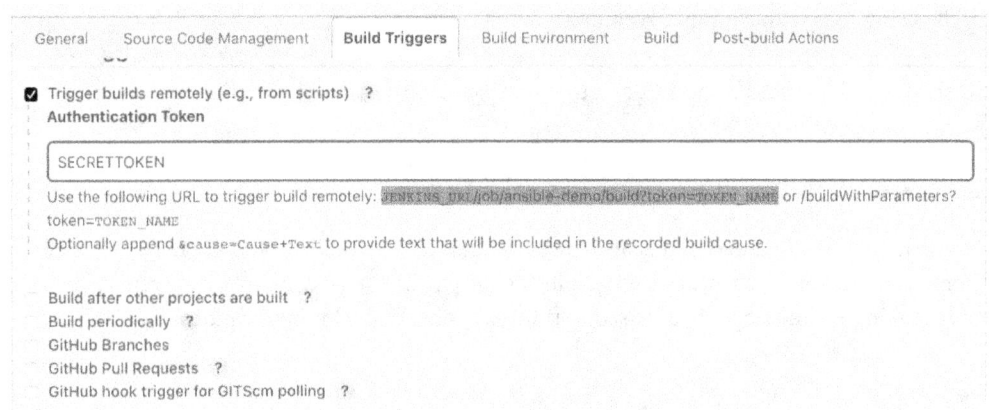

Figure 9.4 – Build trigger configured on the Jenkins pipeline job

4. Whenever there is a change in the application code, you need to trigger the Jenkins pipeline job. To do this, we have configured a webhook in the application repository in GitHub, as shown in the following screenshot:

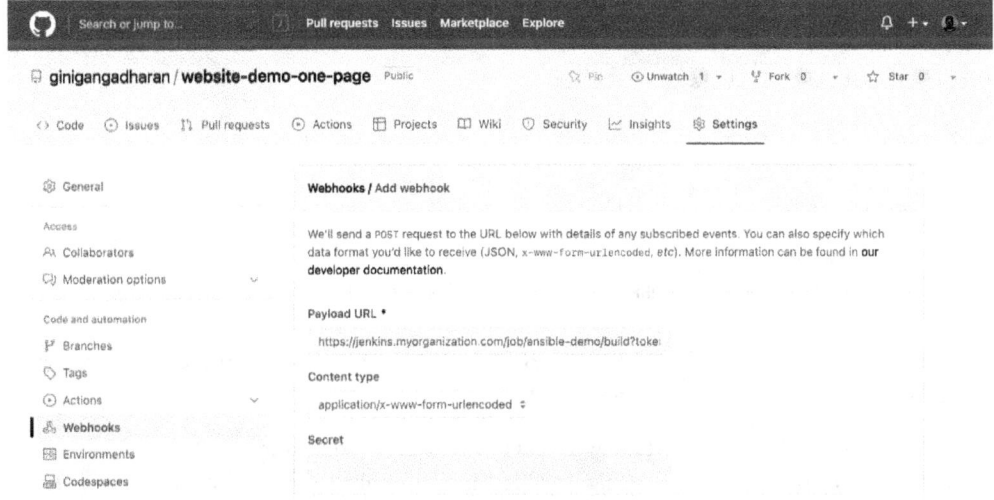

Figure 9.5 – Webhook configuration in the GitHub repository

Additional configurations are available in the GitHub webhook configuration to help you decide on what condition the webhook is to be called. Refer to the GitHub webhook documentation (https://docs.github.com/en/developers/webhooks-and-events/webhooks/about-webhooks) to learn more about webhooks.

> **Triggering Jobs with a Simple Webhook**
>
> To learn more about Jenkins and webhooks, read `https://docs.cloudbees.com/docs/admin-resources/latest/triggering-jobs-simple-webhook`

Using Ansible inside a Jenkins pipeline

Once the Ansible plugin has been installed and configured, an Ansible playbook can be executed from the Jenkins server (or the agent) by calling it inside the pipeline stages. The following screenshot shows sample Jenkin pipeline stages being used to utilize Ansible to deploy applications:

```
pipeline {
    agent any
    stages {
        stage ("Fetch Ansible content") {
            steps {
                git "https://github.com/giniqangadharan/website-demo-one-page.git"
            }
        }
        stage("Deploy application using Ansible") {
            steps {
                ansiblePlaybook credentialsId: 'private-key', disableHostKeyChecking: true, installation: 'Ansible',
inventory: 'produ.inventory', playbook: 'deploy-web.yaml'
            }
        }
    }
}
```

Figure 9.6 – Jenkins pipeline job stages with Ansible tasks to deploy applications

In the following exercise, I will explain how to deploy website content from the source repository using Ansible for continuous deployment tasks.

I will use simple website content (static website) to avoid any complications to help you understand the application deployment concept using Ansible. The playbook can be integrated inside the Jenkins pipeline (or whichever CI/CD tool you are using) to implement the continuous deployment task. Follow these steps:

1. Update the `Chapter-09/hosts` inventory file with `node1` and `node2` as part of the web host group, as follows:

```
● ● ●
[web]
node1 ansible_host=192.168.56.25
node2 ansible_host=192.168.56.24
```

Figure 9.7 – Web hosts in the inventory

2. Create a playbook called `Chapter-09/deploy-web.yaml` and add the following content:

```
● ● ●
---
# Chapter-09/deploy-web.yaml
- name: Deploying Application
  hosts: "{{ NODES }}"
  become: yes
  vars:
    application_repo: 'https://github.com/ginigangadharan/website-demo-one-page'
    application_branch: production
    application_path: /var/www/html

  tasks:
```

Figure 9.8 – Playbook for deploying the web application

The variables can be kept in a separate file or passed from your CI/CD tool as arguments.

3. Add a task that will clean up the application directory and recreate it (this is to ensure any old versions of files are removed from the application path), as follows:

```
● ● ●
# Chapter-09/deploy-web.yaml...

  - name: Delete content & directory if exists
    ansible.builtin.file:
      state: absent
      path: "{{ application_path }}"

  - name: Create application directory
    ansible.builtin.file:
      state: directory
      path: "{{ application_path }}"
      mode: '0755'
```

Figure 9.9 – Tasks to housekeep the application directory

4. Add tasks to the same playbook to install the required packages and dependencies. Even if you are deploying the application on the same server, it is a best practice to install and configure dependencies during every deployment. This can include services, packages, system libraries, Python packages, or other files, depending on your application's type and framework.

5. It is also possible to mention the specific version of the packages, as shown in the following screenshot. Also add tasks to start `firewalld`, open the firewall port for the web service, and start the web service, as follows:

```
# Chapter-09/deploy-website.yml
.
.

  - name: Install httpd, firewalld and Git packages
    ansible.builtin.dnf:
      name:
        - httpd >= 2.4
        - firewalld
        - git
      state: latest

  - name: Enable and Run firewalld service
    ansible.builtin.service:
      name: firewalld
      enabled: true
      state: started

  - name: Permit httpd service in firewall
    ansible.posix.firewalld:
      service: http
      permanent: true
      state: enabled
      immediate: yes

  - name: Enable and start httpd service
    ansible.builtin.service:
      name: httpd
      enabled: true
      state: started
```

Figure 9.10 – Installing the package and starting the necessary services

6. The next step is to deploy the website's content to the application path. (In this exercise, we are using static website content and not a dynamic application.) To identify the servers, update the `index.html` file as follows (we are replacing SERVER_DETAILS with custom text that contains node information in the following task):

```
# Chapter-09/deploy-web.yaml...
.
.
    - name: Git checkout the application or website
      ansible.builtin.git:
        repo: "{{ application_repo }}"
        dest: "{{ application_path }}"
        version: "{{ application_branch }}"

    - name: Update index.html with server details
      ansible.builtin.lineinfile:
        path: "{{ application_path }}/index.html"
        regexp: 'SERVER_DETAILS'
        line: "<h3>(Installed using Ansible. Serving from {{ ansible_hostname }})</h3>"
```

Figure 9.11 – Deploying the application and updating its content

7. Add more tasks as needed, such as configuring the web server with more restrictions or custom SSL certificates.

8. It is important to add the verification step as part of automation. We will add automated website verification here. In this case, this is a simple health check to verify whether the website is working or not. Add a new play (not a task) in the same playbook, as follows:

```
# Chapter-09/deploy-web.yaml...
.
.
- name: Verify deployment
  hosts: "{{ NODES }}"
  become: no
  tasks:
    - name: Verify application health
      ansible.builtin.uri:
        url: http://{{ inventory_hostname }}.lab.local
        status_code: 200
      delegate_to: localhost
```

Figure 9.12 – Adding a play to verify the web service

Note that instead of hardcoding the server names, we are passing the details as Ansible *extra variables*, which will help you pass the server details from your CI/CD tools while executing the Ansible playbook.

Whenever we make changes in the application repository (`https://github.com/ginigangadharan/website-demo-one-page`), GitHub will trigger the build job in the Jenkins server and the playbook will be executed as part of the pipeline tasks. (We will explore the Jenkins job and pipelines using AAP in *Chapter 12, Integrating Ansible with Your Tools*.) For demonstration and testing purposes, let's execute the playbook manually from the console, as follows:

1. Execute the playbook and verify the success of tasks (you will get a similar result in the CI/CD console when you execute the playbook via the CI/CD pipeline):

Figure 9.13 – Ansible playbook output with a health check

The `Verify application health` task is successful, which means the website is working and serving the content.

2. Verify the website's content from a web browser, as shown here:

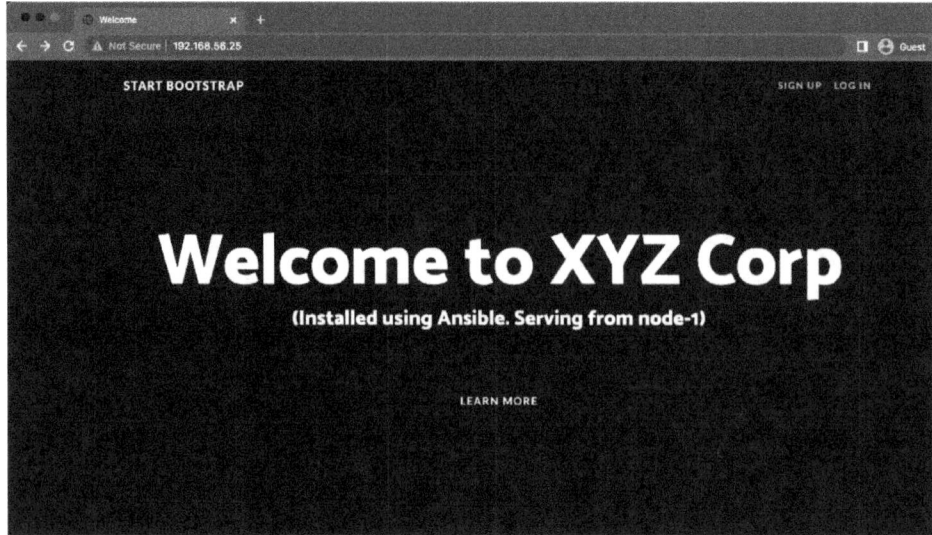

Figure 9.14 – Website deployed using Ansible

For practicing further, expand the deployment use case as follows:

- Deploy other web applications, API applications, or other compressed application files from repositories.

- Include more validations, test cases, scanning tasks, and more as needed as part of the health check.

- Use roles to deploy web services, load balancers, databases, and more.

You will learn about some of these scenarios in *Chapter 12, Integrating Ansible with Your Tools.* In the next section, you will learn how to handle multi-node web server traffic with a load balancer.

Serving applications using a load balancer

So far, you have learned how to deploy applications to multiple servers using Ansible with all the necessary prerequisites, dependencies, and basic health checks. But if the application or website is running on multiple servers, then you will need to tell the end user about multiple servers so that they can access the website. It is a best practice to serve the application from a single entity such as a load balancer, as shown in the following diagram, so that the end user doesn't need to know the actual web or application server IP addresses. It will also help you implement high availability and rolling updates for the application:

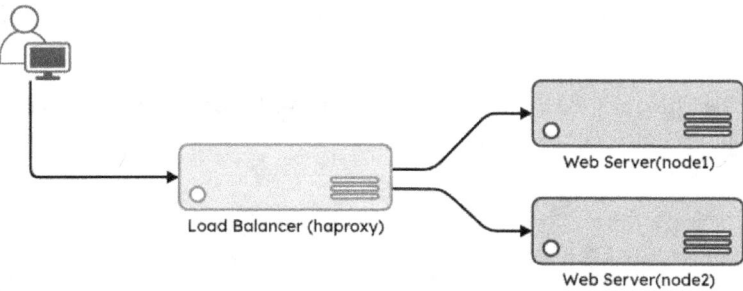

Figure 9.15 – Website hosted on multiple servers with a load balancer

Since we are handling the application deployment using Ansible inside the CI/CD workflow, we can include the load balancer installation and configuration tasks inside the pipeline, as shown in the following diagram:

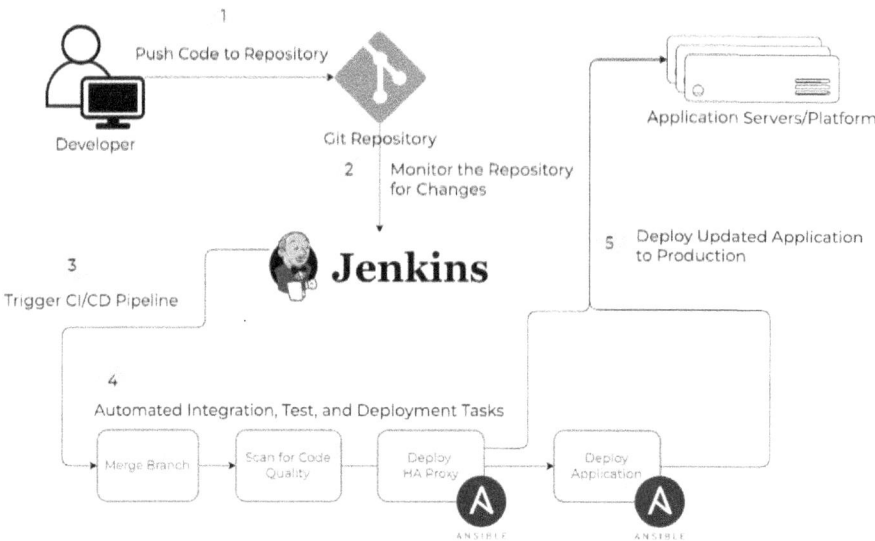

Figure 9.16 – Jenkins pipeline with HA Proxy installation

Since you have deployed website content on node1 and node2, in the following exercise, you will learn how to deploy a simple load balancer using haproxy and then configure node1 and node2 as backends:

1. Update the Chapter-09/hosts inventory file with node3 under the loadbalancer host group:

    ```
    [loadbalancer]
    node3 ansible_host=192.168.56.45
    ```

    Figure 9.17 – Load balancer entry in inventory

2. Instead of creating a playbook from scratch, use the haproxy role by *Jeff Geerling* (https://galaxy.ansible.com/geerlingguy/haproxy) from **Ansible Galaxy**, as follows:

    ● ● ●

    ```
    [ansible@ansible Chapter-09]$ cd roles

    [ansible@ansible roles]$ ansible-galaxy role install geerlingguy.haproxy
    ```

    Figure 9.18 – Installing haproxy role from Ansible Galaxy

3. Create the `Chapter-09/deploy-haproxy.yaml` playbook and include the `geerlingguy.haproxy` role that you installed in the previous step. The following screenshot shows the sample playbook for installing the HAProxy load balancer:

● ● ●

```
---
# Chapter-09/deploy-haproxy.yaml
- name: Deploy Load Balancer using HAProxy
  hosts: loadbalancer
  become: yes
  vars:
    haproxy_frontend_name: 'hafrontend'
    haproxy_backend_name: 'habackend'
    haproxy_backend_servers:
      - name: node1
        address: 192.168.56.25:80
      - name: node2
        address: 192.168.56.24:80
  tasks:
    - name: Install haproxy
      include_role:
        name: geerlingguy.haproxy

    - name: Permit port 80 in firewall
      ansible.posix.firewalld:
        port: 80/tcp
        permanent: true
        state: enabled
        immediate: yes
```

Figure 9.19 – Playbook to install the HAProxy load balancer

4. Include the necessary variables for the `geerlingguy.haproxy` role, as shown in the preceding screenshot. It is possible to customize the execution of the role by referring to the role documentation (`https://galaxy.ansible.com/geerlingguy/haproxy`), but in this demonstration, you will only be adding the load balancer backend and a few other details. (Change the IP address so that it matches your `node1` and `node2` IP addresses, as configured in the inventory.) The role will take care of the `haproxy` installation, configuration, and more. Also, remember to add a task to allow port `80` in the firewall (if you are using a different firewall such as `ufw`, then amend the playbook with the appropriate firewall module).

5. Finally, add a new play in the same playbook (`Chapter-09/deploy-haproxy.yaml`) to automatically validate the load balancer access:

```
# Chapter-09/deploy-haproxy.yaml

- name: Verify load balancer deployment
  hosts: loadbalancer
  become: no
  tasks:
    - name: Verify load balancer health
      ansible.builtin.uri:
        url: http://{{ inventory_hostname }}.lab.local
        status_code: 200
      delegate_to: localhost
```

Figure 9.20 – Adding an Ansible play to verify the load balancer

6. Execute the playbook and verify the output:

```
[ansible@ansible Chapter-09]$ ansible-playbook deploy-haproxy.yaml

TASK [Verify load balancer health] *********************************************************
ok: [node3 -> localhost]
```

Figure 9.21 – The HAProxy playbook with a health check success

The `Verify load balancer health` task is successful, which means the load balancer (`haproxy`) is working and serving the content from the backend web servers – that is, `node1` and `node2`.

7. Verify the website's content using the load balancer IP address (for example, `http://192.168.56.45`) from a web browser, as shown in the following screenshot:

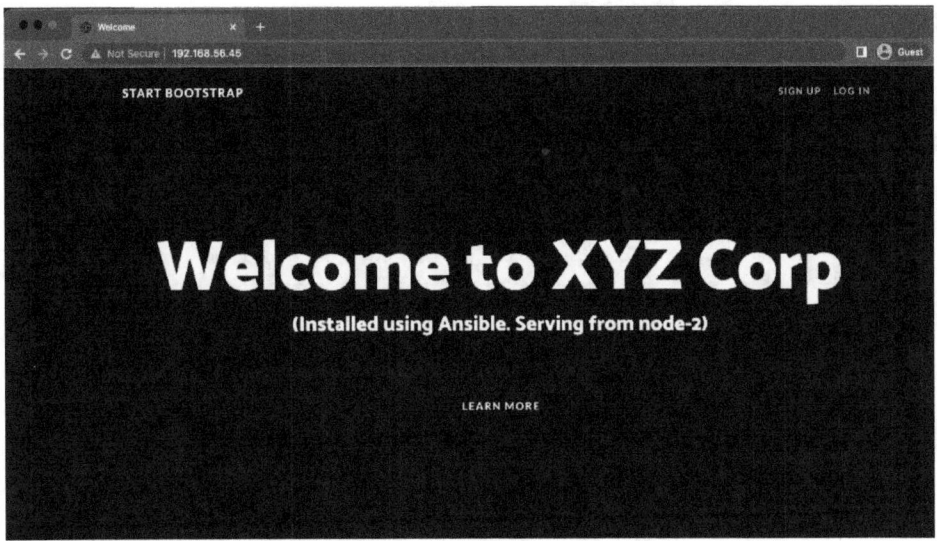

Figure 9.22 – Website accessed using a load balancer IP

Once the load balancer is ready, it is possible to enable the DNS for the load balancer and share it with the end users (for example, `website.example.com`). Users don't need to worry about remembering the IP address of the website or web server. When you have a new version of the website or application, it is possible to update the content without downtime as the load balancer will serve the website from the available web server in the backend.

In the next section, you will learn how to handle rolling updates for websites and applications without downtime using Ansible.

Rolling updates using Ansible

Continuous deployment is a method meant for frequent application deployment (together with CI/CD) and frequent updates of your application or website rather than you having to wait for scheduled downtime and deployment cycles. But you also need to ensure the application will be available during the update process. So far, you have learned that application high availability can be achieved using a load balancer. In this section, you will learn how to update the application on web servers without downtime.

Steps involved in an application update

Depending on your application's type and the components involved, the update process may contain different steps and procedures. The following diagram shows the generic steps involved in the application update process, which is running behind a load balancer:

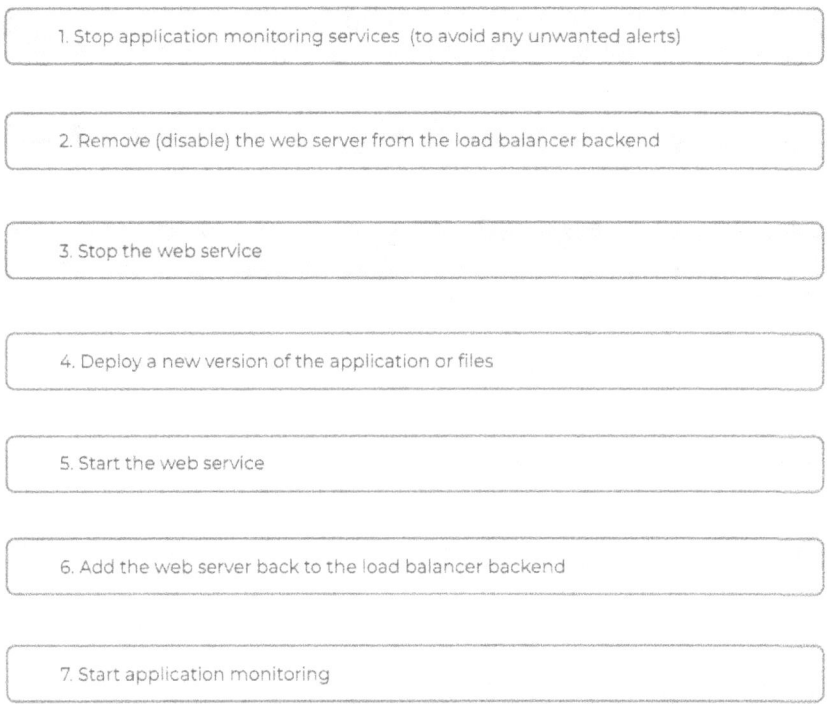

1. Stop application monitoring services (to avoid any unwanted alerts)

2. Remove (disable) the web server from the load balancer backend

3. Stop the web service

4. Deploy a new version of the application or files

5. Start the web service

6. Add the web server back to the load balancer backend

7. Start application monitoring

Figure 9.23 – Steps involved in an application update

It is possible to automate all such tasks using Ansible, including validating and verifying the services, monitoring tasks, and more.

Deploying updates in a batch of managed nodes

If you are running the update tasks on all web servers (for example, `node1` and `node2`) in parallel, then there will not be any servers to serve the requests from the load balancer and the website will be down. This is not the desired behavior, so you need to update the web servers in multiple batches instead of all the servers in a single batch. It is possible to achieve this by passing specific remote node names, such as `extra-variable`, but that is not a best practice or method as you need to execute the playbook multiple times to complete the full website update.

> **CD and Rolling Upgrades Using Ansible**
>
> Ansible orchestration features are very useful for managing multi-tier applications. Refer to the documentation at `https://docs.ansible.com/ansible/latest/user_guide/guide_rolling_upgrade.html` to learn more.

Use the `serial` keyword in Ansible to specify the number of managed nodes to be executed at a time:

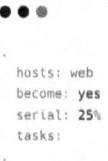

```
hosts: web
become: yes
serial: 25%
tasks:
```

Figure 9.24 – Using serial in Ansible playbook

Here, `serial: 25%` means that the play will be executed for 25% of the total managed nodes at a time.

The following diagram shows the rolling update flow, where only one node will be executed with the update task. The load balancer will still serve the traffic from another node:

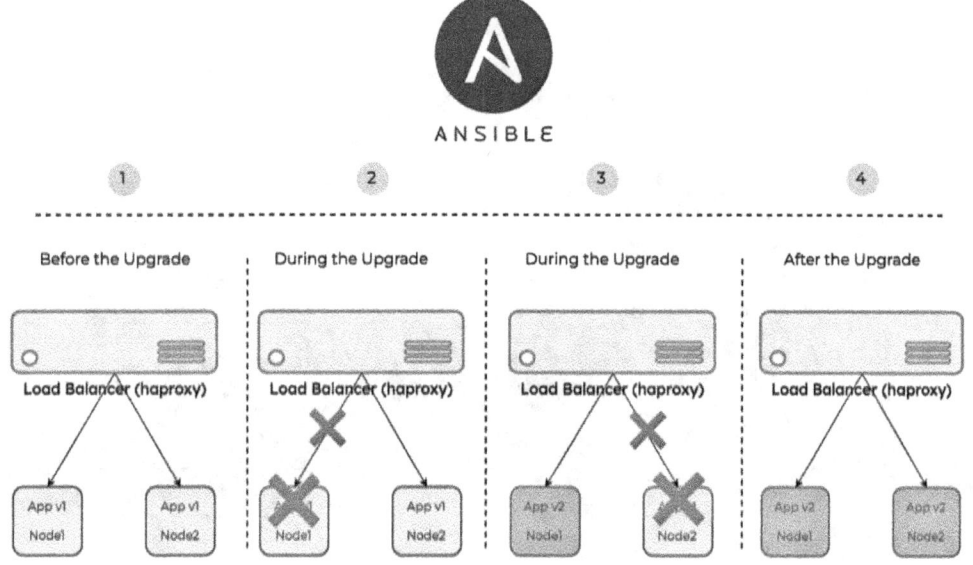

Figure 9.25 – Rolling update using Ansible

It is also possible to mention the exact number of nodes in a batch by using the `serial: 2` or `serial: 5` keyword. The `serial` keyword is very flexible, and you can even control different batches with a different number or managed nodes, as follows:

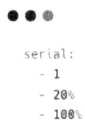

```
serial:
  - 1
  - 20%
  - 100%
```

Figure 9.26 – Using different serial values for host batches

In the next section, you will learn how to use the `serial` keyword to deploy updates on web servers without causing downtime for the application.

Deploying updates on multiple servers without service downtime

Now that the CI/CD pipeline includes more tasks, let's replace the application deployment playbook with the rolling update playbook, as shown in the following diagram:

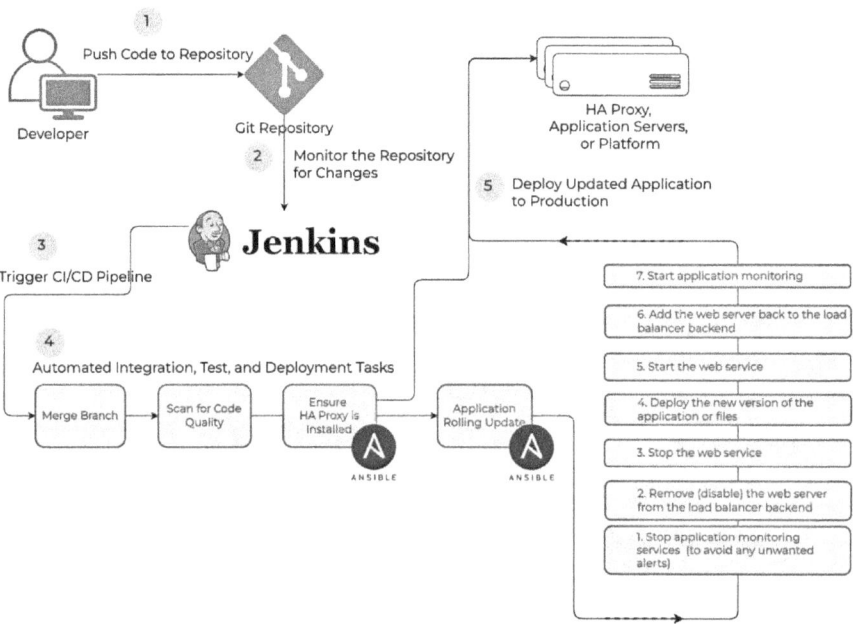

Figure 9.27 – The CI/CD pipeline performing a rolling update of the application

In this exercise, you will use `serial:  1`, which means the operations will be executed only on one managed node at a time. Control this batch size as required based on the number of managed nodes or backend servers:

1. Update the website's content with some changes by creating a new branch in the repository. (Use the `https://github.com/ginigangadharan/website-demo-one-page` repository and make a copy for testing purposes.)

2. Clone the repository to your local machine and switch to the production branch as follows. Also `checkout` a new branch as `v2`:

```
## Clone the repository to your local machine:
[ansible@ansible ~]$ git clone git@github.com:ginigangadharan/website-demo-one-page

## Switch to the repository's directory:
[ansible@ansible ~]$ cd website-demo-one-page

## Switch to the production branch:
[ansible@ansible website-demo-one-page]$ git checkout production
Switched to branch 'production'
Your branch is up to date with 'origin/production'.

## Checkout new branch called v2
[ansible@ansible website-demo-one-page]$ git checkout -b v2
Switched to a new branch 'v2'
```

Figure 9.28 – Clone the repository, checkout to production branch, and create a new branch

3. Update the `index.html` file with some modifications, as follows (for example, add `v2` on the home page to identify the changes):

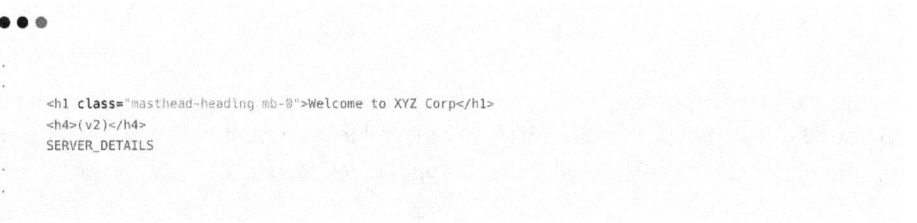

```
<h1 class="masthead-heading mb-0">Welcome to XYZ Corp</h1>
<h4>(v2)</h4>
SERVER_DETAILS
```

Figure 9.29 – Updating index.html in the application repository

After making these changes, save the file.

4. Commit all changes and push the new branch, v2, to the repository:

```
[ansible@ansible website-demo-one-page]$ git add .;git commit -m "v2"

[ansible@ansible website-demo-one-page]$ git push -u origin v2
```

Figure 9.30 – Commit changes and push the v2 branch to origin

Now, the new branch that contains the new version of the application is available in the GitHub repository. This means we can deploy it to the web servers using the Chapter-09/ deploy-web.yaml playbook. But for the rolling update, you will create another playbook called Chapter-09/rolling-update.yaml and add the rolling update tasks inside.

5. Create a new playbook called Chapter-09/rolling-update.yaml and add the serial keyword, as follows:

```
---
# Chapter-09/rolling-update.yaml
- name: Rolling Update
  hosts: "{{ NODES }}"
  become: yes
  serial: 1
  vars:
    haproxy_backend_name: 'habackend'
    application_repo: 'https://github.com/ginigangadharan/website-demo-one-page'
    application_branch: production
    application_path: /var/www/html

  tasks:
```

Figure 9.31 – Rolling update playbook

6. During the update, the node will not be able to serve the web pages. Therefore, you need to inform the load balancer of this. Add a task to disable the host in the haproxy backend that you configured earlier. Also, add a task to stop the web service (httpd) on the server:

```
# Chapter-09/rolling-update.yaml


- name: Disable server in haproxy backend
  community.general.haproxy:
    state: disabled
    host: '{{ inventory_hostname }}'
    wait: yes
    socket: "/var/lib/haproxy/stats"
    backend: "{{ haproxy_backend_name }}"
    fail_on_not_found: yes
  delegate_to: '{{ item }}'
  with_items: '{{ groups.loadbalancer }}'

- name: Stop httpd service
  ansible.builtin.service:
    name: httpd
    state: stopped
```

Figure 9.32 – Removing the host entry from the load balancer

Please note the delegate_to: '{{ item }}' line as this task will be running on the load balancer node. The task will loop through the load balancer nodes by using the with_items: '{{ groups.loadbalancer }}' loop. In our case, there is only one load balancer node.

7. Now, you have similar tasks in the deployment playbook, as shown in the following screenshot:

```
# Chapter-09/rolling-update.yaml


- name: Delete content & directory if exists
  ansible.builtin.file:
    state: absent
    path: "{{ application_path }}"

- name: Create application directory
  ansible.builtin.file:
    state: directory
    path: "{{ application_path }}"
    mode: '0755'

- name: Git checkout - latest application content
  ansible.builtin.git:
    repo: "{{ application_repo }}"
    dest: "{{ application_path }}"
    version: "{{ application_branch }}"

- name: Update index.html with server details
  ansible.builtin.lineinfile:
    path: "{{ application_path }}/index.html"
    regexp: 'SERVER_DETAILS'
    line: "<h3>(Installed using Ansible. Serving from {{ ansible_hostname }})</h3>"
```

Figure 9.33 – Cleaning up the directory and deploying the application

The default branch is set to production (`application_branch: production`) inside the playbook; we will override the branch name using the `extra-variable` later.

8. Once you have the latest application content, start the web service and add the host back to the load balancer backend, as follows:

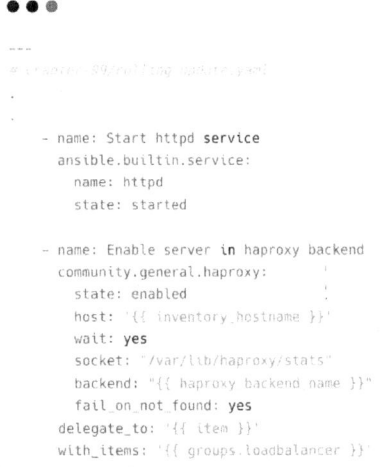

```yaml
---
# <roles-99/rolling-update.yml>
.
.
  - name: Start httpd service
    ansible.builtin.service:
      name: httpd
      state: started

  - name: Enable server in haproxy backend
    community.general.haproxy:
      state: enabled
      host: '{{ inventory_hostname }}'
      wait: yes
      socket: '/var/lib/haproxy/stats'
      backend: "{{ haproxy_backend_name }}"
      fail_on_not_found: yes
    delegate_to: '{{ item }}'
    with_items: '{{ groups.loadbalancer }}'
```

Figure 9.34 – Post-deployment configuration

9. Finally, add another play in the same playbook to verify the web server access via the load balancer:

```yaml
---
# <roles-99/rolling-update.yml>
.
.
- name: Verify load balancer traffic
  hosts: loadbalancer
  become: no
  tasks:
    - name: Verify load balancer traffic
      ansible.builtin.uri:
        url: http://{{ inventory_hostname }}.lab.local
        status_code: 200
      delegate_to: localhost
```

Figure 9.35 – Verifying the play

10. Execute the playbook and check its output. Remember to pass the new branch information as `application_branch=v2`:

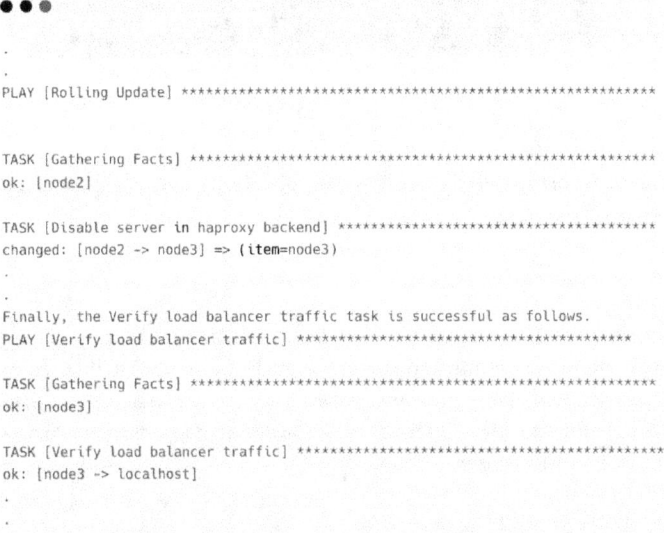

```
[ansible@ansible Chapter-09]$ ansible-playbook rolling-update.yaml -e "NODES=web application_branch=v2"


PLAY [Rolling Update] ************************************************************

TASK [Gathering Facts] **********************************************************
ok: [node1]

TASK [Disable server in haproxy backend] ****************************************
changed: [node1 -> node3] => (item=node3)
```

Figure 9.36 – Ansible rolling update on node1

In the preceding screenshot, the execution is only happening on one node at a time (that is, `node1`) and completes all the tasks for that batch of nodes. Later, the playbook will start the tasks for the new batch (that is, `node2`), as follows:

```
PLAY [Rolling Update] ************************************************************


TASK [Gathering Facts] **********************************************************
ok: [node2]

TASK [Disable server in haproxy backend] ****************************************
changed: [node2 -> node3] => (item=node3)

Finally, the Verify load balancer traffic task is successful as follows.
PLAY [Verify load balancer traffic] *********************************************

TASK [Gathering Facts] **********************************************************
ok: [node3]

TASK [Verify load balancer traffic] *********************************************
ok: [node3 -> localhost]
```

Figure 9.37 – Ansible rolling update on node2

As shown in the preceding outputs, the update is happening in the `rolling` method. This means that at any time, one of the web servers will be available to serve the website in the load balancer backend. It is possible to achieve zero downtime during your application update.

Now, let's verify the website access using the load balancer's IP address and see the changes (notice the **v2** text, which we have changed in the `v2` branch):

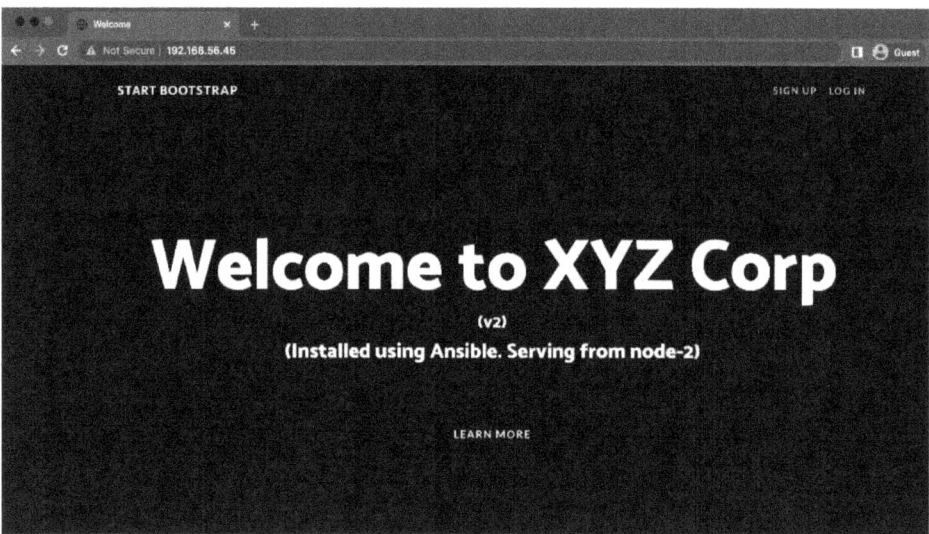

Figure 9.38 – Website after the rolling update (v2)

The process is the same for both static websites and dynamic web applications. Include the exact steps you want to execute during the rolling update.

> **CD and Rolling Upgrades**
>
> Refer to `https://docs.ansible.com/ansible/latest/user_guide/guide_rolling_upgrade.html` for more details about rolling updates using Ansible. Also, check out application deployment (`https://www.ansible.com/use-cases/application-deployment`) and Ansible CD (`https://www.ansible.com/use-cases/continuous-delivery`) use cases.

The flexibility of Ansible makes it suitable for implementing automation in most of your DevOps workflows. Instead of using Ansible alone, it is always possible to integrate Ansible with other tools. With AAP and the Ansible API, it is possible to implement more powerful integration in the CI/CD workflow using webhooks, job templates, and callbacks. You will learn about AAP and integration in *Chapter 12*, *Integrating Ansible with Your Tools*. In the next section, you will learn how to use Ansible with Terraform as a provisioning tool.

Using Ansible as a provisioning tool in Terraform

Ansible can be used as an IaC tool, as you learned in *Chapter 7, Managing Your Virtualization and Cloud Platforms*. At the same time, it is a common practice in the industry to use the right tool for the right task – for example, Terraform for IaC, Ansible for IT automation, Jenkins for CI/CD pipelines, and so on. Instead of comparing similar tools, integrate them in the right place and achieve better results.

Terraform is an open source tool by **HashiCorp** for implementing IaC practices. Terraform can be used to deploy and manage the cloud-based infrastructure and applications using infrastructure code written in a declarative configuration language called **HashiCorp Configuration Language** (**HCL**). Depending on the cloud platform and components, use the provider modules and resources available. Refer to `https://registry.terraform.io/browse/providers` to explore the available and supported providers.

For example, the following Terraform code will provide EC2 instances in the AWS platform with the specified **Amazon Machine Images** (**AMIs**) and other details:

```
resource "aws_instance" "dbnodes" {
  ami             = var.aws_ami_id
  instance_type   = "t2.large"
  key_name        = aws_key_pair.ec2loginkey.key_name
  count           = var.dbnodes_count
  security_groups = ["dbnodes-sg"]
  user_data       = file("user-data-dbnodes.sh")
  tags = {
    Name = "dbnode-${count.index + 1}"
  }
}
```

Figure 9.39 – Terraform code for EC2 provisioning

Terraform is good at handling infrastructure changes and tracking the updates using its state management mechanism. But if you want to configure the operating system-level components, you need to use Terraform provisioners (`https://www.terraform.io/language/resources/provisioners/syntax`) as such configurations cannot be represented in the declarative code. Terraform supports many provisioners such as `file`, `local-exec`, `remote-exec`, and so on. Use either the `local-exec` or `remote-exec` provisioner and use Ansible to configure your system, which has been provisioned by Terraform. Choose either method, depending on your environment and other tool integration options.

Using Terraform's local-exec provisioner with Ansible

To use the `local-exec` provisioner with Ansible, the machine you are running Terraform on should have Ansible installed and configured since the `ansible-playbook` command will be executed on your local machine, as shown in the following diagram:

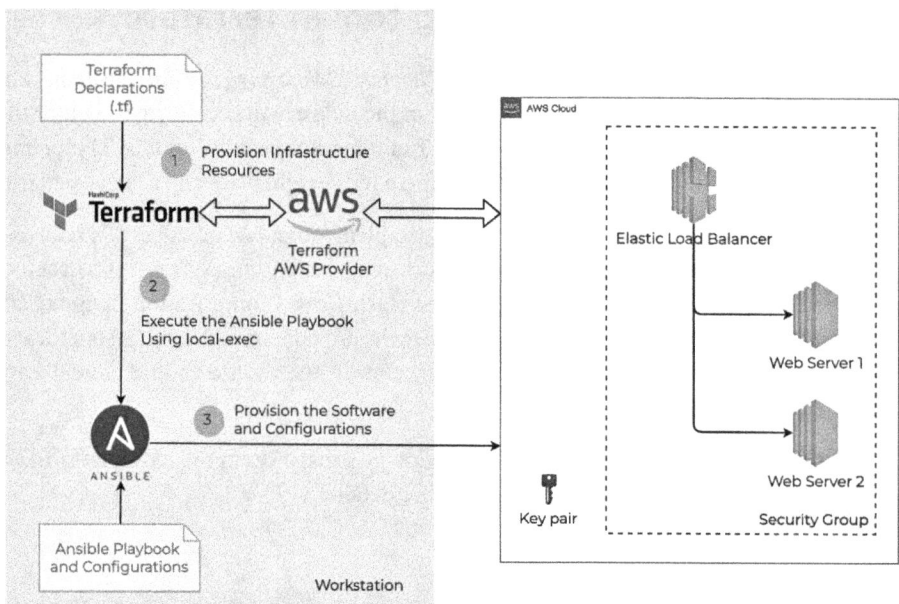

Figure 9.40 – Using Ansible as local-exec with Terraform

You also need to pass the credential details such as the username and SSH private key in the command, as follows:

```
resource "aws_instance" "dbnodes" {
    ami             = var.aws_ami_id
    instance_type   = "t2.large"
    key_name        = aws_key_pair.ec2loginkey.key_name
    count           = var.dbnodes_count
    security_groups = ["dbnodes-sg"]
    user_data       = file("user-data-dbnodes.sh")
    tags = {
        Name = "dbnode-${count.index + 1}"
    }

    provisioner "local-exec" {
        command = "ANSIBLE_HOST_KEY_CHECKING=False ansible-playbook -u ec2-user -i '${self.public_ip}, --private-key
${var.ssh_key_pair} post-configuration.yaml"

    }
}
```

Figure 9.41 – Terraform code for EC2 provisioning with Ansible automation `ec2-user` in the preceding snippet is the default user account in the AWS Linux AMI. This username or credential can be changed, depending on your EC2 AMI or another source image. For example, create a custom base image in AWS, GCP, Azure, or other cloud platforms, and then use that image to create the instances using Terraform and configure them using Ansible.

Using Terraform's remote-exec provisioner with Ansible

If your local machine does not support the installation of Ansible (for example, running Terraform on a Windows machine), then it is possible to use the `remote-exec` provisioner, but you need to ensure that Ansible is installed inside the remote machine provisioned by Terraform. You also need to ensure the playbook and required files have been copied to the target machine before calling the `ansible-playbook` command using the `remote-exec` provisioner. The following diagram shows the high-level workflow in Terraform with Ansible `remote-exec` provisioning:

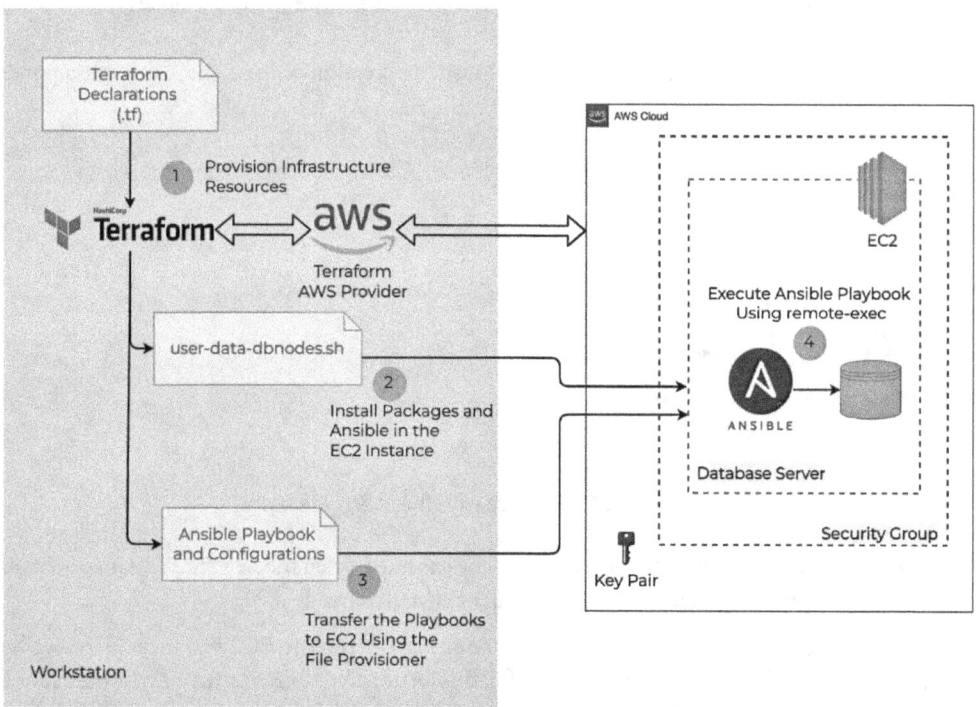

Figure 9.42 – Ansible remote-exec provisioning in Terraform

Refer to `Chapter-09/terraform-aws-ansible-lab` for the Terraform code used in the following explanation. Follow these steps:

1. Use the `user_data` argument to pass the basic commands to be executed during the initialization of the EC2 instance and Ansible installation. These commands can be included in the `user_data` script. A typical user data script can be written as follows:

```
#! /bin/bash
sudo amazon-linux-extras install -y epel
sudo useradd devops
echo -e 'devops\ndevops' | sudo passwd devops
echo 'devops ALL=(ALL) NOPASSWD: ALL' | sudo tee /etc/sudoers.d/devops
sudo sed -i "s/PasswordAuthentication no/PasswordAuthentication yes/g" /etc/ssh/sshd_config
sudo systemctl restart sshd.service
sudo yum install -y python3
sudo yum install -y vim
sudo yum install -y ansible
sudo yum install -y git
```

Figure 9.43 – User data script for installing Ansible inside an EC2 instance

2. As you saw in the previous example, the EC2 instance creation code will have more components now, as follows:

```
resource "aws_instance" "dbnodes" {
  ami           = var.aws_ami_id
  instance_type = "t2.large"
  key_name      = aws_key_pair.ec2loginkey.key_name
  count         = var.dbnodes_count
  security_groups = ["dbnodes-sg"]
  user_data     = file("user-data-dbnodes.sh")
  tags = {
    Name = "dbnode-${count.index + 1}"
  }
}
```

Figure 9.44 – EC2 resource with the user_data script

The `user-data-dbnodes.sh` script will be executed when the new EC2 instance is created. All the components, including Ansible, will be installed inside it.

3. The next step is copying the required playbooks to the remote EC2 instance. Here, we have used the Terraform `file` provisioner for this. After that, execute the playbook using the `remote-exec` provisioner:

```
# copy dbnode-config.yaml
provisioner "file" {
  source      = "dbnode-config.yaml"
  destination = "/home/ec2-user/dbnode-config.yaml"
  connection {
    type        = "ssh"
    user        = "ec2-user"
    private_key = file(pathexpand(var.ssh_key_pair))
    host        = self.public_ip
  }
}

# Execute Ansible Playbook
provisioner "remote-exec" {
  inline = [
    "sleep 120; ansible-playbook dbnode-config.yaml"
  ]
  connection {
    type        = "ssh"
    user        = "ec2-user"
    private_key = file(pathexpand(var.ssh_key_pair))
    host        = self.public_ip
  }
}
```

Figure 9.45 – Copying and executing the playbook inside the EC2 instance using the remote-exec method

For further practicing and learning, include the system configuration tasks inside the dbnode-config.yaml file, such as creating new users, installing database packages, starting services, mounting disk volumes, and more.

Refer to *Using Ansible with Terraform* (https://www.techbeatly.com/using-ansible-with-terraform-ansible-real-life-series) to learn more and understand Terraform and Ansible integration.

Summary

In this chapter, you explored the basic concepts, processes, and technical terms surrounding DevOps, such as CI/CD and continuous deployment. You also learned about some of the possibilities of Ansible integration within the DevOps workflow. Then, you learned how to use Ansible inside a continuous deployment workflow using Jenkins. After that, you learned about how to perform rolling updates without downtime while using Ansible as part of continuous application deployment. Finally, you learned how to integrate Ansible with Terraform for IaC provisioning.

In the next chapter, you will learn how to build, run, and manage containers using Ansible.

Further reading

To learn more about the topics that were covered in this chapter, take a look at the following resources:

- *Use Terraform to Create a FREE Ansible Lab in AWS*: `https://www.techbeatly.com/use-terraform-to-create-a-free-ansible-lab-in-aws`

- *Deep Dive – Automated NetOps – Ansible for Network GitOps*: `https://www.youtube.com/watch?v=JqE13sP2sq8` (Video)

- *Continuous integration vs. delivery vs. deployment*: `https://www.atlassian.com/continuous-delivery/principles/continuous-integration-vs-delivery-vs-deployment`

- *Ansible and HashiCorp: Better Together*: `https://www.hashicorp.com/resources/ansible-terraform-better-together`

- *Manages a Terraform deployment*: `https://docs.ansible.com/ansible/latest/collections/community/general/terraform_module.html`

10
Managing Containers Using Ansible

Since the introduction of containerization, organizations have been able to deploy applications faster and accelerate release cycles with frequent updates and deployments. However, containerizing applications involve more steps compared to traditional server-based deployments. For example, you need to ensure the packaged container image is working as per expectation, security standards are in place, volume mounting is working, secrets are safe inside, and more. When you have more frequent application releases, automating such container build and deployment tasks will help you implement better CI/CD workflows and save time on manual processes.

With the Ansible collections for container management, we can manage the entire life cycle of our containers. This includes building them, pushing them to the registry, scanning them for vulnerabilities, and deploying them.

In this chapter, we will cover the following topics:

- Managing the container host
- Ansible, containers, and CI/CD
- Managing containers using Ansible
- Building container images using Ansible
- Managing multi-container applications using Ansible

First, you will learn how to use Ansible to deploy the container engine to the host machine and run containers inside it. Later, you will learn how to manage the container image build and manage it in the container registry.

Technical requirements

You will need the following technical requirements for this chapter:

- A Linux machine for the Ansible control node (with internet access)
- A Linux machine for installing and configuring Docker
- Access to a Docker container registry (`hub.docker.com`)
- Basic knowledge about containers and container registries (Docker or Podman)

All the Ansible code and playbooks, as well as the commands and snippets, for this chapter can be found in this book's GitHub repository at `https://github.com/PacktPublishing/Ansible-for-Real-life-Automation/tree/main/Chapter-10`.

Managing the container host

Various types of container software are available, such as **Docker** and **Podman**. In this chapter, we will be using Docker to explain and demonstrate container management using Ansible. We will be using Docker **Community Edition** (**CE**), which is free, though you can use Docker **Enterprise Edition** (**EE**) if needed.

Ansible Docker prerequisites

To use the **Ansible Docker** modules, you must install the `docker` library, which you can do using `Python pip` or standard packages managers such as `yum` (`yum install python-docker-py`) of `dnf` if available. If you are using the old version of Python (2.6), then you should install and use the old library called `docker-py`.

Installing Docker on the host using Ansible

Installing Docker software on a host involves multiple steps and configurations. These steps can be completed manually or we can use the Ansible role available in Ansible Galaxy. We will be using the community Ansible role called `geerlingguy.docker` (`https://github.com/geerlingguy/ansible-role-docker`), which was created by the well-known Ansible contributor *Jeff Geerling*.

We will also use `geerlingguy.pip` to install the Docker libraries (for example, `docker`) using Ansible.

Follow these steps to install the Docker software on a Linux machine:

1. Update the inventory with `node1` under the `dockerhost` host group (`chapter-10/hosts`):

```
● ● ●
[dockerhost]
node1 ansible_host=192.168.56.25
```

Figure 10.1 – Configure Docker host in inventory

2. Install `geerlingguy.docker` in the `roles` directory, as shown in the following screenshot:

```
● ● ●
[ansible@ansible Chapter-10]$ ansible-galaxy install geerlingguy.docker -p roles/
[ansible@ansible Chapter-10]$ ansible-galaxy install geerlingguy.pip -p roles/
You can verify the roles installation as follows.
[ansible@ansible Chapter-10]$ ansible-galaxy role list
# /home/ansible/ansible-book-packt/Chapter-10/roles
- geerlingguy.docker, 4.1.3
- geerlingguy.pip, 2.1.0
```

Figure 10.2 – Installing Docker role

3. Create a playbook called `Chapter-10/deploy-docker.yaml` to install the role and libraries:

```
● ● ●
---
# Chapter-10/deploy-docker.yaml
- name: Deploy Docker to Host
  hosts: "{{ NODES }}"
  become: yes
  vars:
    pip_install_packages:
      - name: docker
  tasks:
    - name: Install docker
      include_role:
        name: geerlingguy.docker
    - name: Install Packges
      include_role:
        name: geerlingguy.pip
```

Figure 10.3 – Using deploy-docker.yaml to install Docker on the host

The `pip_install_packages` variable will be used by the `geerlingguy.pip` role to install the listed packages using `pip`.

4. Execute the playbook to deploy Docker on the host:

● ● ●

```
[ansible@ansible Chapter-10]$ ansible-playbook deploy-docker.yaml -e "NODES=dockerhost"
```

Figure 10.4 – Execute the playbook to deploy Docker on the host

5. Verify that it has been installed. Log into `node1` and check the details:

● ● ●

```
[root@node-1 ~]# docker version
Client: Docker Engine - Community
Version:           20.10.14
API version:       1.41
Go version:        go1.16.15
..<output omitted>..
Server: Docker Engine - Community
Engine:
  Version:         20.10.14
  API version:     1.41 (minimum version 1.12)
  Go version:      go1.16.15
..<output omitted>..
containerd:
  Version:         1.5.11
  GitCommit:       3df54a852345ae127d1fa3092b95168e4a88e2f8
..<output omitted>..
```

Figure 10.5 – Verifying the Docker installation

With that, Docker has been installed and configured on the host, which means we can start running containers.

Next, you will learn about the capabilities of Ansible for managing containerized applications.

Ansible, containers, and CI/CD

Containerizing applications will give you more options for integrating, delivering, and deploying them since most of the tools support automated builds, tests, and executions. A typical containerized application workflow can be seen in the following diagram:

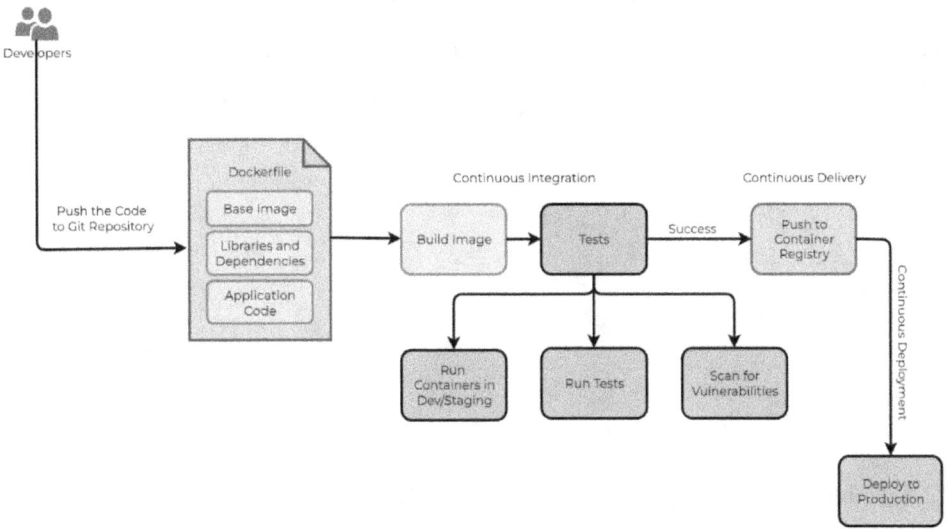

Figure 10.6 – Typical CI/CD tasks in a Docker-based deployment

Most of the tasks in the preceding diagram can be automated using Ansible as the Ansible collection for Docker and Podman contains several modules to support building, running, and managing containers on your container host. Either implement the entire workflow using Ansible or use Ansible with our favourite CI/CD tools and execute the tasks more flexibly. You will learn how to integrate Ansible with Jenkins in *Chapter 12, Integrating Ansible with Your Tools*.

In this next section, you will learn how to manage containers using Ansible and manage the container life cycle.

Managing containers using Ansible

The Ansible collection, `community.docker` (`https://galaxy.ansible.com/community/docker`), contains more than 25 Ansible modules and ~10 plugins for connection, inventory, and more. These modules will help you manage containers, container images, images in the container registry, the Docker network, Docker volumes, Docker swarm, and other container-based operations.

If you are using Podman, then check out the `containers.podman` collection (`https://galaxy.ansible.com/containers/podman`) in Ansible Galaxy.

In the upcoming sections, you will learn how to build, start, and manage containers using Ansible.

Installing the Ansible Docker collection

Installing a collection is straightforward, as you learned in the previous chapters:

1. Update your `ansible.cfg` with the collection path:

```
● ● ●
[defaults]
inventory = ./hosts
remote_user = devops
ask_pass = false

COLLECTIONS_PATHS = ./collections
roles_path = roles
```

Figure 10.7 – ansible.cfg with the collection and role paths

2. Install the `community.docker` Ansible collection:

```
● ● ●
[ansible@ansible Chapter-10]$ ansible-galaxy collection install community.docker
```

Figure 10.8 – Installing community.docker collection

3. Verify that the collection has been installed in the collection path, as shown in the following screenshot:

```
● ● ●
[ansible@ansible Chapter-10]$ ansible-galaxy collection list
...
<output omitted>

Collection        Version
----------------  -------
community.docker  2.3.0
```

Figure 10.9 – Docker collection installed in the collection path

If you are using the Ansible community package, then there might be an old version of the `community.docker` collection in the default path. This is the reason we are installing the latest version of the collection on our project path (`COLLECTIONS_PATHS =`):

● ● ●

```
./collections).
[ansible@ansible Chapter-10]$ ansible-galaxy collection list |grep -i docker
community.docker          1.10.2
community.docker 2.3.0
```

Figure 10.10 – Docker collection from the default Ansible installation

Once the collection is available to use, start using the Docker modules that have been installed.

Installing an Ansible Collection on a Disconnected Ansible Control Node

If you are inside a restricted environment (disconnected or no internet), then follow an alternative method to install an Ansible collection and roles: *How to install an Ansible Collection on a disconnected Ansible control node* (`https://www.techbeatly.com/how-to-install-an-ansible-collection-on-a-disconnected-ansible-control-node`).

Starting a Docker container using Ansible

Use your own container images or use the existing container images from the public Docker registries such as Docker Hub (`https://hub.docker.com`), Quay.io (`https://quay.io/repository`), and GitHub Container Registry (`https://ghcr.io`). It is also possible to use the container images from private repositories, but you will need to authenticate to the container registries (with a username, password, or tokens) to pull or push the container images. You will learn about registry authentication in the *Building container images using Ansible* section.

In this section, you will learn how to run a Docker container using Ansible. To make this demonstration simple, we will be using the default nginx (`https://hub.docker.com/_/nginx`) container image, but always explore using other container images later:

1. Create a playbook called `Chapter-10/container-manage.yaml`, as shown in the following screenshot:

● ● ●

```
---
# Chapter-10/container-manage.yaml
- name: Manage Docker containers
  hosts: "{{ NODES }}"
  become: yes
  vars:
    container_image: nginx
    container_name: web
    container_port: 80
    container_expose_port: 8080
    container_action: 'run'
```

Figure 10.11 – container-manage.yaml

> **Note**
> The variables can be configured in different variable files dynamically, such as `host_vars` or `group_vars`, or via external variables. The variables that have been used inside the playbook have been provided to demonstrate the use case's execution.

2. Add a task to start a container using the variable details, as follows:

● ● ●

```
tasks:
  - name: Create and Start a Docker container
    community.docker.docker_container:
      name: "{{ container_name }}"
      image: "{{ container_image }}"
      state: started
      ports: "{{ container_expose_port }}:{{ container_port }}"
    when: container_action == 'run'
```

Figure 10.12 – container-manage.yaml – part 2

We will use the `container_action` variable later to control other actions for the container.

3. Add another **play** to the same playbook to verify the website access (refer to `Chapter-10/container-manage.yaml` in the repository for the full playbook):

● ● ●

```
- name: Verify web site running inside container
  hosts: "{{ NODES }}"
  become: no
  vars:
    container_expose_port: 8080
    container_action: 'run'
  tasks:
    - name: Verify application health
      ansible.builtin.uri:
        url: http://{{ inventory_hostname }}.lab.local:{{ container_expose_port }}
        status_code: 200
      delegate_to: localhost
      when: container_action == 'run'
```

Figure 10.13 – container-manage.yaml – using a second play to
verify the application running inside the container

(Refer to *Chapter 1, Ansible Automation – Introduction* to see a playbook with multiple plays.)

4. Execute the playbook with NODES set to dockerhost:

●●●

[ansible@ansible Chapter-10]$ ansible-playbook container-manage.yaml -e "NODES=dockerhost"

Figure 10.14 – Execute the playbook on the Docker host

5. Once the playbook has been successfully executed, verify the container from the Docker host (node1), as shown in the following screenshot:

```
[root@node-1 ~]# docker ps
CONTAINER ID   IMAGE    COMMAND              CREATED         STATUS          PORTS                  NAMES
e36fb7419165   nginx    "/docker-entrypoint..."   10 minutes ago   Up 10 minutes   0.0.0.0:8080->80/tcp   web
```

Figure 10.15 – An nginx container running on the Docker host

Here, the nginx container is called web and is exposing the service on port 8080 of the Docker host.

6. Access the website running inside the nginx container. You need to remember to pass port 8080 since the Docker port is exposed on 8080:

●●●

```
[ansible@ansible Chapter-10]$ curl http://node1:8080
<!DOCTYPE html>
<html>
<head>
<title>Welcome to nginx!</title>
..<output omitted>..
<a href="http://nginx.com/">nginx.com</a>.</p>

<p><em>Thank you for using nginx.</em></p>
</body>
</html>
```

Figure 10.16 – The nginx application available after using the curl command

7. Access the website from a web browser, as shown in the following screenshot:

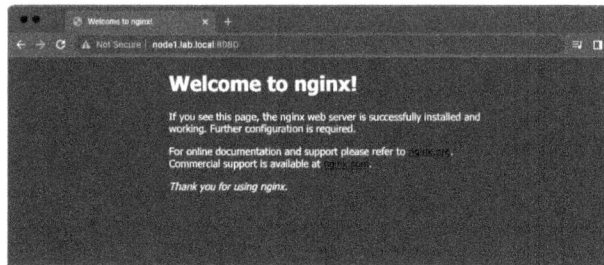

Figure 10.17 – The nginx web server running inside a Docker container deployed using Ansible

Add more complex configurations to the container, such as different Docker networks, mounted volumes, and so on. Refer to the documentation at `https://docs.ansible.com/ansible/latest/collections/community/docker/docker_container_module.html` for various arguments and parameters.

Stopping Docker containers using Ansible

In the CI/CD process, when you build containers for testing, you also need to take care of the cleanup tasks. Once the tests have been completed, you need to stop the container and delete it as part of housekeeping. Use the same Ansible module, `community.docker.docker_container`, to handle the entire container life cycle, such as stopping, deleting, and more.

In this section, you will learn how to stop and remove the container we created in the previous exercise. Follow these steps:

1. Update the previous playbook, `Chapter-10/container-manage.yaml`, and add tasks to the first play, `Manage Docker containers`, as shown in the following screenshot:

```
  - name: Stop Docker container
    community.docker.docker_container:
      name: "{{ container_name }}"
      state: stopped
    when: container_action == 'stop'

  - name: Remove Docker container
    community.docker.docker_container:
      name: "{{ container_name }}"
      state: absent
    when: container_action == 'stop'
```

Figure 10.18 – Adding tasks to stop and remove the container

2. Execute the playbook and pass `container_action=stop` as an extra variable:

●●●

```
[ansible@ansible Chapter-10]$ ansible-playbook container-manage.yaml -e "NODES=dockerhost container_action=stop"
```

Figure 10.19 - Execute the container-manage.yaml file to stop the container

3. On the container host (`node1`), verify whether any containers are running:

●●●

```
[root@node-1 ~]# docker ps -a
CONTAINER ID   IMAGE     COMMAND    CREATED    STATUS    PORTS     NAMES
[root@node-1 ~]#
```

Figure 10.20 – The nginx container has been stopped and removed

Add more tasks to the playbook, such as for verifying the ports, backing up some configurations from container volumes, and accessing API calls as part of testing, as needed.

In this section, you tested simple container executions using the public container image from Docker Registry. In the next section, you will learn how to build a custom container image with all the necessary dependencies using Ansible and run containers using custom images.

Managing container images using Ansible

As we learned from *Figure 10.6*, your integration stage will begin when the developers push the code or merge the branches in a Git repository. Call the container build commands directly from your CI/CD tools, such as **Jenkins** or **GitHub Actions**. However, commands and pipeline tasks are unpredictable, so you will not have much control over the output and results. This is where you can utilize Ansible playbooks as you have more flexibility and control over the build processes and outputs.

In the next few sections, you will learn how to create Docker container registry access, build container images using Ansible, and save the container images in the container registry.

Configuring Docker Registry access

Before pushing the latest images to the container registries, you need to log into the registry with your credentials. Access Docker Registry using a username and password, but it is a best practice to use **Access Tokens** instead of passwords. The following diagram shows how Ansible accesses the container registry to manage container images:

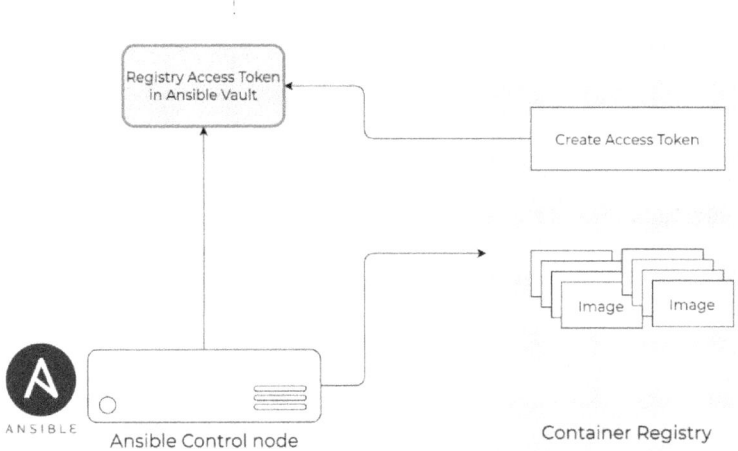

Figure 10.21 – Ansible to Container Registry access

For this demonstration, you will be using Docker Registry. Check out the documentation (refer to the *Container Registry Access Tokens* information box) for other registries. Follow these steps:

1. Log in to Docker Hub at hub.docker.com.

2. At the top right, click on your profile name and select **Account Settings** from the menu.

3. Select the **Security** tab.

4. Click on the **New Access Token** button and enter a name for your token. After that, select the **Read, Write, Delete** permission under **Access permissions** and click **Generate**:

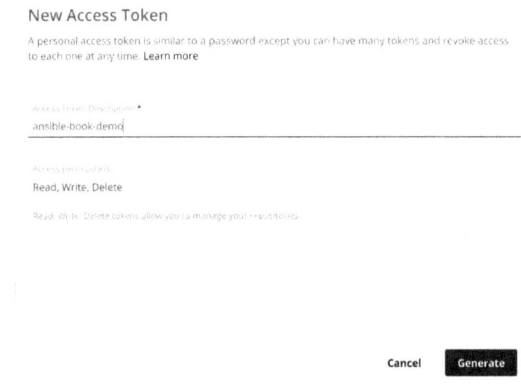

Figure 10.22 – Creating a new access token in Docker Hub

5. You will see the instructions and token text on the next screen. Remember to copy and keep the token safe as this token text will not be visible later.

> **Container Registry Access Tokens**
>
> To learn more about Access Tokens, please refer to `https://docs.docker.com/docker-hub/access-tokens/` (Docker Hub) and `https://docs.quay.io/glossary/access-token.html` (Quay).

6. Create an Ansible Vault file to keep the Docker Registry credentials in, as follows:

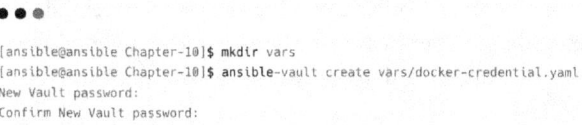

```
[ansible@ansible Chapter-10]$ mkdir vars
[ansible@ansible Chapter-10]$ ansible-vault create vars/docker-credential.yaml
New Vault password:
Confirm New Vault password:
```

Figure 10.23 – Using an Ansible Vault file for your Docker Registry credentials

7. Add your Docker username and Access Token value to the file and save it:

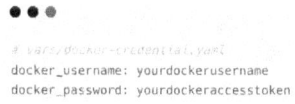

```
# vars/docker-credential.yaml
docker_username: yourdockerusername
docker_password: yourdockeraccesstoken
```

Figure 10.24 – Adding your Docker username and password to the Ansible Vault file

These variables will be used to access Docker Registry in the next sections. It is possible to keep this sensitive information in environment variables or the built-in secret management features of your CI/CD software (such as credentials in Jenkins).

Building container images using Ansible

As you may recall, you can use the existing available Docker container images from public registries such as Docker Hub or Quay. But for our application, we need to build container images and use them to deploy the application.

With the help of `community.docker.docker_image` and other modules, we can easily build container images and push those images to container registries.

Create applications or find sample applications on the internet and use them for practicing further. The following are some options:

- `https://github.com/spring-projects/spring-petclinic`

- `https://github.com/docker/getting-started/tree/master/app`

- `https://github.com/dockersamples/example-voting-app`

- `https://github.com/dockersamples`

In this section, you will containerize a simple Node.js application (`https://github.com/ginigangadharan/nodejs-todo-demo-app`) using a Dockerfile. A Dockerfile (`https://docs.docker.com/engine/reference/builder`) is a simple plain text file containing instructions for building the container image. After that, you will use Ansible to build the container image and push it to Docker Registry. The following diagram shows the steps involved:

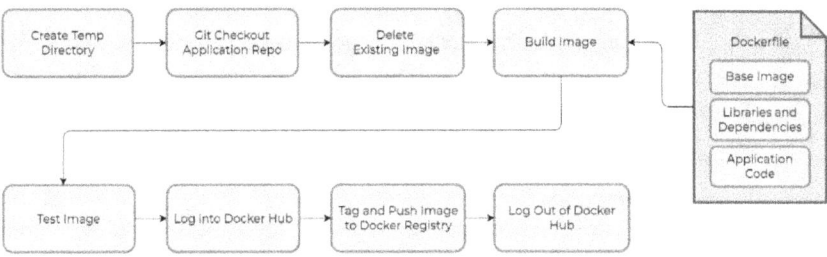

Figure 10.25 – Building and managing a container image using Ansible

Podman and Containerfiles

If you are using a different container engine, such as Podman, then check out the respective module documentation and use it accordingly. Podman modules for Ansible can be found at `https://docs.ansible.com/ansible/latest/collections/containers/podman/index.html`, while the Podman build documentation is available at `https://docs.podman.io/en/latest/markdown/podman-build.1.html`. This will help you learn more about Containerfiles.

Access the repository and perform the following steps:

1. Verify the Dockerfile inside the application repository (`https://github.com/ginigangadharan/nodejs-todo-demo-app`):

```
• • •
# syntax=docker/dockerfile:1
FROM node:12-alpine
RUN apk add --no-cache python2 g++ make
WORKDIR /app
COPY . .
RUN yarn install --production
CMD ["node", "src/index.js"]
EXPOSE 3000
```

Figure 10.26 – Verifying the Dockerfile to build the container image

The Dockerfile contains instructions for building the container image and exposing the application on port 3000.

The repository also contains a simple **ToDo** application written in **Node.js** with supported files and directories:

```
• • •
$ ls -l
total 344
-rw-r-xr-x  1 gini  staff     182  3 Apr 15:33 Dockerfile
-rw-r--r--  1 gini  staff     204  3 Apr 14:07 README.md
-rw-rw-r--@ 1 gini  staff     646 10 Feb 16:59 package.json
drwxrwxr-x@ 5 gini  staff     160  3 Apr 14:06 spec
drwxrwxr-x@ 7 gini  staff     224  3 Apr 14:06 src
-rw-rw-r--@ 1 gini  staff  162208 10 Feb 16:59 yarn.lock
```

Figure 10.27 – Application repository content

2. Fork this repository and make changes as needed.

3. Create a playbook called `Chapter-10/container-build.yaml` and add the required variables to build the container image as follows:

● ● ●

```
---
# ./chapter-10/build-dockerimage-10.yaml
- name: Building Container Images
  hosts: "{{ NODES }}"
  become: yes
  vars:
    application_repo: https://github.com/ginigangadharan/nodejs-todo-demo-app
    application_branch: main
    application_name: todo-app
    application_version: v1
    container_image_repository: ginigangadharan
    container_registry_url: https://index.docker.io/v1/

  vars_files:
    - vars/docker-credential.yaml
    .
    .
```

Figure 10.28 – Playbook to build the container image

Using variables will help you dynamically pass the values to the same playbook for different image build tasks.

Docker Registry URL

Note that `https://index.docker.io/v1/` is the default URL for the Docker Hub registry. If you are using a different registry or other private container registries, then find the correct registry URL and use it as `container_registry_url`.

Also check the included variable file `vars/docker-credential.yaml`, which contains the Docker Registry username and Access Token (refer to the *Configuring Docker Registry access* section in this chapter for more details).

4. Add a task that will create a temporary working directory on the host and check out the application repository. This is to avoid using the default directory names and overwriting issues when the same playbook executes in parallel. You also need to delete the directory at the end of the play as part of housekeeping:

```
# Chapter-10/container-build.yaml
..
  tasks:
    - name: Create temporary location
      ansible.builtin.tempfile:
        state: directory
        prefix: "container_build_"
      register: temp_location

    - debug:
        msg: "{{ temp_location.path }}"

    - name: Git checkout the application
      ansible.builtin.git:
        repo: "{{ application_repo }}"
        dest: "{{ temp_location.path }}"
        version: "{{ application_branch }}"
```

Figure 10.29 – Tasks to build the container image

5. Add a task to delete the image if it already exists with the same name and tag before creating
    the new container image. Also add the task for building the container image by providing the
    working directory path – that is, `temp_location.path`:

```
# Chapter-10/container-build.yaml
..
    - name: Delete existing container image with same name and tag
      community.docker.docker_image:
        name: "{{ application_name }}"
        tag: "{{ application_version }}"
        state: absent

    - name: Build container image
      community.docker.docker_image:
        name: "{{ application_name }}"
        tag: "{{ application_version }}"
        build:
          path: "{{ temp_location.path }}"
        source: build
        state: present
```

Figure 10.30 – Delete and create new container image

6. If you want to include tasks such as scanning the image, testing the vulnerabilities, and more, they can be included at this stage (or include this as part of the post-build stage in your CI/CD pipeline):

```
---
# Chapter-10/container-build.yaml
...
    - name: Integration and other tests
      debug:
        msg: "Your tests can be included here..."
...
```

Figure 10.31 – Include scanning or testing tasks

7. Now, we need to authenticate to Docker Registry before pushing the image to the repository. Once authenticated, push the image to Docker Registry, as follows:

```
---
# Chapter-10/container-build.yaml
...
    - name: Log into DockerHub
      community.docker.docker_login:
        registry_url: "{{ container_registry_url }}"
        username: "{{ docker_username }}"
        password: "{{ docker_password }}"

    - name: Push container image to registry
      community.docker.docker_image:
        name: "{{ application_name }}"
        tag: "{{ application_version }}"
        repository: "{{ container_image_repository }}/{{ application_name }}:{{ application_version }}"
        source: local
        push: yes
...
```

Figure 10.32 – Authenticate to Docker Hub and push the image to the container registry

8. Optionally, add the latest tag to the image that will be used when you don't mention any tag while pulling the image:

```
● ● ●
---
# Chapter-10/container-build.yaml
..
  - name: Add tag latest to image
    community.docker.docker_image:
      name: "{{ application_name }}:{{ application_version }}"
      repository: "{{ container_image_repository }}/{{ application_name }}:latest"
      force_tag: yes
      push: yes
      source: local

  - name: Log out of DockerHub
    community.docker.docker_login:
      registry_url: "{{ container_registry_url }}"
      state: absent
```

Figure 10.33 – Add the latest tag to the image and log out from Docker Hub

Also, notice the last task to log out from the container registry. For security reasons, always log out of Docker Registry once the image has been pushed (or pulled).

9. Finally, delete the temporary working directory as part of the cleanup process:

```
● ● ●
---
# Chapter-10/container-build.yaml
..
  - name: Delete temporary location
    ansible.builtin.file:
      path: "{{ temp_location.path }}"
      state: absent
```

Figure 10.34 – Delete the temporary working directory

10. Execute the playbooks and verify the result. Remember to include --ask-vault-password in the command since you have included the Docker credential using Ansible Vault:

```
● ● ●
[ansible@ansible Chapter-10]$ ansible-playbook container-build.yaml -e "NODES=dockerhost" --ask-vault-password
Vault password:
```

Figure 10.35 – Execute the playbook to build and push the container image

11. Now, verify the image from multiple places, such as from a Docker host and the Docker Registry GUI (Docker Hub: https://hub.docker.com/repositories).

12. Check the image on the Docker host (`node1`):

```
[root@node-1 ~]#
todo-app                        v1      8408ad9523d3    2 minutes ago    407MB
ginigangadharan/todo-app    v1      83c58c775765    26 hours ago     407MB
```

Figure 10.36 – Container image built using Ansible

13. Also, verify the image in the Docker Hub GUI (`https://hub.docker.com/repositories`), as shown in the following screenshot:

Figure 10.37 – Docker image in Docker Hub

14. Click on the container image entry to view more details about the image:

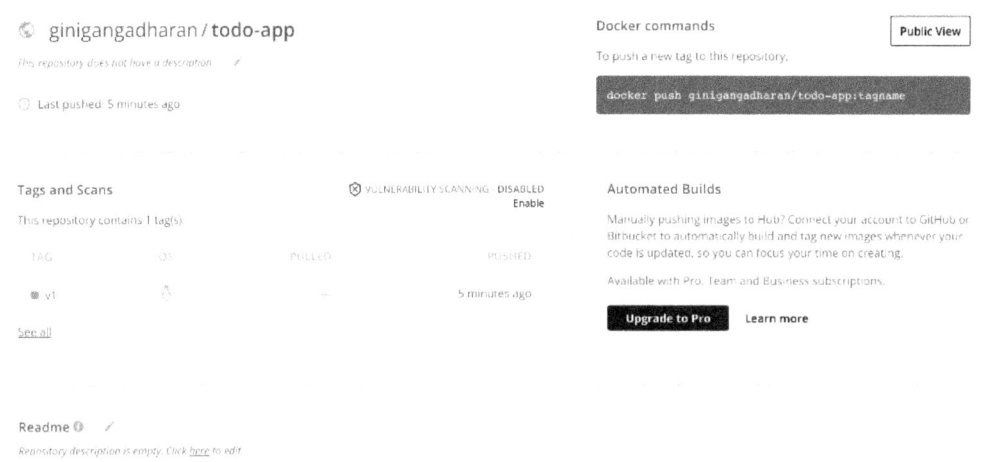

Figure 10.38 – Docker image details

Add more details to the image, such as `README` information about the container image, supported tags, documentation links, and so on.

With that, we have the latest container image ready with our application in the container registry. We can test it with the same `Chapter-10/container-manage.yaml` playbook.

Follow these steps to run container with new image:

1. Run a new container using the `Chapter-10/container-manage.yaml` playbook but pass appropriate extra variables such as `container_image`, `container_name`, `container_port`, and `container_expose_port`, as follows:

    ```
    [ansible@ansible Chapter-10]$ ansible-playbook container-manage.yaml -e "NODES=dockerhost
    container_image=ginigangadharan/todo-app container_name=todo-app container_port=3000 container_expose_port=8081"
    ```

    Figure 10.39 – Run container with different image

2. Once the container has been created, verify it on the Docker host (`node1`), as follows:

    ```
    [root@node-1 ~]# docker ps |grep todo
    0e158f5710bf   ginigangadharan/todo-app   "docker-entrypoint.s…"   3 minutes ago   Up 3 minutes   0.0.0.0:8081-
    >3000/tcp   todo-app
    ```

    Figure 10.40 – The ToDo application container deployed using Ansible

3. Verify the application from a web browser. Remember to add port `8081`, as shown in the following screenshot, since we used port `8081` while running the container:

    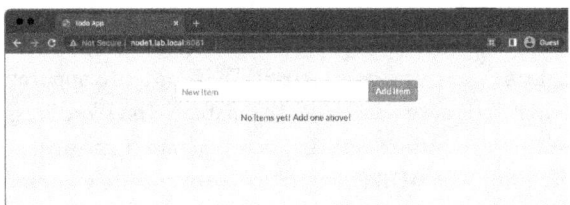

    Figure 10.41 – Accessing the ToDo app from a web browser

4. Add some entries and test the application, as shown in the following screenshot:

    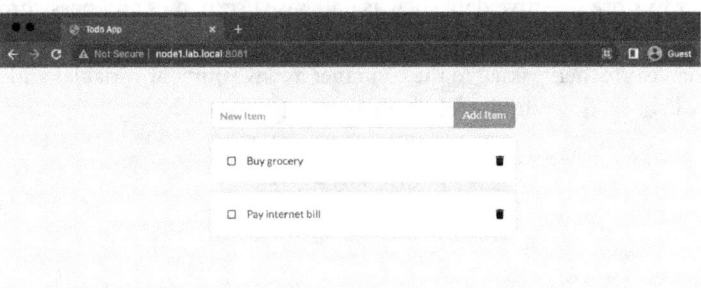

    Figure 10.42 – Testing the ToDo application with sample entries

5. Stop the container as part of housekeeping using the same playbook but by passing the `container_action=stop` action:

```
[ansible@ansible Chapter-10]$ ansible-playbook container-manage.yaml -e "NODES=dockerhost
container_image=githigangadharan/todo-app container_name=todo-app container_port=3800 container_expose_port=8081
container_action=stop"
```

Figure 10.43 – Stop container using playbooks

> **Note**
>
> The data you have stored will be lost when you stop and kill the container as you did not mount any volumes for data persistence. Add more configurations and volume details to `Chapter-10/container-manage.yaml` and enhance your playbook.

In this section, you learned how to handle single containers and container images. But it is possible to handle any number of images and containers using Ansible based on your application stack. In the next section, you will learn how to handle multiple containers using Ansible.

Managing multi-container applications using Ansible

In this section, you will use the well-known **Content Management System** (**CMS**) application stack known as WordPress (`https://wordpress.org`). The WordPress application is based on multiple application stacks, including PHP, a web server, and a database. The WordPress application is available as a container image (`https://hub.docker.com/_/wordpress`). For the database, we will deploy another container using MariaDB (`https://hub.docker.com/_/mariadb`).

Please refer to the `Chapter-10/deploy-wordpress-on-docker.yaml` file to see the Ansible playbook for deploying the WordPress CMS using Ansible. Follow these steps:

1. We declared the essential parameters on top of the playbook, as shown in the following screenshot. Remember to store sensitive data such as database usernames and passwords using Ansible Vault (or Credential in Ansible Automation Controller) or other secret management services. These variables are then passed to the container as environment variables and Docker volumes will be created, as shown in the following screenshot:

```
---
- name: Deploy wordpress stack on Docker
  hosts: "{{ NODES }}"
  become: yes
  vars:
    db_volume: 'mariadb'
    wordpress: 'wordpress'
    mysql_root_password: 'secretrootpassword'
    mysql_username: 'wordpressuser'
    mysql_password: 'secretpassword'
    mysql_database: 'wordpressdb'
    container_port: 8082
  tasks:
```

Figure 10.44 – Deploying WordPress using Ansible

2. There are two tasks, as shown in the following screenshot. The first task will deploy the MariaDB container, while the second task will deploy the WordPress container:

```
- name: Deploy MariaDB server for Database
  community.docker.docker_container:
    state: started
    image: mariadb
    name: mariadb
    volumes:
      - "{{db_volume}}:/var/lib/mysql"
    env:
      MYSQL_ROOT_PASSWORD: "{{ mysql_root_password }}"
      MYSQL_PASSWORD: "{{ mysql_password }}"
      MYSQL_DATABASE: "{{ mysql_database }}"
      MYSQL_USER: "{{ mysql_username }}"

- name: Deploy WordPress
  community.docker.docker_container:
    state: started
    image: wordpress
    name: wordpress
    restart_policy: always
    ports:
      - "{{ container_port }}:80"
    links:
      - "{{ db_volume }}:/var/lib/mysql"
    volumes:
      - "{{ wordpress }}:/var/www/html"
    env:
      MYSQL_PASSWORD: "{{ mysql_password }}"
      MYSQL_DATABASE: "{{ mysql_database }}"
      MYSQL_USER: "{{ mysql_username }}"
      MYSQL_HOST: mariadb
```

Figure 10.45 – Ansible tasks for deploying the WordPress and MariaDB containers

3. Execute the playbook to deploy the WordPress stack with the MariaDB database on `node1`:

Figure 10.46 – Deploy WordPress using Ansible

4. On `node1`, verify the Docker containers and Docker volumes, as shown in the following screenshot:

```
[devops@node-1 ~]$ sudo docker ps
CONTAINER ID   IMAGE       COMMAND              CREATED         STATUS          PORTS                    NAMES
d5253f49d1c9   wordpress   "docker-entrypoint.s…" 15 minutes ago  Up 15 minutes   0.0.0.0:8082->80/tcp
wordpress
74eb2db91a52   mariadb     "docker-entrypoint.s…" 15 minutes ago  Up 15 minutes   3306/tcp
mariadb

[devops@node-1 ~]$ sudo docker volume ls
DRIVER    VOLUME NAME
local     mariadb
local     wordpress
```

Figure 10.47 – The WordPress and MariaDB containers running on the Docker host

5. Verify the WordPress application from a browser using port `8082`, which we have configured to expose. The initial configuration for WordPress will be visible, as shown in the following screenshot. Now, we can configure the WordPress CMS application (`https://wordpress.org/support/article/how-to-install-wordpress/#setup-configuration-file`):

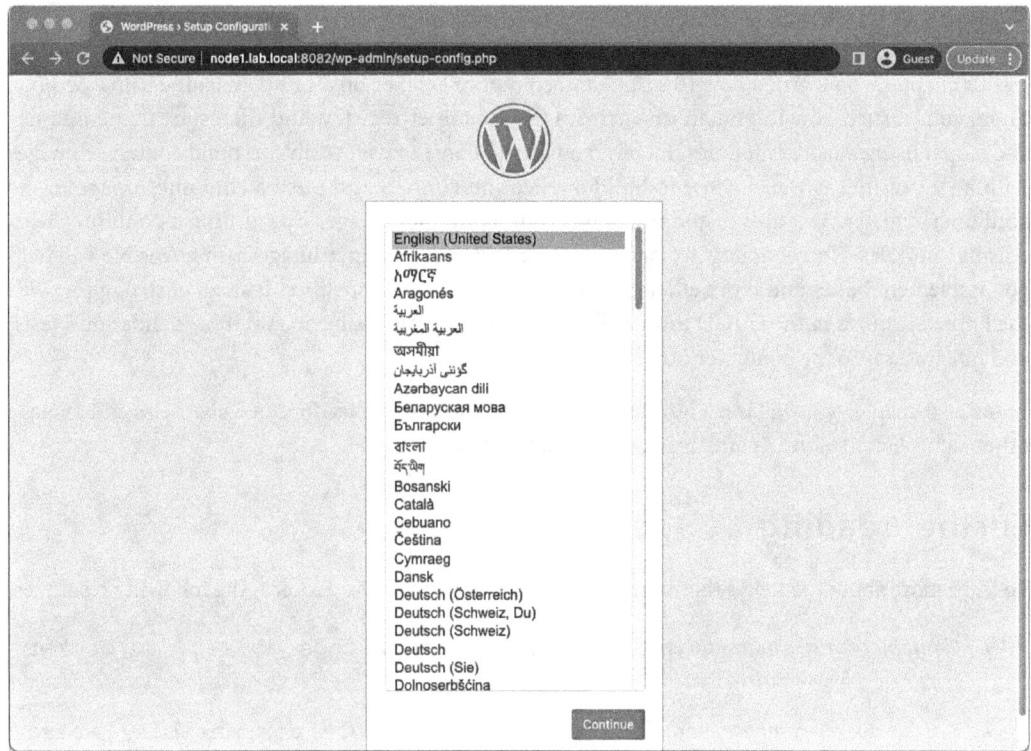

Figure 10.48 – The initial WordPress configuration screen

The playbook can be enhanced by configuring additional volumes and configurations; please refer to the WordPress installation documentation to implement more automation steps for such use cases (`https://wordpress.org/support/article/how-to-install-wordpress/#setup-configuration-file`).

With that, you have learned that the entire container image life cycle can be automated using Ansible in different stages of your CI/CD pipelines. This will give you more control over building and testing compared to using the native container management features in the available CI/CD tools.

Summary

In this chapter, you learned how to install and configure Docker on a Linux machine using Ansible. Then, you learned how to pull an image from the container registry, and then start that container and stop it using Ansible modules. Finally, you learned how to use Ansible to build container images with a Dockerfile, as well as how to build application content and push a container image to the container registry. You also tested the newly built container images by running a container with Ansible modules. Knowing how to manage containers and container images using Ansible will help you implement better and more efficient CI/CD workflows and pipelines. Instead of struggling with the limited features in the CI/CD tools, utilize the flexibility of Ansible to add more validations, tests, and integrations to the container build process.

In the next chapter, you will learn how to manage containerized applications in Kubernetes and manage other Kubernetes resources and applications using Ansible.

Further reading

To learn more about the topics that were covered in this chapter, take a look at the following resources:

- *Container registry access tokens for Docker Hub*: `https://docs.docker.com/docker-hub/access-tokens`

- *Ansible docker-compose module*: `https://docs.ansible.com/ansible/latest/collections/community/docker/docker_compose_module.html`

- *Docker RUN and environment variables*: `https://docs.docker.com/engine/reference/commandline/run/#set-environment-variables--e---env--env-file`

- *Dockerfile documentation*: `https://docs.docker.com/engine/reference/builder/`

- *Top 5 Free Resources to Learn Docker*: `https://www.techbeatly.com/top-5-free-resources-to-learn-docker/`

11

Managing Kubernetes Using Ansible

Due to the containerization of applications and the revolution in microservices, Kubernetes-based platforms have become popular. The containerization of applications and container orchestration using Kubernetes provide additional layers and complexity to infrastructure that requires automated solutions for managing a large number of components.

In the previous chapter, you learned about the capabilities of Ansible to build and manage container images and containers. When it comes to container orchestration tools, such as Kubernetes or Red Hat OpenShift, there are Ansible collections available with modules and plugins for supporting and managing your Kubernetes and Red Hat OpenShift clusters and resources.

Using Ansible for Kubernetes resource management will help you to implement more integrations in your DevOps workflow and **Continuous Integration/Continuous Deployment (CI/CD)** pipelines to deploy your applications very flexibly.

In this chapter, we will cover the following topics:

- An introduction to Kubernetes

- Managing Kubernetes clusters using Ansible

- Configuring Ansible for Kubernetes

- Deploying applications to Kubernetes using Ansible

- Scaling Kubernetes applications

- Executing commands inside a Kubernetes Pod

We will learn about using the Ansible collection for Kubernetes management and automating Kubernetes cluster operations and resource management with it.

Technical requirements

The following are the technical requirements to proceed with this chapter:

- One Linux machine for the Ansible control node

- A working Kubernetes cluster with API access (refer to `https://minikube.sigs.k8s.io/docs/start` to spin up a local Kubernetes cluster)

- Basic knowledge about containers and Kubernetes

All the Ansible code, Ansible playbooks, commands, and snippets for this chapter can be found in the GitHub repository at `https://github.com/PacktPublishing/Ansible-for-Real-life-Automation/tree/main/Chapter-11`.

An introduction to Kubernetes

Kubernetes is an open source container orchestration platform where we can deploy and manage our containerized applications without worrying about the underlying layers. This model of service is known as **Platform as a Service** (**PaaS**), where developers have the freedom to deploy their applications and other required resources, such as storage, network, and secrets, without assistance from the platform team.

The Kubernetes platform contains many components to manage container deployment and orchestration, as shown in *Figure 11.1*:

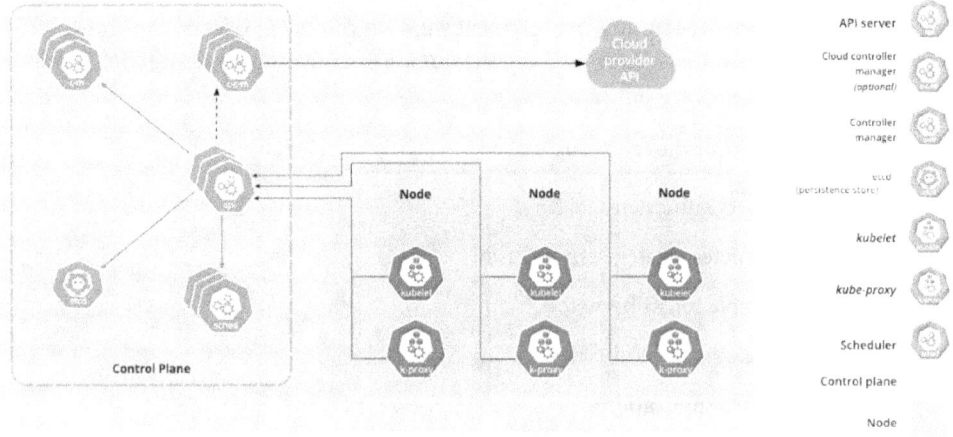

Figure 11.1 – The components of a Kubernetes cluster (source: `https://kubernetes.io/docs/concepts/overview/components/`)

Let's briefly have a look at these components in the following sections.

The Kubernetes control plane

The control plane is responsible for making decisions on behalf of the cluster and application, such as scheduling the application Pods, detecting and responding to Pod failures, and managing cluster nodes. The control plane has multiple components to handle these operations, as follows:

- **kube-apiserver** – Exposes the Kubernetes cluster API that will take care of all central management and communications. Each and every task inside a Kubernetes cluster is operated via the Kubernetes API server.

- **kube-controller-manager** – Consists of several controller processes, such as a Node controller, Job controller, Service account controller, and Endpoints controller.

- **etcd** – A **high-availability (HA)** key-value store used for storing Kubernetes cluster data.

- **kube-scheduler** – Helps to select nodes for deploying Pods.

Components on the nodes

These are the Kubernetes components running on every node in the cluster, managing the application Pods and its network:

- **Container runtime** – The actual software that is running the containers in the backend, such as **containerd** and **CRI-O**

- **kubelet** – Takes care of the running containers, as per specifications

- **kube-proxy** – Helps to implement Kubernetes Services by maintaining network rules on the nodes

Like any other open source project, Kubernetes is also free to use and supported by the Kubernetes user and developer community (`https://kubernetes.io/community`). If an organization is looking for enterprise Kubernetes distributions and support, there are different Kubernetes distributions available on the market as turnkey solutions, such as Red Hat OpenShift, Rancher, and VMware Tanzu.

Explaining the Kubernetes platform, concepts, and architecture is beyond the scope of this book. You can find more details in the *Further reading* section at the end of this chapter.

For using Ansible with Kubernetes, we can use **minikube**, which is a local Kubernetes cluster for learning about and developing Kubernetes Deployments. Refer to the documentation at `https://minikube.sigs.k8s.io/docs/start`, where you will find installation instructions for Linux, Windows, and macOS platforms.

> **The Kubernetes Documentation**
>
> Refer to the Kubernetes documentation at `https://kubernetes.io/docs/home` to learn more about Kubernetes. Also refer to *Top 15 Free Kubernetes Courses* (`https://www.techbeatly.com/kubernetes-free-courses`) to learn the basics about Kubernetes.

In the next section, we will learn about the Ansible method for managing Kubernetes clusters.

Managing Kubernetes clusters using Ansible

Deploying a Kubernetes cluster involves many steps, including preparing nodes, installing container runtime packages, and configuring networking. There are multiple methods we can use for deploying Kubernetes clusters within testing or production environments. The installation method for a Kubernetes cluster also depends on your requirements, whether you are using single-node clusters or multi-node clusters with HA or you require the option to scale the cluster whenever needed, for example.

Kubespray is a production-grade Kubernetes cluster deployment method that uses Ansible as its foundation for provisioning and orchestration. Using Kubespray, it is possible deploy a Kubernetes cluster on top of bare-metal servers, virtual machines, and private cloud or public cloud platforms (for example, AWS, GCE, Azure, and OpenStack).

Kubespray is highly customizable and you can configure the cluster with different Kubernetes components of your choice, as follows:

- A supported **Container Network Interface** (**CNI**) – Calico, Flannel, Kube-router, Kube-OVN, Weave, or Multus

- A supported **Container Runtime Interface** (**CRI**) – containerd, Docker, CRI-O, Kata Containers, or gVisor

- Supported cloud providers – AWS, Azure, OpenStack, or vSphere

- A supported **Ingress** – Kube-vip, ALB Ingress, MetalLB, or Nginx

- Supported operating systems – Debian, Fedora CoreOS, Flatcar Container Linux, openSUSE, Red Hat Enterprise Linux, CentOS, or Amazon Linux 2

Kubernetes deployment using Kubespray is explained in the documentation (`https://kubespray.io`) and other online guides.

Once provisioned, the Kubernetes cluster can be scaled as needed (by adding or removing nodes) using Kubespray without worrying about the manual configurations of the new nodes and cluster-joining steps. Refer to the following information box and the *Further reading* section at the end of this chapter for links to further information on Kubespray.

> **Kubespray**
>
> Refer to the documentation (`https://kubespray.io`) and project repository (`https://github.com/kubernetes-sigs/kubespray`) to learn more about the different options available with Kubespray. Also, check out `https://www.techbeatly.com/deploying-kubernetes-with-kubespray` for a detailed guide on how to deploy a multi-node Kubernetes cluster using Kubespray.

For the demonstration of Ansible use cases, we have used a minikube Kubernetes environment, as follows:

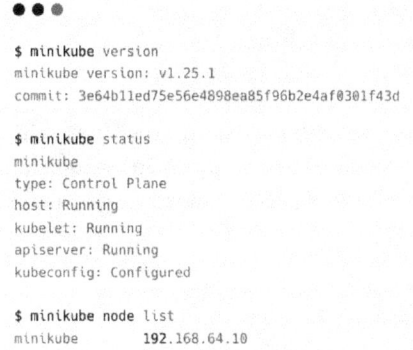

```
$ minikube version
minikube version: v1.25.1
commit: 3e64b11ed75e56e4898ea85f96b2e4af0301f43d

$ minikube status
minikube
type: Control Plane
host: Running
kubelet: Running
apiserver: Running
kubeconfig: Configured

$ minikube node list
minikube          192.168.64.10
```

Figure 11.2 – minikube Kubernetes cluster details

Refer to the documentation (`https://minikube.sigs.k8s.io/docs/start`) for more details on how to create a minikube environment.

We will learn how to configure Ansible for Kubernetes cluster access and check cluster resources in the next section.

Configuring Ansible for Kubernetes

Ansible can communicate with Kubernetes clusters using the Kubernetes Python libraries or directly via the Kubernetes API, as shown in *Figure 11.3*:

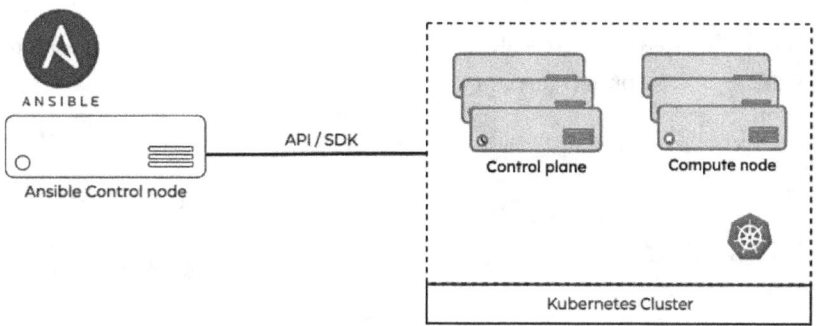

Figure 11.3 – Communication between Ansible and Kubernetes

Ansible modules and plugins for managing Kubernetes are available in the `kubernetes.core` Ansible collection. (The Ansible Kubernetes collection was released as `community.kubernetes` prior to the release of `kubernetes.core` 1.1.) We will install, configure, and use the `kubernetes.core` collection in the following sections.

Python requirements

To communicate with the Kubernetes or OpenShift API, use the **Python client for the OpenShift API** (`https://github.com/openshift/openshift-restclient-python`) Python library. Before using any of the Kubernetes modules, you need to install the required Python libraries, as follows:

```
$ pip install openshift
$ pip install PyYAML
```

If you are using Ansible inside a Python virtual environment, then remember to activate it and install the libraries within it.

> **The Python Library for Kubernetes**
>
> The OpenShift REST client depends on the Kubernetes Python client (`https://github.com/kubernetes-client/python`) and this Kubernetes Python client will be installed as part of the dependencies. Also check the Ansible Content Collection for Red Hat OpenShift (Ansible blog) – `https://www.ansible.com/blog/introducing-the-ansible-content-collection-for-red-hat-openshift`

Installing the Ansible Kubernetes collection

Install the Ansible Kubernetes collection from Ansible Galaxy (`https://galaxy.ansible.com/kubernetes/core`), as follows:

1. Configure the `ansible.cfg` object type with the collection path if you want to install the collection inside the project directory:

```
[defaults]

COLLECTIONS_PATHS = ./collections
```

Figure 11.4 - Configure collection path in ansible.cfg

2. Install the `kubernetes.core` collection:

```
● ● ●
[ansible@ansible Chapter-11]$ ansible-galaxy collection install kubernetes.core
```

Figure 11.5 - Install kubernetes.core collection

The kubernetes.core collection will be installed inside the collection directory (COLLECTIONS_ PATHS = ./collections) as you mentioned in the ansible.cfg file.

Connecting Ansible to Kubernetes

Ansible will try to use the $HOME/.kube/config file, which is the default Kubernetes configuration file (kubeconfig) containing Kubernetes cluster details, credential details, and connection contexts. If your kubeconfig file is residing in a different path or has a different filename, then specify this in the Kubernetes module parameter inside the Ansible playbook.

As we mentioned earlier, we have a Kubernetes cluster running on minikube. The kubeconfig file and certificate credentials are copied to the Ansible control node, as shown in the following figure:

```
● ● ●
[ansible@ansible Chapter-11]$ ls -l ~/.kube/
total 16
-rw-r--r--. 1 ansible ansible 1111 Apr 25 14:03 ca.crt
-rw-r--r--. 1 ansible ansible 1147 Apr 25 14:03 client.crt
-rw-------. 1 ansible ansible 1675 Apr 25 14:03 client.key
-rw-rw-r--. 1 ansible ansible  824 Apr 25 13:58 minikube-config
```

Figure 11.6 – The kubeconfig file and certificates

If you are using basic authentication (a username and password) for accessing the cluster, then specify these details inside the Kubernetes module itself.

For production environments, remember to follow best practices for storing the credentials, certificates, and access keys, using Ansible Vault or any other secret management system.

Installing the kubectl CLI tool

Since we are working from the control node, install the kubectl CLI tool on the machine and execute the kubectl command to verify the resources that we will be creating using Ansible.

If you have a machine installed with kubectl and access to the Kubernetes cluster, then skip this step.

Follow the kubectl installation document (https://kubernetes.io/docs/tasks/tools/ install-kubectl-linux) and use the appropriate method suitable for your environment (since we are running the Ansible control node on a Red Hat Enterprise Linux machine, the following commands are based on Red Hat Enterprise Linux/Fedora distributions):

1. Add the repository for the `kubectl` package.

● ● ●

```
[ansible@ansible Chapter-11]$ cat <<EOF | sudo tee /etc/yum.repos.d/kubernetes.repo
[kubernetes]
name=Kubernetes
baseurl=https://packages.cloud.google.com/yum/repos/kubernetes-el7-x86_64
enabled=1
gpgcheck=1
repo_gpgcheck=1
gpgkey=https://packages.cloud.google.com/yum/doc/yum-key.gpg https://packages.cloud.google.com/yum/doc/rpm-package-
key.gpg
EOF
```

Figure 11.7 – Add the repository for the kubectl package

2. Install the `kubectl` package:

● ● ●

```
[ansible@ansible Chapter-11]$ sudo yum install -y kubectl
```

Figure 11.8 – Installing the kubectl utility

3. Configure the KUBECONFIG environment variable, as our `kubeconfig` filename is different (`/home/ansible/.kube/minikube-config`) from the default filename (`/home/ansible/.kube/config`):

● ● ●

```
[ansible@ansible Chapter-11]$ export KUBECONFIG=$KUBECONFIG:/home/ansible/.kube/minikube-config
```

Figure 11.9 – Configure the KUBECONFIG environment variable

4. Verify the `kubectl` CLI version and cluster details.

● ● ●

```
[ansible@ansible Chapter-11]$ kubectl version
Client Version: version.Info{Major:"1", Minor:"23", GitVersion:"v1.23.6",
GitCommit:"ad3338546da947756e8a88aa6822e9c11e7eac22", GitTreeState:"clean", BuildDate:"2022-04-14T08:49:13Z",
GoVersion:"go1.17.9", Compiler:"gc", Platform:"linux/amd64"}
Server Version: version.Info{Major:"1", Minor:"23", GitVersion:"v1.23.1",
GitCommit:"86ec240af8cbd1b60bcc4c03c28da9b98005b92e", GitTreeState:"clean", BuildDate:"2021-12-16T11:34:54Z",
GoVersion:"go1.17.5", Compiler:"gc", Platform:"linux/amd64"}
```

Figure 11.10 – kubectl version information

The kubectl CLI is able to access the cluster and we will go on to use it in the following sections to verify the resources in the Kubernetes cluster.

Verifying the Kubernetes cluster details using Ansible

Collecting the cluster details and resource information is an important step. The Kubernetes cluster details can be collected using the kubectl CLI, as shown in *Figure 11.11*:

```
[ansible@ansible Chapter-11]$ kubectl get po -n kube-system
NAME                                    READY   STATUS    RESTARTS   AGE
coredns-64897985d-msdjx                 1/1     Running   18         164d
etcd-minikube                           1/1     Running   21         164d
kube-apiserver-minikube                 1/1     Running   22         164d
kube-controller-manager-minikube        1/1     Running   21         164d
kube-proxy-bh9wj                        1/1     Running   19         164d
kube-scheduler-minikube                 1/1     Running   21         164d
metrics-server-6b76bd68b6-4lcww         1/1     Running   69         164d
storage-provisioner                     1/1     Running   80         164d
```

Figure 11.11 – Kubernetes cluster details, collected using the kubectl CLI

Use the kubernetes.core.k8s_info module to collect similar details about the Kubernetes cluster using Ansible (prior to Ansible 2.9, the module was called k8s_facts).

Follow these steps to create our first Ansible playbook to interact with a Kubernetes cluster:

1. Create Chapter-11/k8s-details.yaml, as follows:

```
# Chapter-11/k8s-details.yaml
---
- name: Ansible Kubernetes Info
  hosts: localhost
  tasks:
    - name: Get a list of all pods from any namespace
      kubernetes.core.k8s_info:
        kubeconfig: /home/ansible/.kube/minikube-config
        kind: Pod
        namespace: kube-system
      register: pod_list

    - name: Display Pod Details
      debug:
        msg: "{{ pod_list }}"
```

Figure 11.12 – The task for fetching Pod details from the kube-system namespace

We are fetching the Pod details from the `kube-system` namespace in the Kubernetes cluster. Note the `hosts: localhost` line as the execution needs to happen on `localhost`. The Python libraries will take care of the Kubernetes operations.

2. Execute the playbook and verify the output. You will see a lot of details about the Pods running inside the `kube-system` namespace, as shown in *Figure 11.13*:

```
[ansible@ansible Chapter-11]$ ansible-playbook k8s-details.yaml |more

...<output omitted for brevity>...

TASK [Display Pod Details] ********************************************************
ok: [localhost] => {
    "msg": {
        "api_found": true,
        "changed": false,
        "failed": false,
        "resources": [
            {
                "apiVersion": "v1",
                "kind": "Pod",
                "metadata": {
                    "creationTimestamp": "2022-02-01T06:57:46Z",
                    "generateName": "coredns-64897985d-",
                    "labels": {
                        "k8s-app": "kube-dns",
                        "pod-template-hash": "64897985d"
                    },
                    "managedFields": [
                        {
                            "apiVersion": "v1",
                            "fieldsType": "FieldsV1",

...<output omitted for brevity>...
```

Figure 11.13 – Pod details from a Kubernetes namespace

3. Add one more task in the same playbook to collect the Kubernetes cluster node details.

```
- name: Get a list of Nodes
  kubernetes.core.k8s_info:
    kubeconfig: /home/ansible/.kube/minikube-config
    kind: Node
  register: node_list

- name: Display Pod Details
  debug:
    msg: "{{ item.metadata.labels['kubernetes.io/hostname'] }}"
  loop: "{{ node_list.resources }}"
```

Figure 11.14 – The task to get the Kubernetes node details

4. Execute the playbook again and verify the results.

See the node name (in this case, `minikube`), as in the output in *Figure 11.15*. If you have multiple nodes, you will see multiple entries here.

● ● ●

```
[ansible@ansible Chapter-11]$ ansible-playbook k8s-details.yaml
...<output omitted>..
ames': ['k8s.gcr.io/kube-scheduler@sha256:8be4eb1593cf9ff2d91b44596633b7815a3753696031a1eb4273d1b39427fa8c',
'k8s.gcr.io/kube-scheduler:v1.23.1'], 'sizeBytes': 53488305}, {'names': ['k8s.gcr.io/ingress-nginx/kube-webhook-
certgen@sha256:64d8c73dca984af206adf9d6d7e46aa550362b1d7a01f3a0a91b20cc67868660'], 'sizeBytes': 47736388}, {'names':
['k8s.gcr.io/coredns/coredns@sha256:5b6ec0d6de9baaf3e92d0f66cd96a25b9edbce8716f5f15dcd1a616b3abd590e',
'k8s.gcr.io/coredns/coredns:v1.8.6'], 'sizeBytes': 46829283}, {'names': ['kubernetesui/metrics-
scraper@sha256:36d5b3f60e1a144cc5ada820910535074bdf5cf73fb70d1ff1681537eef4e172', 'kubernetesui/metrics-
scraper:v1.0.7'], 'sizeBytes': 34446077}, {'names': ['gcr.io/k8s-minikube/storage-
provisioner@sha256:18eb69d1418e854ad5a19e399310e52808a8321e4c441c1dddad8977a0d7a944', 'gcr.io/k8s-minikube/storage-
provisioner:v5'], 'sizeBytes': 31465472}]}, 'apiVersion': 'v1', 'kind': 'Node'}] => {
    "msg": "minikube"
}

PLAY RECAP ***************************************************************************************************************
localhost                  : ok=5    changed=0    unreachable=0    failed=0    skipped=0    rescued=0    ignored=0
```

Figure 11.15 – Fetching the Kubernetes cluster node details

Customize the playbook for different clusters by using different `kubeconfig` files or providing the credential details in the module itself, such as `api_key`, `client_cert`, and `client_key`. Refer to the `kubernetes.core.k8s_info` module (https://docs.ansible.com/ansible/latest/collections/kubernetes/core/k8s_info_module.html) for more details.

Now, you have learned how to communicate with Kubernetes clusters using Ansible modules and how to fetch cluster details. In the next section, we will learn more about Ansible for Kubernetes automation by creating resources and objects in Kubernetes clusters.

Deploying applications to Kubernetes using Ansible

Containerized applications can be deployed inside Kubernetes via the Kubernetes dashboard (web UI) or using the `kubectl` CLI (https://kubernetes.io/docs/reference/kubectl). By using Ansible, we can automate most of the deployment operations that take place inside our Kubernetes clusters. Since Ansible can easily integrate within CI/CD pipelines, it is possible to achieve more control over your application deployments in a containerized environment such as Kubernetes.

Applications are deployed inside logical isolated groups called Kubernetes **namespaces**. There can be default namespaces and Kubernetes cluster-related namespaces, and we can also create additional namespaces as required to deploy applications. *Figure 11.16* demonstrates the relation between Deployments, Pods, Services, and namespaces in a Kubernetes cluster:

Figure 11.16 – Kubernetes Deployments and namespaces

In the following exercise, we will deploy an application in Kubernetes by creating a dedicated namespace, Deployment configuration, and Service:

1. Prepare the `Chapter-11/todo-app-deploy.yaml` Deployment definition file, which will be used to create a Kubernetes Deployment resource in the next steps. We can create **Jinja2 templates** to automate the definition files to handle complex Deployments, but here we will use a simple Deployment definition, using the `ginigangadharan/todo-app:latest` image that we created in *Chapter 10*, *Managing Containers Using Ansible*.

● ● ●

```
---
apiVersion: apps/v1
kind: Deployment
metadata:
  name: todo-app
  labels:
    app: todo
spec:
  replicas: 1
  selector:
    matchLabels:
      app: todo
  template:
    metadata:
      labels:
        app: todo
    spec:
      containers:
      - name: todoapp
        image: ginigangadharan/todo-app:latest
        ports:
        - containerPort: 3000
```

Figure 11.17 – The Kubernetes definition file for the to-do app Deployment

2. Prepare a `Chapter-11/todo-app-service.yaml` Kubernetes Service definition file to expose the application using `NodePort 300080`. Use different `NodePort`, `ClusterIP`, or `LoadBalancer` type Services depending on the Kubernetes cluster that you are using.

```
---
apiVersion: v1
kind: Service
metadata:
  name: todoapp-svc
spec:
  type: NodePort
  ports:
    - targetPort: 3000
      port: 3000
      nodePort: 30080
  selector:
    app: todo
```

Figure 11.18 – The Kubernetes definition file for the to-do app Service

3. Create a `Chapter-11/k8s-app-deploy.yaml` playbook and add contents, as follows:

```
---
# Chapter-11/k8s-app-deploy.yaml
- name: Deploying Application to Kubernetes
  hosts: localhost
  gather_facts: false
  vars:
    kubeconfig_file: /home/ansible/.kube/minikube-config
    namespace_name: todoapp-ns
  tasks:
    - name: Create a k8s namespace
      kubernetes.core.k8s:
        kubeconfig: "{{ kubeconfig_file }}"
        name: "{{ namespace_name }}"
        api_version: v1
        kind: Namespace
        state: present
```

Figure 11.19 – A playbook to deploy the application in Kubernetes

4. Execute the playbook:

```
[ansible@ansible Chapter-11]$ ansible-playbook k8s-app-deploy.yaml
```

Figure 11.20 – Execute the playbook to deploy todo-app

5. Verify that the namespace was created using the `kubectl` CLI, as shown in *Figure 11.21*:

```
[ansible@ansible Chapter-11]$ kubectl get namespace todoapp-ns
NAME         STATUS  AGE
todoapp-ns   Active  6s
```

Figure 11.21 – A Kubernetes namespace created using Ansible

6. Add a task in the same playbook to create the Deployment using the `todo-app-deploy.yaml` Deployment definition file and to create the Kubernetes Service resource using the `todo-app-service.yaml` file, as follows:

```
# Chapter 11/k8s_app_deploy.yaml - Tasks for deployment and service

 - name: Create Deployment
   kubernetes.core.k8s:
     kubeconfig: {{ kubeconfig_file }}
     state: present
     src: todo-app-deploy.yaml
     namespace: {{ namespace_name }}

 - name: Expose application on NodePort
   kubernetes.core.k8s:
     kubeconfig: {{ kubeconfig_file }}
     state: present
     src: todo-app-service.yaml
     namespace: {{ namespace_name }}
```

Figure 11.22 – Tasks to create Deployment and Service resources in Kubernetes

7. Execute the playbook again to create the Deployment.

8. Verify that the Deployment, Pods, ReplicaSet, and Service resources were created, as shown in *Figure 11.23*:

```
[ansible@ansible Chapter-11]$ kubectl -n todoapp-ns get all
NAME                              READY   STATUS    RESTARTS   AGE
pod/todo-app-546b5b58d-bhhnz     1/1     Running   0          5m36s

NAME                  TYPE       CLUSTER-IP     EXTERNAL-IP   PORT(S)          AGE
service/todoapp-svc   NodePort   10.98.213.33   <none>        3000:30080/TCP   5m35s

NAME                        READY   UP-TO-DATE   AVAILABLE   AGE
deployment.apps/todo-app    1/1     1            1           5m36s

NAME                                  DESIRED   CURRENT   READY   AGE
replicaset.apps/todo-app-546b5b58d    1         1         1       5m36s
```

Figure 11.23 – Deployment, Pod, and Service created by Ansible

9. As we are using a minikube cluster, we can get the exposed Service details using the `minikube service list` command from the machine running the minikube cluster (not from the Ansible control node), as shown in *Figure 11.24*:

```
$ minikube service list
|---------------|---------------------------------|---------------|----------------------------|
|   NAMESPACE   |              NAME               |  TARGET PORT  |            URL             |
|---------------|---------------------------------|---------------|----------------------------|
| default       | kubernetes                      | No node port  |                            |
| ingress-nginx | ingress-nginx-controller        | http/80       | http://192.168.64.10:31729 |
|               |                                 | https/443     | http://192.168.64.10:30711 |
| ingress-nginx | ingress-nginx-controller-admission | No node port |                            |
| kube-system   | kube-dns                        | No node port  |                            |
| kube-system   | metrics-server                  | No node port  |                            |
| todoapp-ns    | todoapp-svc                     |          3000 | http://192.168.64.10:30080 |
|---------------|---------------------------------|---------------|----------------------------|
```

Figure 11.24 – Exposed Service details in a minikube cluster

Find the URL for the `todo-app` application (`http://192.168.64.10:30080`, in this case) based on the name and use it in the next step.

10. Access the application from a web browser using the URL `http://192.168.64.10:30080`, as shown in *Figure 11.25*:

Figure 11.25 – Accessing the to-do app deployed in a Kubernetes cluster

11. Expand the playbook with more resources and configurations as part of your application deployment. For example, we can create an Ingress resource with the Ingress definition, as follows:

```
apiVersion: networking.k8s.io/v1
kind: Ingress
metadata:
  name: todoapp-ingress
  annotations:
    nginx.ingress.kubernetes.io/rewrite-target: /$1
spec:
  rules:
    - host: todoapp.local
      http:
        paths:
          - path: /
            pathType: Prefix
            backend:
              service
                name: todoapp-svc
                port:
                  number: 3000
```

Figure 11.26 – Ingress resource definition for the to-do app

12. Use the Ingress definition in the Ansible task to create an Ingress resource, as follows:

```
- name: Create ingress resource
  kubernetes.core.k8s:
    kubeconfig: "{{ kubeconfig_file }}"
    state: present
    src: todo-app-ingress.yaml
    namespace: "{{ namespace_name }}"
```

Figure 11.27 – Create an Ingress resource using Ansible

We can also pass the Deployment information in the form of variables and integrate them into our CD workflow.

You have learned how to deploy applications and Service resources in Kubernetes using Ansible. Most of the resources in Kubernetes can be created using the `kubernetes.core.k8s` module with the resource definition files. However, there are other useful modules in the `kubernetes.core` collection and we will go on to learn how to use a few of them, such as scaling applications and running commands inside a Pod.

Scaling Kubernetes applications

The **ReplicaSet** resource in Kubernetes ensures that a specified number of application Pod replicas are running as part of the Deployment. This mechanism will help to scale the application horizontally whenever needed and without additional resource configurations. A ReplicaSet resource will be created when you create a deployment resource in Kubernetes, as shown in *Figure 11.28*:

```
[ansible@ansible Chapter-11]$ kubectl -n todoapp-ns get all
NAME                          READY    STATUS      RESTARTS   AGE
pod/todo-app-546b5b58d-bhhnz  1/1      Running     0          5m36s

NAME                  TYPE        CLUSTER-IP      EXTERNAL-IP   PORT(S)           AGE
service/todoapp-svc   NodePort    10.98.213.33    <none>        3000:30080/TCP    5m35s

NAME                        READY    UP-TO-DATE   AVAILABLE   AGE
deployment.apps/todo-app    1/1      1            1           5m36s

NAME                                  DESIRED   CURRENT   READY   AGE
replicaset.apps/todo-app-546b5b58d    1         1         1       5m36s
```

Figure 11.28 – A ReplicaSet resource created as part of Deployment

Specify the initial number of replicas inside the Deployment definition file as `replicas: 1`. ReplicaSet will scale the number of Pods based on the replica number.

When there is extra traffic on the application Pods, scale the application using the `kubectl scale` command, as follows (modify the Deployment, not the ReplicaSet):

```
$ kubectl -n todoapp-ns scale deployment/todo-app --replicas=3
deployment.apps/todo-app scaled
```

Figure 11.29 – Scaling an application using kubectl

Wait for the replication changes to take effect and check the resource details again. You will find multiple Pods (three) now running, as shown in *Figure 11.30*:

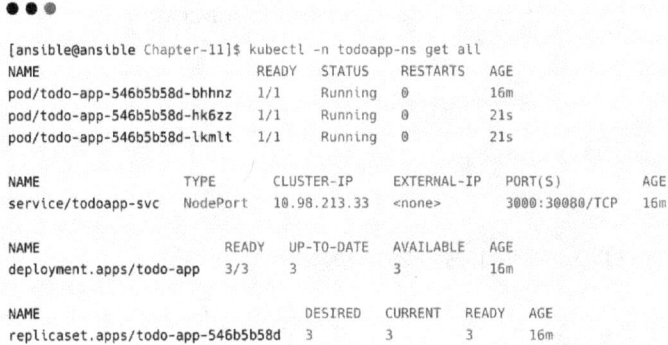

```
[ansible@ansible Chapter-11]$ kubectl -n todoapp-ns get all
NAME                          READY    STATUS      RESTARTS   AGE
pod/todo-app-546b5b58d-bhhnz  1/1      Running     0          16m
pod/todo-app-546b5b58d-hk6zz  1/1      Running     0          21s
pod/todo-app-546b5b58d-lkmlt  1/1      Running     0          21s

NAME                  TYPE        CLUSTER-IP      EXTERNAL-IP   PORT(S)           AGE
service/todoapp-svc   NodePort    10.98.213.33    <none>        3000:30080/TCP    16m

NAME                        READY    UP-TO-DATE   AVAILABLE   AGE
deployment.apps/todo-app    3/3      3            3           16m

NAME                                  DESIRED   CURRENT   READY   AGE
replicaset.apps/todo-app-546b5b58d    3         3         3       16m
```

Figure 11.30 – An application scaled up using a ReplicaSet resource

The traffic will be distributed to all Pods using the `service/todoapp-svc` module as a load balancer. Refer to *Figure 11.31*, which highlights the multiple *endpoints*, which are Pod IP addresses:

```
[ansible@ansible Chapter-11]$ kubectl -n todoapp-ns describe service/todoapp-svc
Name:                   todoapp-svc
Namespace:              todoapp-ns
Labels:                 <none>
Annotations:            <none>
Selector:               app=todo
Type:                   NodePort
IP Family Policy:       SingleStack
IP Families:            IPv4
IP:                     10.98.213.33
IPs:                    10.98.213.33
Port:                   <unset>  3000/TCP
TargetPort:             3000/TCP
NodePort:               <unset>  30080/TCP
Endpoints:              172.17.0.10:3000,172.17.0.11:3000,172.17.0.9:3000
Session Affinity:       None
External Traffic Policy: Cluster
Events:                 <none>
```

Figure 11.31 – A Kubernetes Service with multiple Pod replicas as endpoints

We can manage the scaling of the Kubernetes application based on traffic and conditions automatically using Ansible. We will learn more about the `kubernetes.core.k8s_scale` module and practice using it in the next section.

Scaling Kubernetes Deployments using Ansible

The `k8s_scale` module is part of the `kubernetes.core` collection and we can use it for scaling up or scaling down Kubernetes Deployments.

Before proceeding with the exercise, remember to scale down the application to one replica (`--replicas=1`), as follows:

```
$ kubectl -n todoapp-ns scale deployment/todo-app --replicas=1
deployment.apps/todo-app scaled
```

Figure 11.32 – Scale down todo-app replica

Follow these steps to create Ansible artifacts for scaling a Kubernetes application:

1. Create a `Chapter-11/k8s-app-scale.yaml` playbook and add content, as follows:

```
●●●
---
# Chapter-11/k8s-app-scale.yaml
- name: Scaling Applications in Kubernetes
  hosts: localhost
  gather_facts: false
  vars:
    kubeconfig_file: /home/ansible/.kube/minikube-config
    namespace_name: todoapp-ns
  tasks:
    - name: Scale deployment
      kubernetes.core.k8s_scale:
        kubeconfig: "{{ kubeconfig_file }}"
        api_version: v1
        kind: Deployment
        name: todo-app
        namespace: "{{ namespace_name }}"
        replicas: 4
        wait_timeout: 30
```

Figure 11.33 – An Ansible playbook to scale a Kubernetes Deployment

Adjust wait_timeout: 30 depending on your next task in the workflow. For example, increase the value to ensure the Pod replicas are created and running successfully before proceeding with the next task.

2. Execute the playbook:

```
[ansible@ansible Chapter-11]$ ansible-playbook k8s-app-scale.yaml
```

Figure 11.34 – Execute Kubernetes Deployment scaling playbook

3. Verify the Pod replicas using the kubectl command, as shown in *Figure 11.35*:

```
●●●
[ansible@ansible Chapter-11]$ kubectl -n todoapp-ns get pods
NAME                      READY   STATUS    RESTARTS   AGE
todo-app-546b5b58d-5j8nj  1/1     Running   0          28s
todo-app-546b5b58d-7sr8j  1/1     Running   0          28s
todo-app-546b5b58d-bhhnz  1/1     Running   0          24m
todo-app-546b5b58d-r9nmz  1/1     Running   0          28s
```

Figure 11.35 – Pod replicas after scaling

Customize the scaling based on several conditions, as follows:

*   current_replicas: x: This will change the replicas only if the current number of replicas matches the current_replicas value.

- `src: deployment.yml`: This will read the Deployment and the replicas definition from a file.

- `wait: no`: This will not wait for the scaling to complete.

Refer to the `k8s_scale` module documentation for more details (`https://docs.ansible.com/ansible/latest/collections/kubernetes/core/k8s_scale_module.html`).

In the next section, we will learn about the `k8s_exec` module and how to use it to troubleshoot Pods.

Executing commands inside a Kubernetes Pod

In a normal situation, we do not need to log in to a Pod or container, as the application is exposed on some ports and Services are talking over these exposed ports. However, when there are issues, we need to access the containers and check what is happening inside, by checking logs, accessing other Pods, or running any necessary troubleshooting commands.

Use the `kubectl exec` command if you are doing this troubleshooting or information gathering manually:

```
$ kubectl exec --stdin --tty POD_NAME -- /bin/bash
```

Figure 11.36 – Execute commands inside a Pod using the kubectl utility

However, when we automate Kubernetes operations using Ansible, use the `k8s_exec` module and automate the verification tasks or validation tasks as well.

For such scenarios, we can deploy debug Pods using suitable images (for example, images with required utilities, such as `ping`, `curl`, or `netstat`) and execute validation commands from these Pods. A typical deployment scenario with test Pods (`curl-pod`) is shown in *Figure 11.37*, as follows:

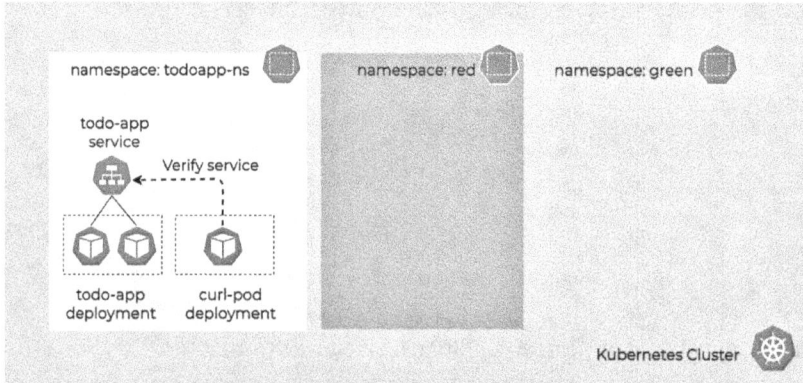

Figure 11.37– A debug Pod to validate and verify applications

In the following scenario, we will create a playbook to deploy a `curl-pod` Pod using a `busyboxplus` image (https://hub.docker.com/r/radial/busyboxplus) and verify the `todoapp-svc` Service from the `curl-pod` Pod:

1. Create the `curl-app-pod.yaml` definition file for the `curl-pod` Pod, as follows:

```
● ● ●

---
apiVersion: v1
kind: Pod
metadata:
  name: curl-pod
  namespace: todoapp-ns
  labels:
    app: curl-pod
spec:
  containers:
    - name: curl
      image: radial/busyboxplus:curl
      command:
        - "sleep"
        - "50000"
```

Figure 11.38 – Kubernetes Deployment definition for a curl-pod Pod

We have added a `sleep 50000` command; otherwise, the Pod will exit without running any processes.

2. Create a `Chapter-11/curl-app-deploy.yaml` playbook and add content, as follows:

```
● ● ●

---
# Chapter-11/curl-app-deploy.yaml
- name: Deploying curl Pod
  hosts: localhost
  gather_facts: false
  vars:
    kubeconfig_file: /home/ansible/.kube/minikube-config
    namespace_name: todoapp-ns
  tasks:

    - name: Create a Pod with curl image
      kubernetes.core.k8s:
        kubeconfig: "{{ kubeconfig_file }}"
        state: present
        src: curl-app-pod.yaml
        namespace: "{{ namespace_name }}"
```

Figure 11.39 – Ansible playbook for deploying curl-pod

3. Add a task to execute the `curl` command inside the `curl-pod` Pod to verify whether the `todoapp-svc` Service is accessible or not. Then, add tasks for displaying the output of the `curl` command, and validate the success and fail status (`curl_output.failed == true`), as follows:

● ● ●

```
# chapter-11/curl-app-deploy.yaml - tasks for curl command and status

- name: Verify todo-app sevice
  ignore_errors: yes
  kubernetes.core.k8s_exec:
    kubeconfig: "{{ kubeconfig_file }}"
    namespace: "{{ namespace_name }}"
    pod: curl-pod
    command: curl todoapp-svc:3000
  register: curl_output

- name: Display service check output
  debug:
    msg: "{{ curl_output.stdout_lines }}"
  when: curl_output.failed == false

- name: Display service check output
  debug:
    msg: "Service (todoapp-svc) is not reachable !"
  when: curl_output.failed == true
```

Figure 11.40– Tasks to execute the curl command inside the curl-pod Pod and to display its status

The last task is optional but adding more validations and messages will help you to implement a better workflow.

4. Once the verification is complete, delete the `curl-pod` Pod, as we do not require it anymore.

● ● ●

```
# chapter-11/curl-app-config.yaml - tasks to delete the curl pod after testing

- name: Delete curl pod
  kubernetes.core.k8s:
    kubeconfig: "{{ kubeconfig_file }}"
    state: absent
    src: curl-app-pod.yaml
    namespace: "{{ namespace_name }}"
```

Figure 11.41 – Remove the curl-pod Pod after testing

5. Execute the playbook and verify the output. We can see the `curl` output as shown in *Figure 11.42* if the `todoapp-svc` Service is reachable from the `curl-pod` Pod:

```
● ● ●
[ansible@ansible Chapter-11]$ ansible-playbook curl-app-deploy.yaml
...<ouput omitted>...

TASK [Display service check output] ***************************************************************
ok: [localhost] => {
    "msg": [
        "",
        "<!DOCTYPE html>",
        "<html>",
        "<head>",
        "    <meta charset=\"utf-8\" />",
        "    <meta name=\"viewport\" content=\"width=device-width, initial-scale=1, shrink-to-fit=no, maximum-
scale=1.0, user-scalable=0\" />",
        "    <link rel=\"stylesheet\" href=\"css/bootstrap.min.css\" crossorigin=\"anonymous\" />",
        "    <link rel=\"stylesheet\' href=\"css/font-awesome/all.min.css\" crossorigin=\"anonymous\" />",
        "    <link href=\"https://fonts.googleapis.com/css?family=Lato&display=swap\" rel=\"stylesheet\" />",
        "    <link rel=\"stylesheet\" href=\"css/styles.css\" />",
        "    <title>Todo App</title>",
        "</head>",
        "<body>",
        "    <div id=\"root\"></div>",
        "    <script src=\"js/react.production.min.js\"></script>",
        "    <script src=\"js/react-dom.production.min.js\"></script>",
        "    <script src=\"js/react-bootstrap.js\"></script>",
        "    <script src=\"js/babel.min.js\"></script>",
        "    <script type=\"text/babel\" src=\"js/app.js\"></script>",
        "</body>",
        "</html>"
    ]
}

...<ouput omitted>...
```

Figure 11.42 – The curl command output for todoapp-svc

The `kubernetes.core.k8s_exec` module is very useful for the validation and verification of Kubernetes applications and cluster management using Ansible. The command can be executed inside a debug Pod, as we learned in the preceding exercise, or even inside the application Pod for verifying facts.

Explore the `kubernetes.core` collection and find other useful modules and plugins to automate your Kubernetes Deployments and resources.

Summary

In this chapter, we have learned about the Ansible collection for Kubernetes cluster and resource management. We started by covering the basics of Kubernetes components and discussed how to use Kubespray to deploy and manage Kubernetes clusters and their supported features.

After that, we learned the method of connecting a Kubernetes cluster to Ansible to automate cluster operations. We have used the Kubernetes Ansible collection to deploy applications and scale

Deployments. We have also learned how to execute commands inside a running Kubernetes Pod using Ansible, which can be utilized for validation and troubleshooting purposes. This chapter has provided a brief introduction to Kubernetes automation using Ansible and other important information, such as Kubernetes content collection and methods of connecting Ansible to Kubernetes.

In the next chapter, you will learn about the different available methods of integrating your CI/CD and communication tools using Ansible. We will learn more about the enterprise version of Ansible, which is called Ansible Automation Platform, and its flexible integration features.

Further reading

For more information on the topics covered in this chapter, please refer to the following links:

- *Introduction to Ansible for Kubernetes guide* – `https://docs.ansible.com/ansible/latest/collections/kubernetes/core/docsite/scenario_guide.html`

- *Play with Kubernetes – free hands-on labs* – `https://labs.play-with-k8s.com/`

- *CRI-O* – `https://cri-o.io`

- *containerd* – `https://containerd.io/`

- *Top 15 Free Kubernetes Courses* – `https://www.techbeatly.com/kubernetes-free-courses/`

- *What is Kubernetes?* – `https://www.redhat.com/en/topics/containers/what-is-kubernetes`

- *Free OpenShift labs* – `learn.openshift.com`

- *The Ansible kubernetes.core collection* – `https://galaxy.ansible.com/kubernetes/core` or `https://console.redhat.com/ansible/automation-hub/repo/published/kubernetes/core`

- *The Ansible community.kubernetes collection* – `https://galaxy.ansible.com/community/kubernetes`

- *Templating (Jinja2)* – `https://docs.ansible.com/ansible/latest/user_guide/playbooks_templating.html`

- *What is PaaS?* – `https://www.redhat.com/en/topics/cloud-computing/what-is-paas`

- *Microk8s* – `https://microk8s.io`

- *Red Hat OpenShift* – `https://www.redhat.com/en/technologies/cloud-computing/openshift`

- *Kubespray* – `https://kubespray.io`, `https://github.com/kubernetes-sigs/kubespray`

12

Integrating Ansible with Your Tools

As an organization grows its IT infrastructure, more and more tools are often needed to solve the technical challenges. Instead of these tools working alone in silos, it is desirable to implement integration between these tools to increase efficiency and scalability. As an example, the **IT Service Management (ITSM)** tool can send an alert to approvers, or the container platform can trigger a new deployment of the application when a new version has been developed by the team. There are an immeasurable number of opportunities in terms of integrating multiple siloed tools in IT infrastructure.

The same goes for automation as well; Ansible can be used as the key automation tool for implementing integration between multiple infrastructure and application support tools. In the previous chapters, you learned about the Ansible automation and integration opportunities for the infrastructure (public and private cloud), DevOps, networks, applications, and more. In this chapter, you will learn more about the enterprise automation solution called **Ansible Automation Platform (AAP)** and the integration methods between other tools in the IT infrastructure environment.

We will start by introducing Red Hat AAP and its components, features, and benefits. Then, you will learn how to use AAP by creating various automation resources in automation controller, such as projects, job templates, and credentials.

In this chapter, we will cover the following topics:

- Introduction to Red Hat AAP
- Red Hat AAP components
- Database management using Red Hat AAP
- Integrating Jenkins with AAP
- Integrating an automation controller with Slack and notification services

Let's start learning about AAP, the enterprise IT automation tool.

Technical requirements

You will need the following technical requirements for this chapter:

- One or more Linux machines with Red Hat repositories configured. If you are using other Linux operating systems instead of **Red Hat Enterprise Linux** (**RHEL**) machines, then make sure you have the appropriate repositories configured to get packages and updates.

- Access to Red Hat AAP.

- A GitHub account.

- A Jenkins server and basic knowledge about Jenkins pipelines.

- A Slack application account and basic knowledge about Slack usage.

A single 60-day self-supported subscription to Red Hat AAP is available for testing AAP and its features. Please refer to `https://www.redhat.com/en/technologies/management/ansible/trial` to learn more about the AAP trial subscription.

All the Ansible artifacts, commands, and snippets for this chapter can be found in this book's GitHub repository at `https://github.com/PacktPublishing/Ansible-for-Real-life-Automation/tree/main/Chapter-12`.

Introduction to Red Hat AAP

So far, you have learned how to use Ansible, develop playbooks, create roles, use content collections, and more for your use cases. Anyone can install and use Ansible in their workstation or some random servers in an environment and use them for their automation use cases. However, there won't be any standardization, traces, or accountability as each person works on their methods and practices. This will result in the following challenges in the organization:

- Individuals work in silos, which will result in no collaboration in the workplace.

- Automation artifacts (playbooks, roles, and collections) are not shared between individuals or teams.

- No logging or auditing options will be available since the automation is running on an individual's workstation or some random servers.

- Less control over who can execute the playbooks or automated jobs.

- Difficulty keeping secrets and credentials.

- Lack of job scheduling and monitoring features.

- Complexity in managing managed nodes information.

When it comes to enterprise automation, the solution must be able to implement governance, standardization, collaboration, accountability and auditing.

Red Hat AAP helps organizations cover most of the previously mentioned challenges by implementing an enterprise automation solution that can scale the automation and orchestrate the infrastructure and applications. AAP includes all the tools and features required for enterprise automation, such as **Graphical User Interface (GUI)** and **Text User Interface (TUI)**-based tools, analytics, dashboards, auditing, and more.

Note that unlike Ansible (the `ansible-core` or `ansible` package), Red Hat AAP is not free and is delivered as a subscription model (`https://www.ansible.com/products/pricing`), like any other Red Hat product. For testing and **proof of concept** (**POC**) purposes, request a 60-day trial subscription and refer to `https://www.redhat.com/en/technologies/management/ansible/try-it` to learn more. Please refer to `https://access.redhat.com/documentation/en-us/red_hat_ansible_automation_platform/2.1/html/red_hat_ansible_automation_platform_installation_guide/index` to learn how to install and configure Red Hat AAP.

It is also possible to get a no-cost RHEL individual developer subscription for testing and development purposes. Refer to `https://developers.redhat.com/articles/faqs-no-cost-red-hat-enterprise-linux` to learn more about this free RHEL subscription.

Red Hat AAP Managed Service in Microsoft Azure

Red Hat offers AAP as a managed service on Microsoft Azure, by which the organization can utilize the AAP solution just like any other Microsoft Azure cloud service. This is fully supported by Red Hat and the billing will be handled inside the same cloud service bill. Refer to `https://www.redhat.com/en/technologies/management/ansible/azure` to learn more about this managed service.

Now, let's explore the various features of Red Hat AAP.

Features of Red Hat AAP

The following important features are included in Red Hat AAP:

- **WebUI**: The web-based GUI can help administrators and developers manage the entire automation solution from a web browser. Most of the configurations in AAP can be executed from the WebUI itself, as shown in the following screenshot:

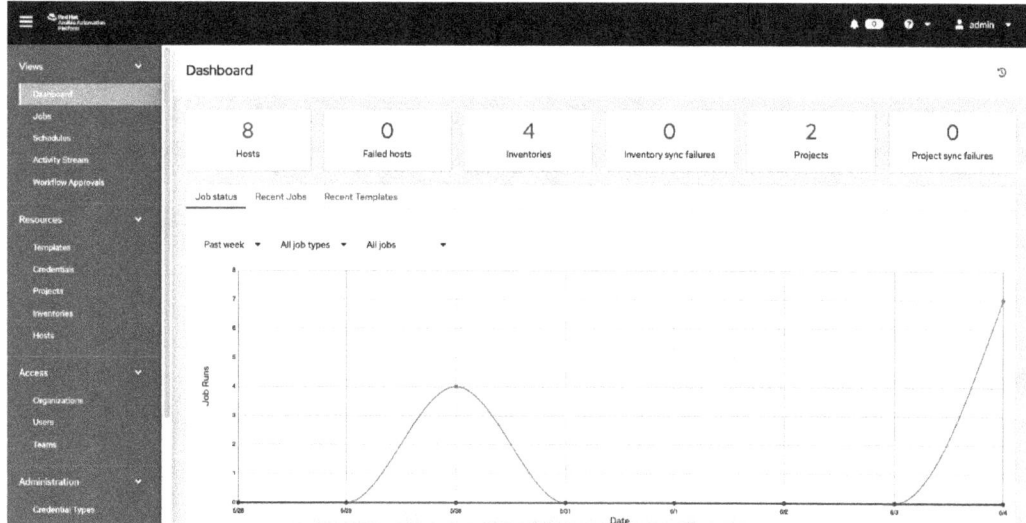

Figure 12.1 – The Red Hat AAP dashboard

- **Role-Based Access Control (RBAC)**: Administrators can create teams and roles in AAP and assign the roles with adequate permissions to the users or teams. These permissions can be configured for every component in AAP, such as projects, job templates, inventories or credentials.

- **Logging and auditing**: AAP will store the activity logs, including job execution history, and it is possible to access the old job details any time from the WebUI. AAP also includes the option to integrate with logging aggregators such as Logstash, Splunk, Loggly, Sumo Logic, and others so that we can keep your logs in a central system.

- **REST API (also known as the RESTful API)**: The powerful and well-documented REST API will help you integrate AAP with the existing tools and applications in your environment.

Automation Controller API Guide

The automation controller REST API also helps you manage automation controller operations such as job template creation, credential management or configuring authentication. Refer to the API guide `https://docs.ansible.com/automation-controller/latest/html/controllerapi/index.html` for more details.

- **Job templates and workflows**: It is possible to predefine and configure the job templates and workflow templates to quickly execute complex automation jobs (`https://docs.ansible.com/automation-controller/latest/html/userguide/workflows.html`).

- **Credential management**: With credentials, you store secrets and sensitive information such as passwords, API keys, tokens, and so on. There are several predefined credential types available in AAP and it is also possible to create custom credential types if required (`https://docs.ansible.com/automation-controller/latest/html/userguide/credential_types.html`):

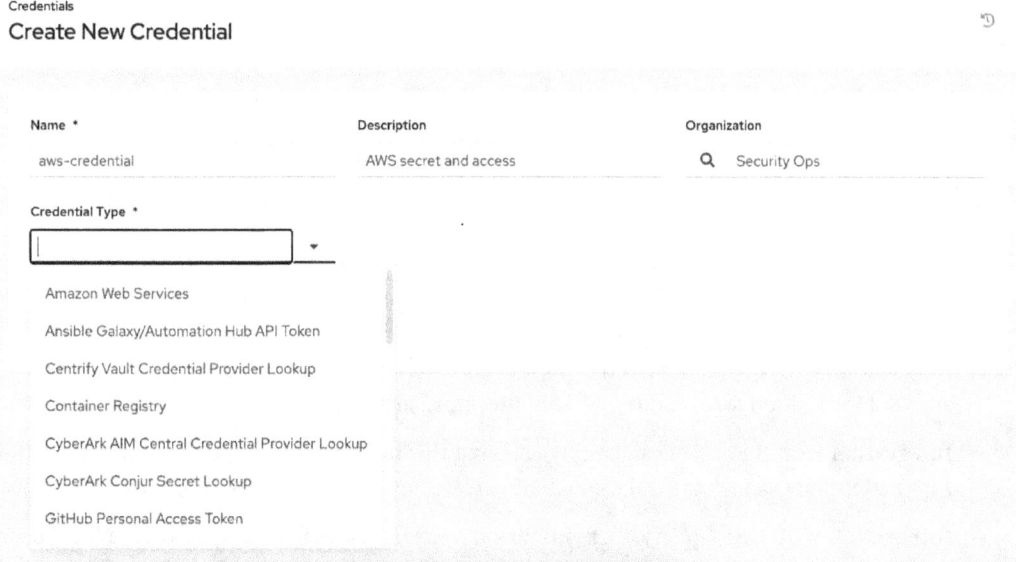

Figure 12.2 – Automation controller credentials

- **Job scheduling**: Automation jobs can be scheduled and executed automatically without user intervention, as shown in the following screenshot. This is useful for many automation jobs that must be executed in a specific period and without manual intervention. For example, we can schedule a weekly or monthly reboot job (as we learned in *Chapter 3, Automating Your Daily Jobs*) using an automation controller:

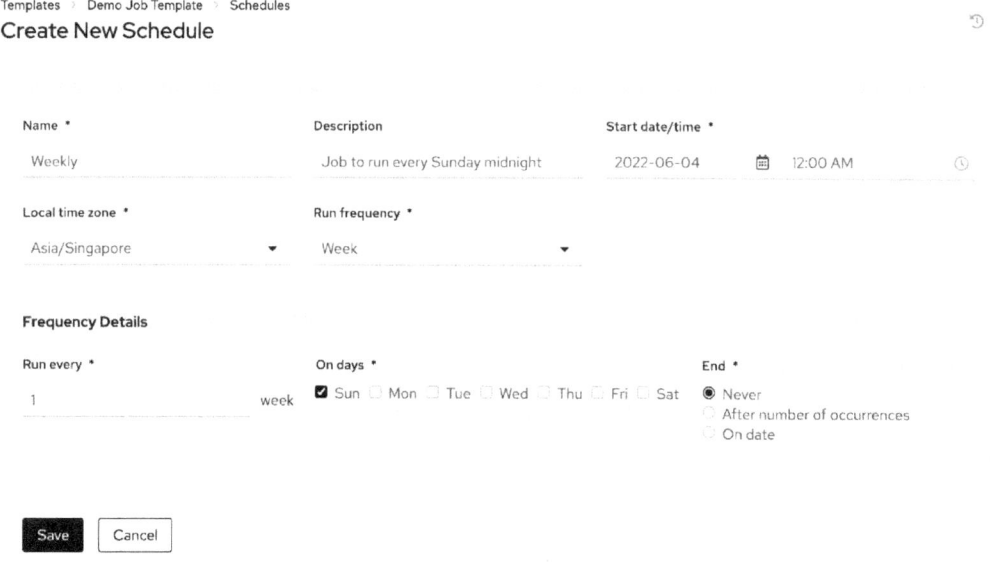

Figure 12.3 – Job scheduling in an automation controller

- **Notifications**: AAP supports multiple types of notifications, such as email, IRC, and webhook, and tool integration for Grafana, Slack, Mattermost, and others.

- **Integration with Red Hat Insights**: By enabling this integration, it is possible to analyze the automation status and data for the automation platforms (AAP clusters) in your environments.

- **Integration with third-party authentication systems**: Instead of managing the local user accounts in AAP, integrate with the existing authentication providers, such as Active Directory, Google OAuth2, LDAP, RADIUS, SAMPLE, or TACACS.

> **Red Hat AAP Features and Benefits**
>
> Please go to `https://www.redhat.com/en/technologies/management/ansible/features` to learn more about the features of AAP and their benefits.

In the next section, you will learn more about the components of Red Hat AAP.

Red Hat AAP components

Red Hat AAP is an automation suite that contains multiple components, as shown in the following diagram:

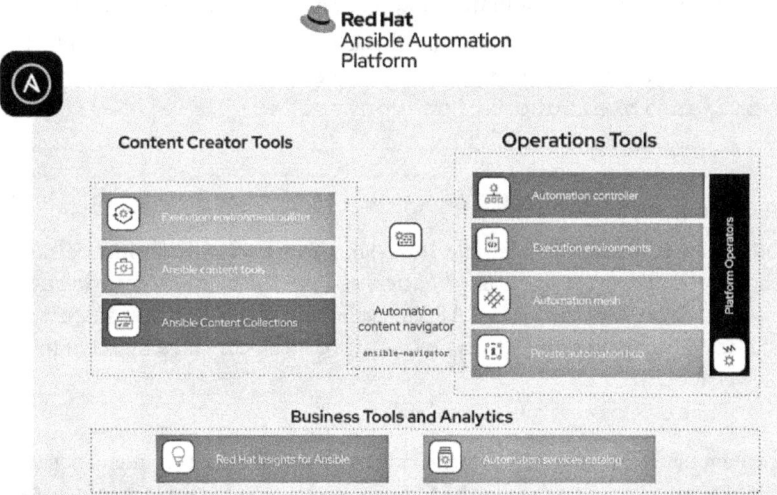

Figure 12.4 – High-level diagram of Red Hat AAP 2.1 components (source: https://
www.ansible.com/blog/introducing-red-hat-ansible-automation-platform-2.1)

In the following sections, you will learn about the different components of AAP, such as the automation controller, execution environments, and automation mesh.

Ansible automation controller

The **automation controller** was referred to as **Ansible Tower** previously. It is the control plane and the core component of AAP. With the introduction of the Ansible automation controller, the control plane components (WebUI and API) become decoupled from the **execution environment** (EE), which also helps the solution with additional execution nodes.

With the automation controller, we can manage the AAP operations from the WebUI, such as managing remote nodes (inventory), credentials, projects, job templates, and other operations. The automation controller contains many components, as follows:

- A REST API for inventory, credential, and job management
- A database for storing the resource details, including the automation job's history
- Am automation mesh connector and receptor
- A message queue and caching
- A task scheduler

Based on the RBAC configuration, the user will have different permissions on the automation controller interface.

> **What is the Red Hat Ansible Automation Platform Automation Controller?**
>
> To learn more about automation controller, check *What is the Red Hat Ansible Automation Platform automation controller?* (`https://www.redhat.com/en/technologies/management/ansible/automation-controller`).

> **Ansible Tower and AAP Life Cycle**
>
> Ansible automation controller 3.8 (Ansible Tower) has been part of AAP since version 1.2. Refer to the Ansible Tower life cycle (`https://access.redhat.com/support/policy/updates/ansible-tower`) and AAP life cycle (`https://access.redhat.com/support/policy/updates/ansible-automation-platform`) to learn more about these versions.

Automation content navigator (`ansible-navigator`) is a TUI utility for interacting with automation execution environments. It can be seen in the following screenshot. `ansible-navigator` is also helpful for developing and testing automation content:

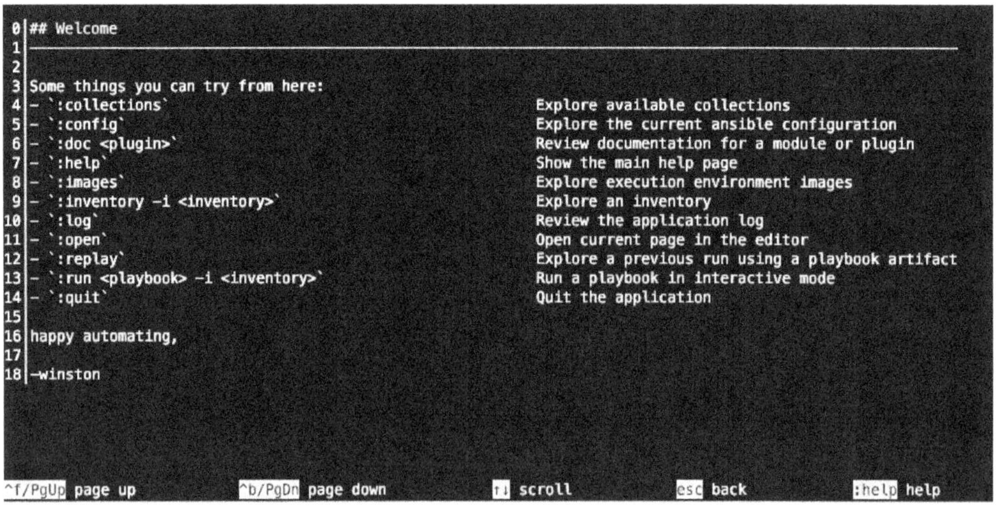

```
0 | ## Welcome
1 |
2 |
3 | Some things you can try from here:
4 | - `:collections`                          Explore available collections
5 | - `:config`                               Explore the current ansible configuration
6 | - `:doc <plugin>`                         Review documentation for a module or plugin
7 | - `:help`                                 Show the main help page
8 | - `:images`                               Explore execution environment images
9 | - `:inventory -i <inventory>`             Explore an inventory
10| - `:log`                                  Review the application log
11| - `:open`                                 Open current page in the editor
12| - `:replay`                               Explore a previous run using a playbook artifact
13| - `:run <playbook> -i <inventory>`        Run a playbook in interactive mode
14| - `:quit`                                 Quit the application
15|
16| happy automating,
17|
18| -winston
```

`^f/PgUp` page up   `^b/PgDn` page down   `↑↓` scroll   `esc` back   `:help` help

Figure 12.5 – The ansible-navigator TUI

Instead of using `ansible-playbook` and other `ansible-*` commands, we can use `ansible-navigator` to manage all command-line operations of Ansible, such as executing playbooks, managing collections, executing environments, and so on.

Refer to the Ansible navigator documentation (`https://ansible-navigator.readthedocs.io`) and Ansible navigator cheat sheet (`https://www.techbeatly.com/ansible-navigator-cheat-sheet`) to learn more.

Automation execution environments

The automaton execution environment provides a standard and portable mechanism for executing Ansible playbooks. Execution environments are consistent by design and deliver container images that contain the following components:

- Ansible
- Ansible Runner
- Required Ansible collections
- Other dependencies for the playbook's execution (for example, Python libraries, system packages)

Previously, Python virtual environments (for example, `/var/lib/awx/venv/ansible`) were used to achieve this functionality. This was not easy to manage and maintain in terms of consistency as you needed to manage the Python virtual environments on all nodes in the AAP or Ansible Tower cluster. By containerizing the executables and dependencies, it is possible to distribute the same image to multiple cluster nodes consistently:

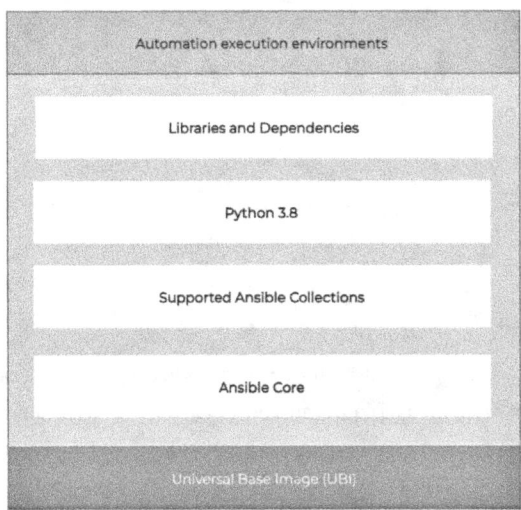

Figure 12.6 – Components inside the automation execution environment

Execution environment builder (`ansible-builder`) is a command-line utility that can be used to build and manage container images for the automation execution environment. Refer to the automation execution environment documentation (`https://docs.ansible.com/automation-controller/latest/html/userguide/execution_environments.html`) to learn more about building and distributing automation execution environments.

The automation mesh

The automation mesh is a service mesh concept for Ansible automation and was introduced in AAP 2.1. The automation mesh replaces the isolated node concept in older versions of AAP (version 1.2 and below) and provides the flexibility to scale the automation landscape. The automation mesh replaces the need for jump servers (or bastion hosts), and it is possible to place the execution nodes near the managed nodes, as shown in the following diagram:

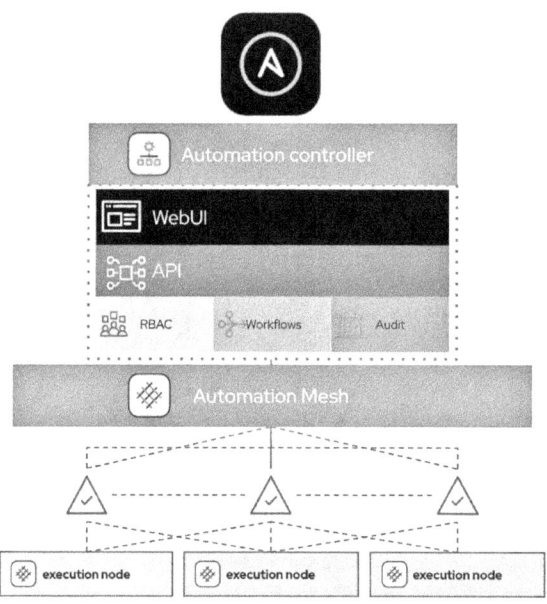

Figure 12.7 – AAP and the automation mesh (source: https://www.ansible.com/
blog/whats-new-in-ansible-automation-platform-2.1-automation-mesh)

Refer to the documentation (`https://www.ansible.com/products/automation-mesh`) to learn more about the automation mesh.

Automation Hub

Ansible Automation Hub is the official location for downloading the supported and certified Ansible collections. The hosted Automation Hub is available at `https://console.redhat.com/ansible/automation-hub`, as shown in the following screenshot:

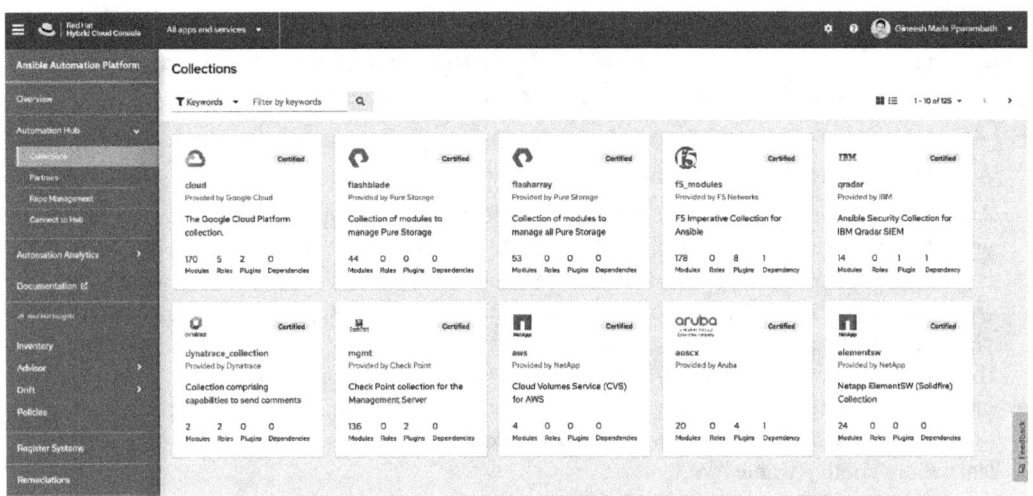

Figure 12.8 – Ansible Automation Hub

To manage your content collection and other supported collections, it is possible to use a **Private Automation Hub (PAH)**. This concept was introduced in AAP 1.2. Content from other sources (such as Red Hat Ansible Automation Hub or Ansible Galaxy) can be synced to private automation hubs, as shown in the following diagram:

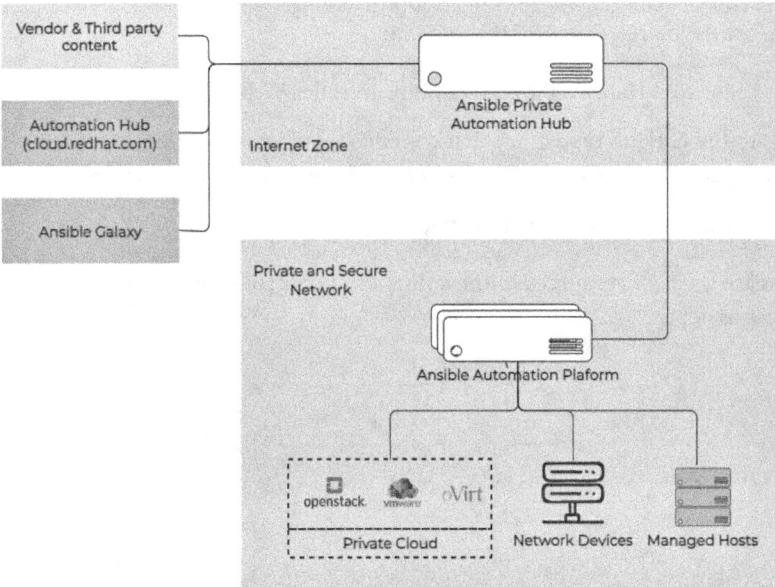

Figure 12.9 – Private Automation Hub with AAP

Refer to the documentation (`https://docs.ansible.com/ansible/latest/reference_appendices/automationhub.html`) and product page (`https://www.ansible.com/products/automation-hub`) to learn more about Automation Hub.

Ansible AWX and the Ansible Automation Controller

Ansible AWX is an open source community project sponsored by Red Hat. It is an upstream project for the automation controller component in Red Hat AAP. (Red Hat Ansible Tower 3.2 is the first version based on the Ansible AWX project.) Ansible AWX is a fast-moving project (`https://github.com/ansible/awx`) that is primarily supported by the community via IRC (`https://web.libera.chat/#ansible-awx`) and the AWX mailing list (`https://groups.google.com/forum/#!forum/awx-project`). Refer to the project page (`https://www.ansible.com/community/awx-project`) and frequently asked questions (`https://www.ansible.com/products/awx-project/faq`) to learn more about Ansible AWX.

In the next section, you will learn how to start using Red Hat AAP for automation use cases.

Database management using Red Hat AAP

In *Chapter 8*, *Helping the Database Team with Automation*, you learned how to use Ansible to automate database creation and user management operations. This section will reuse the same Ansible artifacts but in a different repository and execute the job from Red Hat AAP.

The following are the prerequisites for this section:

- Access to the Red Hat Ansible automation controller WebUI
- Access to the GitHub repository that contains the necessary Ansible artifacts (`https://github.com/ginigangadharan/ansible-database-demo`)
- Access to the target nodes (database node)

Job templates are the core resources in the automation controller, but a job template requires few other resources as dependencies.

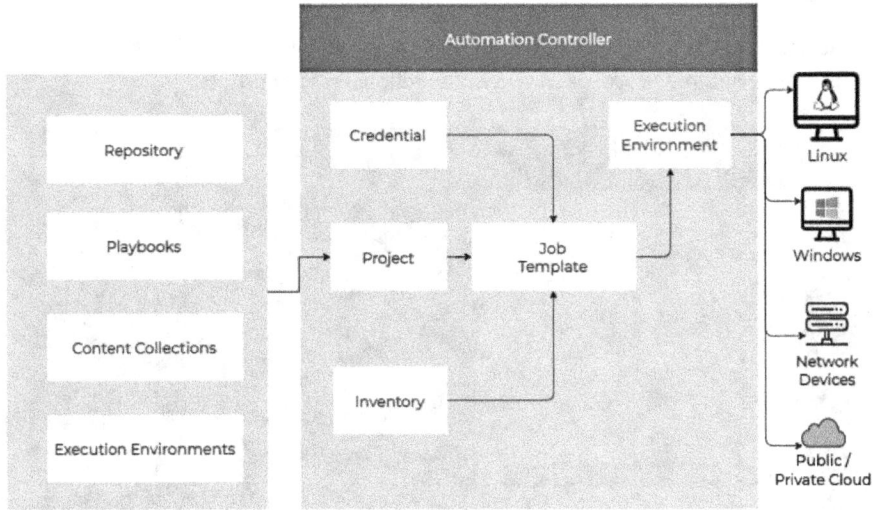

Figure 12.10 – Components and dependencies in Ansible Automation Platform

You will learn the basic operations in the Ansible automation controller and create various resources, as follows:

- Organizations
- Projects
- Inventory, managed nodes, and groups
- Credentials
- A job template with survey forms and extra variables

In the following section, we will learn how to access the Ansible automation controller and create the resources inside.

Accessing the Ansible automation controller

Access the automation controller IP address or hostname from a supported web browser and log in with your username and password, as shown in the following screenshot:

Figure 12.11 – Ansible automation controller WebUI login page

Based on the RBAC configurations and permissions, you will find allowed menus and configuration items on the dashboard:

Figure 12.12 – Ansible automation controller dashboard

To manage different projects, users, and other resources, create a new *organization*; see the example shown in the following screenshot. An organization is a logical collection of projects, inventories, teams and users (this is an optional step; use the **Default** organization that already exists in the automation controller):

Figure 12.13 – Creating a new organization

Once you have created a new organization, create a new project.

Creating a project in automation controller

Click on the **Projects** menu and then click **Add** to create a new project, as shown in the following screenshot. For **Source Control URL**, use https://github.com/ginigangadharan/ansible-database-demo:

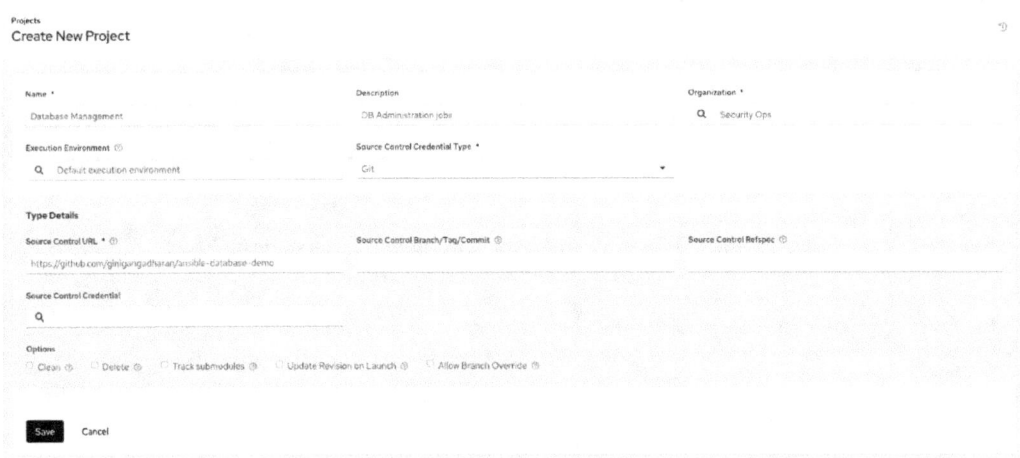

Figure 12.14 – Creating a new project in the automation controller

Since the GitHub repository is public, we do not require a credential to access it. For private GitHub repositories, you need to provide one; you will learn how to create credentials in the next section when we add the inventory and managed nodes.

Verify the project's sync status, as shown in the following screenshot:

Figure 12.15 – Project details and job status

As you can see, the automation controller has finished syncing the content from the GitHub repository to a local directory (`Playbook Directory: _15__database_management`).

If you come across any issues while syncing content, check your internet connection or connection to the Git server (for private repositories) and ensure the necessary access is in place.

Creating the inventory and managed nodes

In the previous chapters, you created Ansible inventories using static files (and dynamic inventory plugins), as shown in the following screenshot:

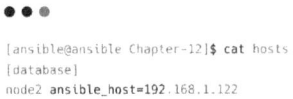

```
[ansible@ansible Chapter-12]$ cat hosts
[database]
node2 ansible_host=192.168.1.122
```

Figure 12.16 – Creating an Ansible inventory using a static file

Now, let's create a new inventory, managed nodes, and host groups in the automation controller using the WebUI:

1. Open the **Inventories** tab on the left-hand side and **Add** a new inventory, as shown in the following screenshot:

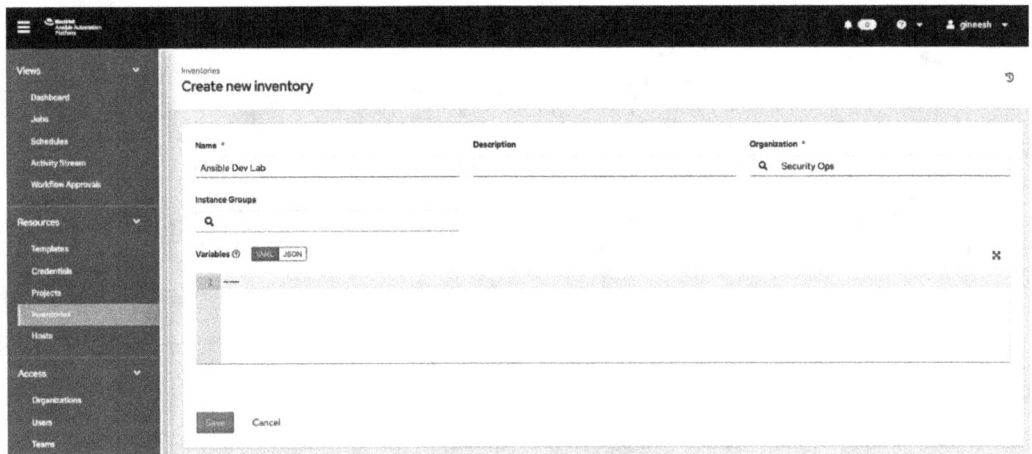

Figure 12.17 – Creating an Ansible inventory in the automation controller

A new inventory will be created and its status will be displayed:

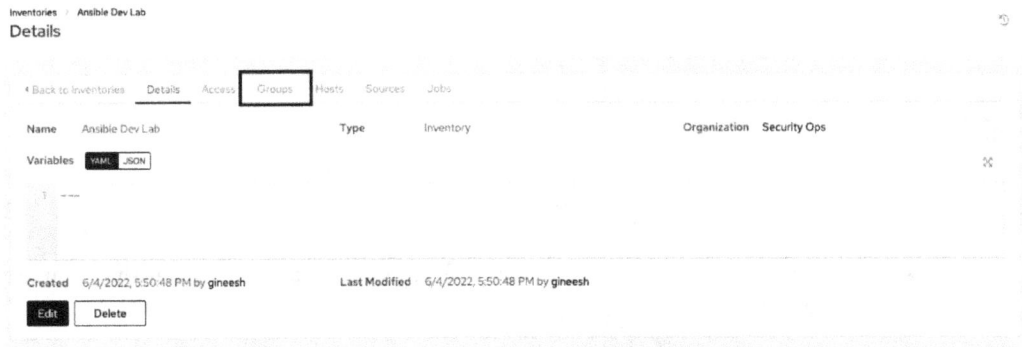

Figure 12.18 – The Ansible inventory on the automation controller

2.   Now, select the **Groups** tab at the top and click **Add** to create a new host group:

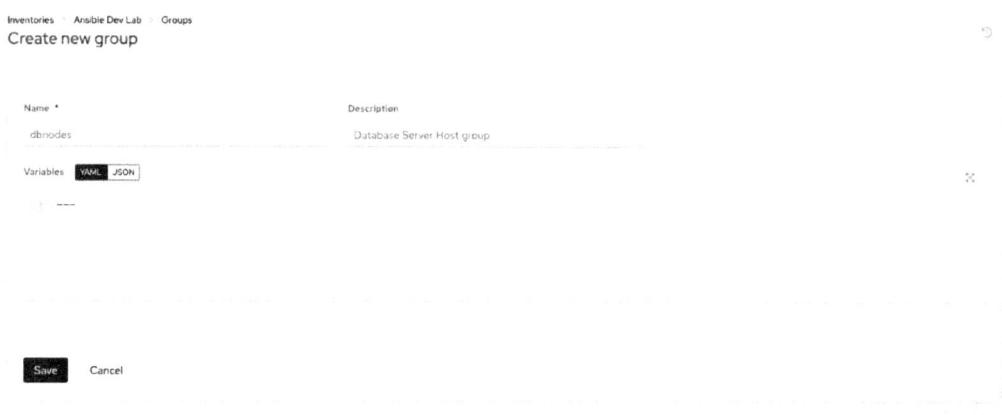

Figure 12.19 – Creating a new host group

New host group details will be displayed, as shown in the following screenshot:

Figure 12.20 – New host group details

3.   Click on the **Hosts** tab at the top. Then, click **Add** and select **Add new host**, as shown in the following screenshot:

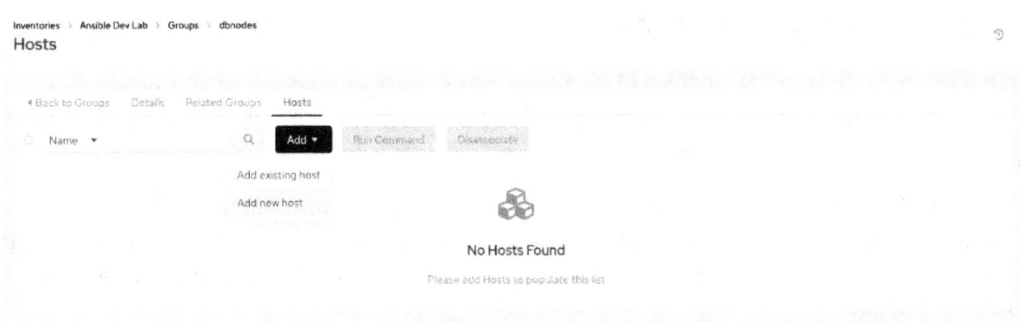

Figure 12.21 – Adding new hosts to the host group

If you have already added the hosts directly, then add the host to this host group by selecting **Add existing host**.

4. Create a new host with a hostname and `ansible_host` information, as shown here:

Figure 12.22 – Creating a new host

More variables can be added for host groups as hosts under the **Variables** section, as shown in *Figures 12.22* and *12.20*.

Importing Your Inventory into the Automation Controller

Instead of adding managed nodes one by one, it is also possible to import your inventory from the project source, public cloud, or other inventory management systems. Refer to the documentation at `https://docs.ansible.com/automation-controller/latest/html/administration/scm-inv-source.html` to learn more.

Creating credentials in the automation controller

In the previous chapters, we used SSH keys to access the managed nodes. The SSH private key was stored in the Ansible control node (`/home/ansible/.ssh/id_rsa`), but in this case, we must pass the private key using a credential.

Open the **Credential** tab on the left-hand side of the dashboard and click on **Add** to add a new credential. Fill in its details, as shown in the following screenshot. Select **Machine** for **Credential Type** and enter a username (for example, `devops`) and the SSH private key (you can also use `password` as the credential):

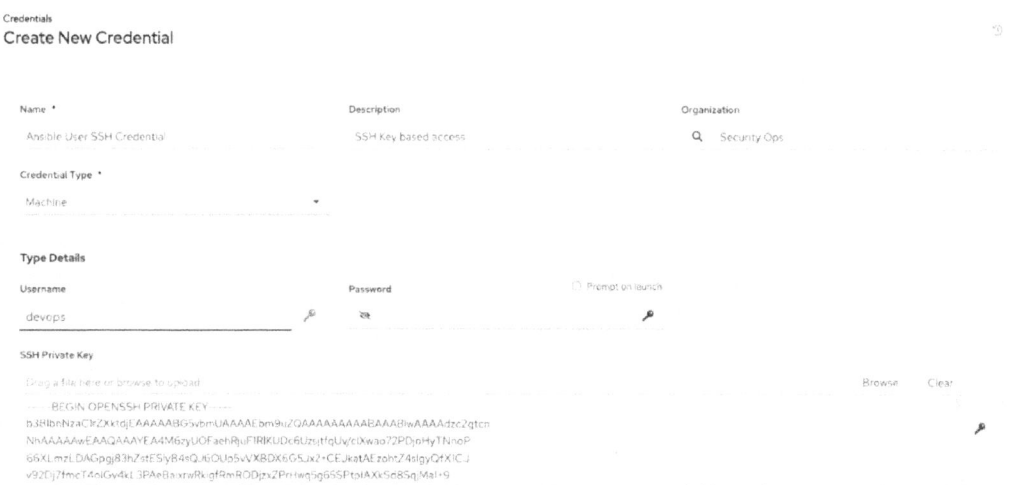

Figure 12.23 – Adding the Machine credential

Since we have enabled `NOPASSWD` (refer to *Chapter 1, Ansible Automation – Introduction*, the *Configuring your managed nodes* section) for the `sudo` access, we do not need to enter any data in the **Privilege Escalation** fields:

Figure 12.24 – Machine credential privilege escalation fields

Save the credential and verify its details. The passwords and SSH private key content will be encrypted and will not be visible to anyone. You can only reset the SSH private key content or password in the credential, as shown in the following screenshot:

Credentials > Ansible User SSH Credential
Edit Details

Name *
Ansible User SSH Credential

Description
SSH Key based access

Organization
Q   Security Ops

Credential Type *
Machine

Type Details

Username
ansible

Password

Prompt on launch

SSH Private Key
Drag a file here or browse to upload
$encrypted$

Browse...   Clear

Figure 12.25 – Encrypted SS private key content

With that, you have created a project, an inventory with a managed node (dbnode1), and a credential to access the node. Now, you must create a job template using the Chapter-08/postgres-manage-database.yaml playbook, as explained in the next section.

> **Ansible Automation Controller – Credentials**
>
> Learn more about the credentials in the automation controller at https://docs.ansible.com/automation-controller/latest/html/userguide/credentials.html. Please go to https://docs.ansible.com/automation-controller/latest/html/userguide/credential_types.html to learn more about custom credentials.

Creating a new job template

A job template in the automation controller is a preconfigured definition with a set of parameters and a playbook for executing an Ansible automation job. To execute the same jobs several times, use a job template and pass the parameter as needed. Follow these steps:

1. Open the **Templates** tab on the left-hand side of the dashboard. Then, click on **Add** and select **Add job template**, as shown here:

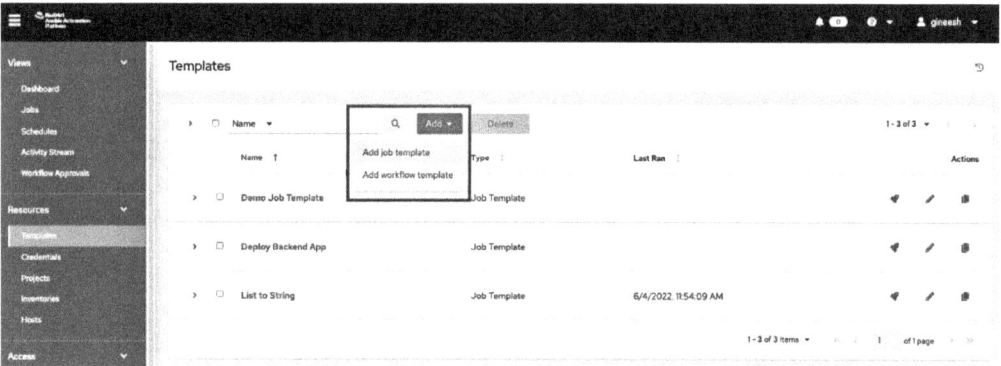

Figure 12.26 – Creating a job template

2. Enter the details of the job template:

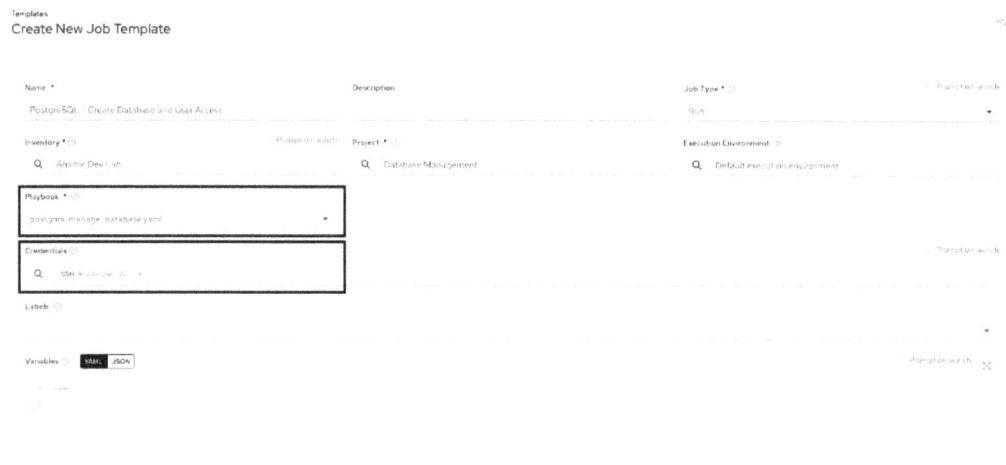

Figure 12.27 – Job template details

> **Job Templates in the Automation Controller**
>
> Refer to the documentation at `https://docs.ansible.com/automation-controller/latest/html/userguide/job_templates.html` to learn more about how to customize job templates with additional options.

We have used variables inside the playbook (*Figure 12.28*) to demonstrate the operations without complexity. In the actual environment, the variables can be fetched from additional variable files, group variables, host variables, or extra variables. Now, we must pass the variables and values to the job template instead of using the hardcoded variables in the Ansible playbook:

```
# postgres-manage-database.yaml
---
- name: Managing PostgreSQL Database Server
  hosts: "{{ NODES }}"
  vars:
    ansible_become_user: postgres
    postgres_user: postgres
    postgres_password: 'PassWord'
    postgres_host: localhost
    postgres_database: db_sales
    postgres_table: demo_table
    postgres_new_user_name: devteam
    postgres_new_user_password: 'DevPassword'
  tasks:
    .
    .
```

Figure 12.28 – Hardcoded variables in the Ansible playbook

It is possible to override these values by passing the parameters to the playbook using the **Survey form** option in the job template (also use the **Extra variables** option in the job template, but you need to edit the job template and update it every time you want to pass different values). The survey questions will be passed to the playbooks as extra_vars automatically.

3. Click on the **Survey** tab of **Job template**, as follows:

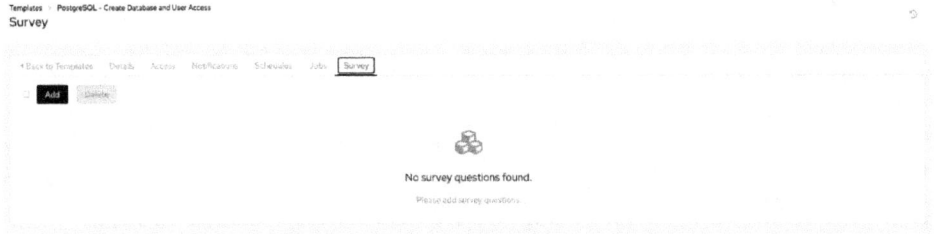

Figure 12.29 – Job template survey

4.   Create a new survey question by clicking the **Add** button and filling in the details:

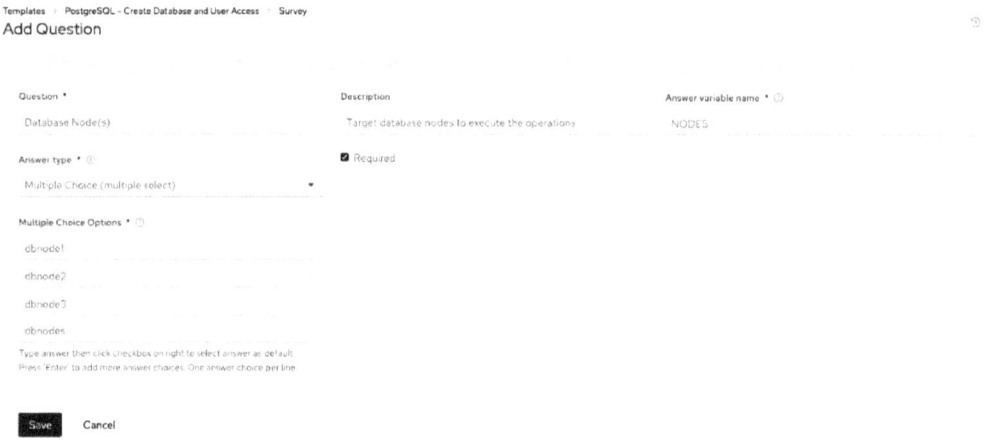

Figure 12.30 – Job template survey form for collecting the NODES information

In the preceding screenshot, we can see the following details:

- The answer variable's name is NODES, which we will pass to the playbook for hosts (hosts: "{{ NODES }}").

- A multiple-choice menu will be displayed with options.

Proceed to create survey questions for other variables as needed. Also, remember to enable the survey, as highlighted in the following screenshot:

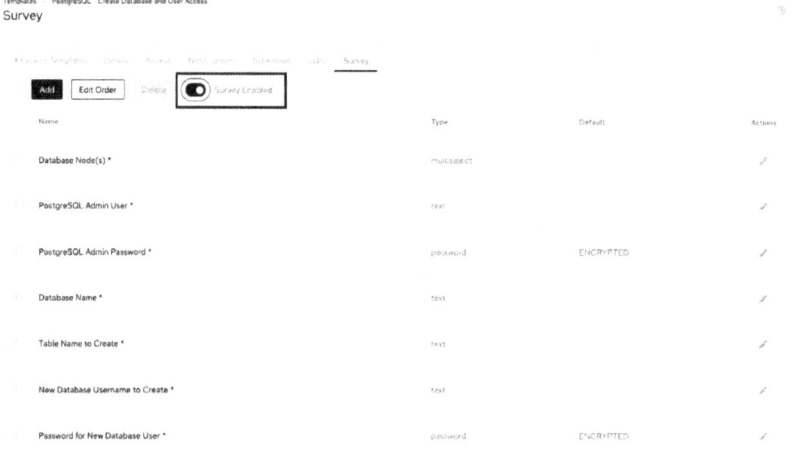

Figure 12.31 – Ansible survey variables for the job template

Remember to use the `password` type for the variables that carry sensitive information such as passwords or secrets.

> **Surveys in the Automation Controller**
>
> A survey can be used to set extra variables for the playbook with user-friendly forms and validations. Refer to the documentation at `https://docs.ansible.com/automation-controller/latest/html/userguide/job_templates.html#surveys` to learn more about survey forms.

We assume that the following variables will not change every time and do not need to be inside the survey form:

```
ansible_become_user: postgres
postgres_host: localhost
```

Instead, add them as extra variables and save the job template:

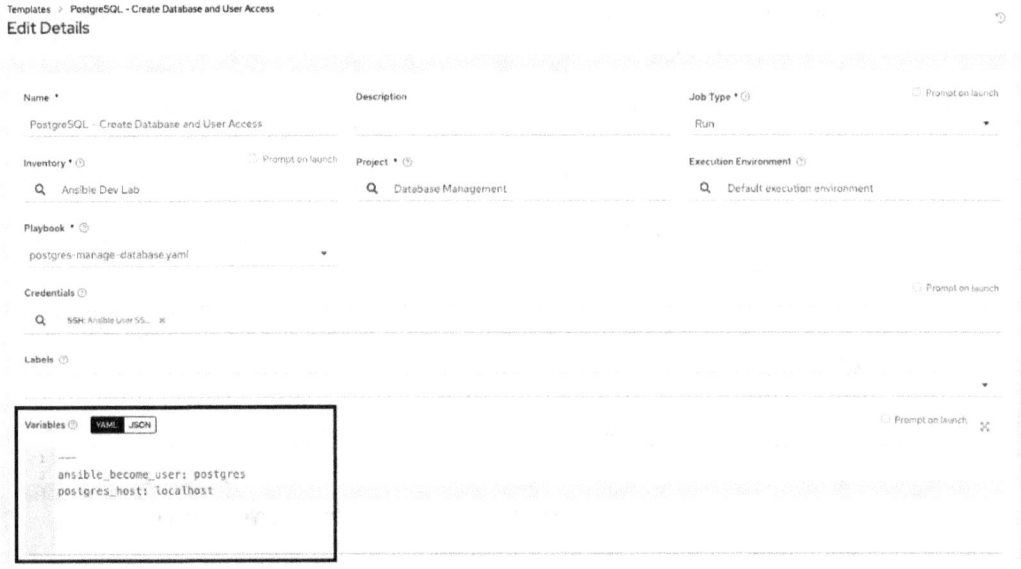

Figure 12.32 – Extra variables in the Ansible job template

Also pass variables such as `ansible_become_user` in the host variable or group variable section, but this will impact all the jobs on that node, which is not the desired configuration in this case.

In the next section, you will learn how to execute the automation job that we configured previously.

Executing an automation job from the automation controller

In the previous sections, you configured many resources in the automation controller, such as the organization, projects, inventory, managed nodes, credentials, job templates, and more. Now, let's test the job template by launching the automation job.

Ansible Collection and Execution Environment

Use the collection from the execution environment or the project repository (`COLLECTIONS_PATHS = ./collections`). Not all Ansible collections (or latest versions) will be available in the execution environment, and it is possible to create a custom execution environment with the required collections and libraries. Refer to the documentation at `https://docs.ansible.com/automation-controller/latest/html/userguide/execution_environments.html` to learn more about creating an execution environment. In this demonstration, we will be using the `community.postgresql` Ansible collection, which is already available in the project repository (refer to *Chapter 8, Helping the Database Team with Automation*).

Open the **Templates** tab from the automation controller dashboard and find the PostgreSQL job template that we created in the previous section. Click on the **Launch Template** button (the small rocket icon) and wait for the survey template to pop up:

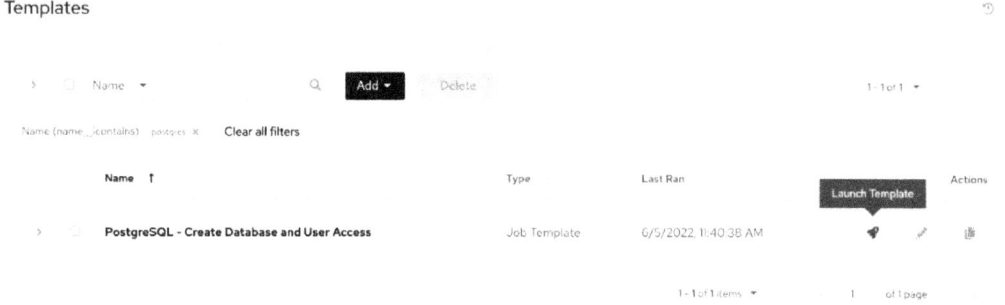

Figure 12.33 – Launching the job template from the automation controller

Enter the details in the survey's form and click **Next**:

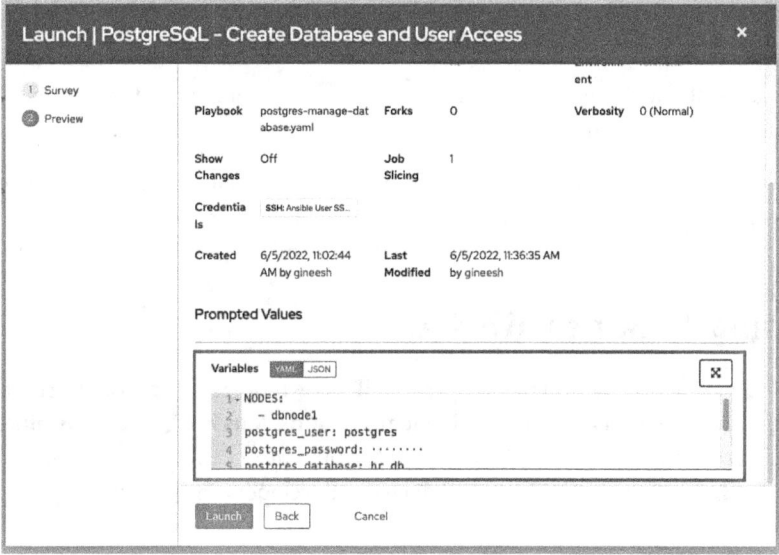

Figure 12.34 – Entering the details in the survey form

The details will be displayed on the preview screen. Notice that the survey variables are passing as extra variables. Click on the **Launch** button and wait for the job screen to load:

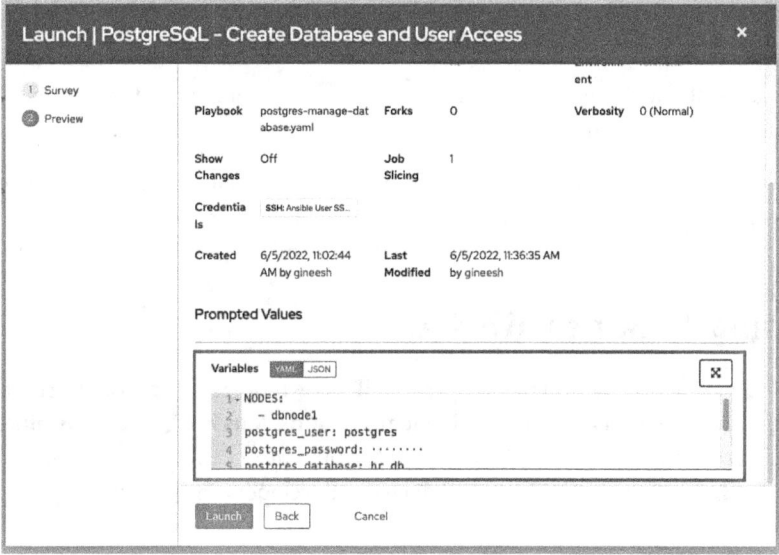

Figure 12.35 – Details of the job in preview

The job will be executed on the target nodes. Every detail about the job will be displayed on the **Output** tab, including the playbook execution output, as shown in the following screenshot:

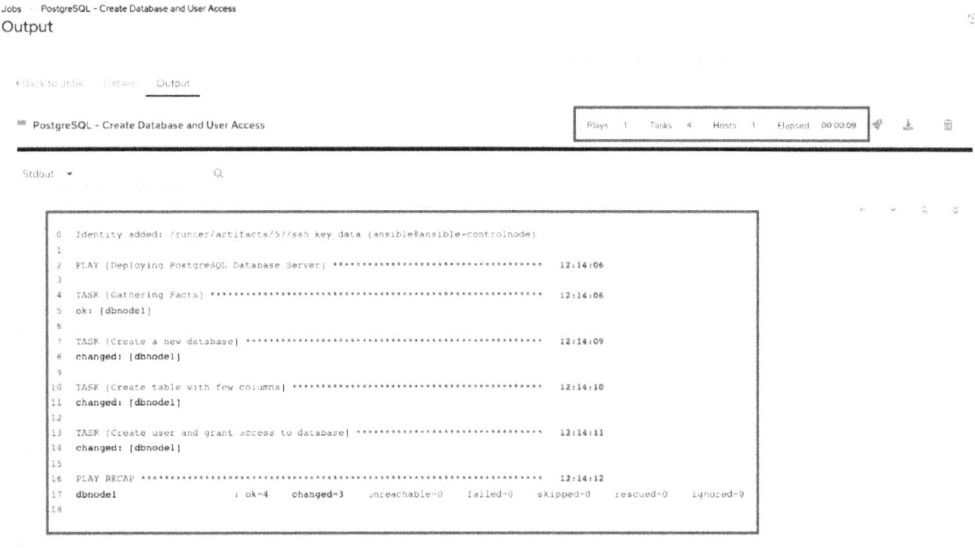

Figure 12.36 – Job execution output on the automation controller

Verify the database and user creation on the database nodes (dbnode1) as needed.

The same job template can be used to execute the operations on different database nodes or databases as the variables are dynamic and do not need to be modified in the playbook.

In this section, you learned how to create basic resources and launch the automation job in the automation controller. In the next section, you will learn how to integrate other applications, such as the Jenkins CI/CD server, with the automation controller and trigger automated executions.

Integrating Jenkins with AAP

Jenkins (https://www.jenkins.io) is a well-known open source tool (written in the Java programming language) that can be used to implement **continuous integration/continuous delivery (CI/CD)** and deployment solutions. Automating the build and deployment is the key to effective DevOps practices. As shown in the following diagram, developers and testers can offload such tasks to CI/CD tools such as Jenkins:

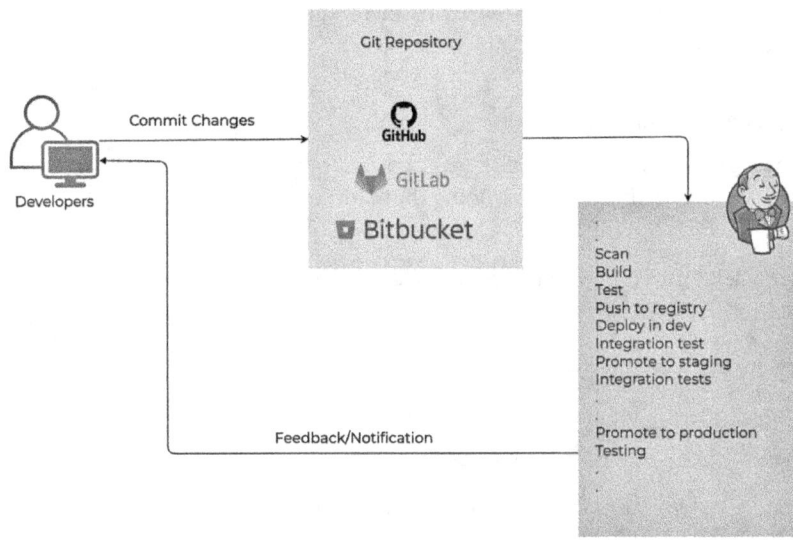

Figure 12.37 – CI/CD workflow using Jenkins

Jenkins can execute many tasks natively or use plugins but for complex tasks, Jenkins can utilize the appropriate tools. For example, instead of calling complex scripts or commands inside the Jenkins pipeline, a specific job can be offloaded to the Ansible automation controller, as shown in the following diagram:

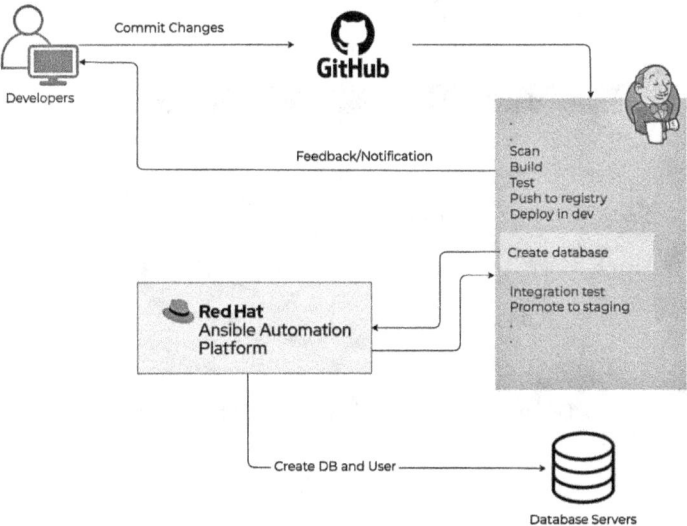

Figure 12.38 – Jenkins integration with AAP for database operations

The automation controller will execute the job based on the parameters passed and return the result to Jenkins as feedback (success/fail and other messages).

> **GitOps and CI/CD Courses**
>
> Refer to `https://www.techbeatly.com/gitops-cicd/` to find free GitOps and CI/CD courses and certifications, including those for Jenkins.

In the following demonstration, you will reuse the database creation job template as part of deploying the ToDo application.

The following are the prerequisites:

- The Jenkins server must have been installed with the Ansible Tower plugin (`https://plugins.jenkins.io/ansible-tower`).

- Knowledge of Jenkins operations and configurations.

- The `PostgreSQL - Create Database and User Access` job template must have been configured on the automation controller.

- Application repository (`https://github.com/ginigangadharan/nodejs-todo-demo-app`).

Ansible Tower plugin for Jenkins

The Ansible Tower plugin in Jenkins helps you execute automation jobs on Ansible Tower (or the automation controller) by passing the appropriate parameters:

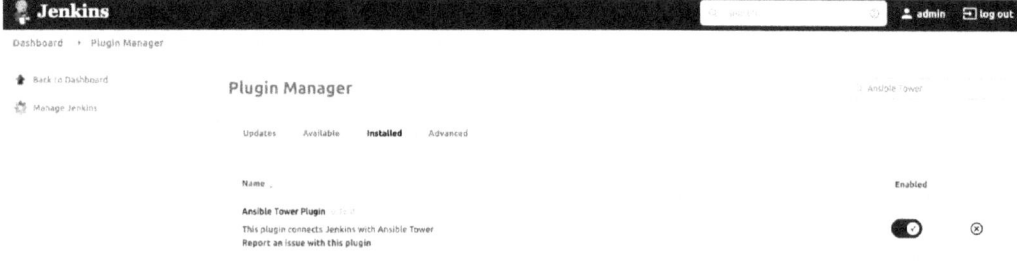

Figure 12.39 – Ansible Tower plugin for Jenkins

Install the plugin if you haven't done so yet and create an Ansible Tower (automation controller) connection (add the appropriate credentials needed) under **Manage Jenkins | Configure System**:

Ansible Tower

Tower Installation

Name ?

AAP-Demo

URL ?

https://192.168.1.103/

Credentials

admin/****** ∨   Add ∨

☑ Force Trust Cert ?

☐ Enable Debugging ?

Test Connection

Save   Apply

Figure 12.40 – Automation controller configuration for the Ansible Tower plugin in Jenkins

Verify the connection from Jenkins to the Ansible automation controller by tapping the **Test Connection** button, as shown in the preceding screenshot.

In the next section, you will learn how to create and configure multibranch pipelines in Jenkins.

Multibranch pipelines in Jenkins

Create a multibranch pipeline in Jenkins using the ToDo app repository (`https://github.com/ginigangadharan/nodejs-todo-demo-app`), as shown in the following screenshot:

Branch Sources

Git
Project Repository ?

https://github.com/ginigangadharan/nodejs-todo-demo-app

Credentials ?

- none - ∨   Add ∨

Behaviours

Figure 12.41 – A multibranch pipeline in Jenkins

Also, configure the automated pipeline trigger by scanning the source repository every minute. By enabling this, Jenkins will scan the repository for changes every minute and trigger the build process automatically:

Scan Multibranch Pipeline Triggers

☑ Periodically if not otherwise run   ?
  Interval   ?

  1 minute   ⌄

Figure 12.42 – Scanning multibranch pipeline triggers

Creating a Jenkinsfile

A `Jenkinsfile` is a **Pipeline-as-Code** mechanism for Jenkins. Instead of creating the build stages and steps manually in the GUI, each step can be written in a `Jenkinsfile`. The following screenshot shows the available `Jenkinsfile` in the `todo` application source repository:

```
● ● ●

pipeline {
    agent any

    environment {
        // Git Repo
        GIT_URL = "https://github.com/ginigangadharan/nodejs-todo-demo-app"
        // Database variables
        DATABASE_SERVER = "dbnode1"
        POSTGRES_USER = "postgres"
        POSTGRES_PASSWORD = "PassWord"
        POSTGRES_DATABASE = "app2_db"
        POSTGRES_TABLE = "data_table"
        POSTGRES_NEW_USER_NAME = "devteam"
        POSTGRES_NEW_USER_PASSWORD = "DevPassword"
    }
    .
    .
    .
```

Figure 12.43 – A Jenkinsfile with environment variables

In a production environment, sensitive variables need to be considered when storing vault services or credentials.

The following screenshot shows calling the database creation job in the automation controller:

```
● ● ●
    stage("Creating Database") {
        steps {
            echo "Create database and user access using Ansible Automation Controller"
            script {
                // Trigger Ansible controller job
                ansible_controller_job();
            }
        }
    }
```

Figure 12.44 – Database creation stage in the Jenkinsfile

`ansible_controller_job()` is written at the bottom of the `Jenkinsfile`, as shown here:

```
● ● ●
def ansible_controller_job(){
    ansibleTower(
        towerServer: 'AAP-Demo',
        templateType: 'job',
        jobTemplate: 'PostgreSQL - Create Database and User Access',
        importTowerLogs: true,
        inventory: 'Ansible Dev Lab',
        jobTags: '',
        skipJobTags: '',
        limit: '',
        removeColor: false,
        verbose: true,
        credential: '',

        extraVars: '''---
NODES: ["$DATABASE_SERVER"]
postgres_user: "$POSTGRES_USER"
postgres_password: "$POSTGRES_PASSWORD"
postgres_database: "$POSTGRES_DATABASE"
postgres_table: "$POSTGRES_TABLE"
postgres_new_user_name: "$POSTGRES_NEW_USER_NAME"
postgres_new_user_password: "$POSTGRES_NEW_USER_PASSWORD"
'''
    )
}
```

Figure 12.45 – ansible_controller_job in the Jenkinsfile

Notice the `extraVars` parameter we are passing since those variables are mandatory and need to pass via extra variables or a survey form (which is not applicable in this case).

In the next section, you will explore the automated build trigger and automated database provisioning that can be done via the Ansible automation controller.

Triggering a build in the Jenkins pipeline

A new build will be triggered automatically whenever there is a change (new commit) in the repository or whenever the build is initiated manually. The following screenshot shows the build history in the job; notice that the **Creating Database** stage is highlighted:

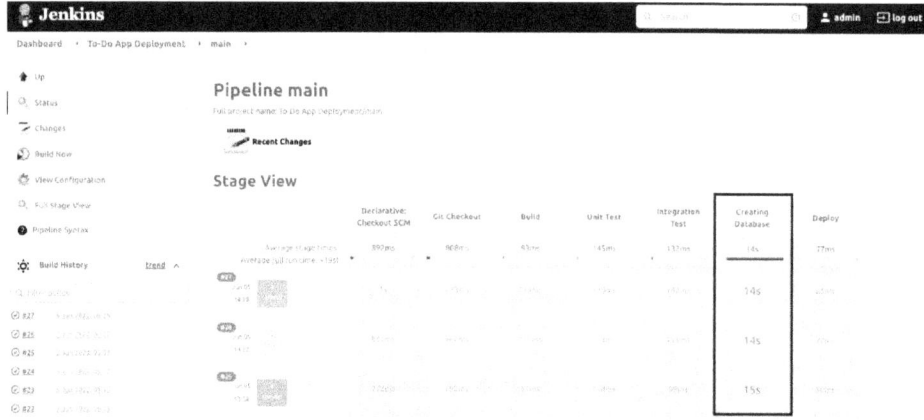

Figure 12.46 – Build jobs in the Jenkins pipeline

Check the build console logs and find the Ansible automation controller job execution with detailed output, as shown in the following screenshot:

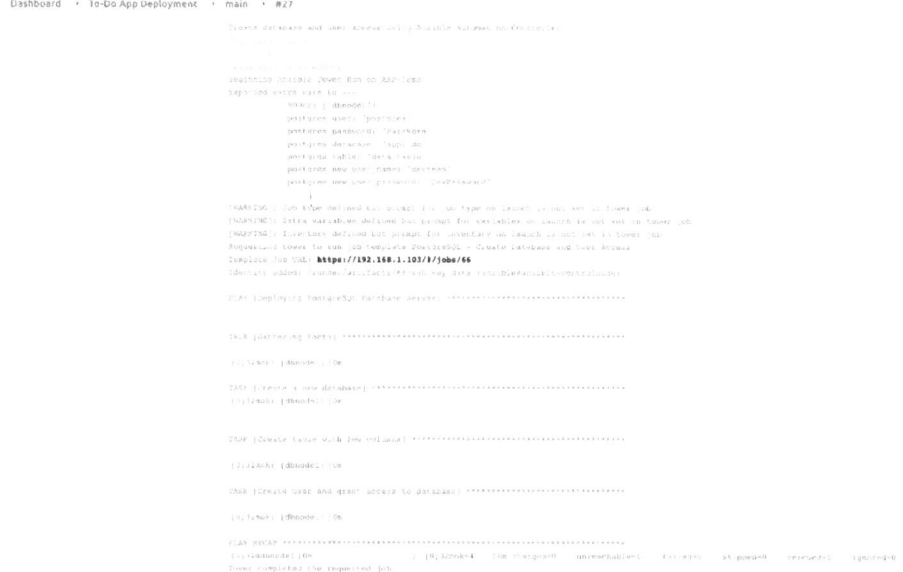

Figure 12.47 – The Jenkins build console with Ansible job execution details

With that, you have learned how to call automation controller jobs from a Jenkins pipeline and automate complex build steps.

Further enhancements to the Jenkins pipeline

Explore the integration between Jenkins and the Ansible automation controller by adding more Ansible jobs to the pipeline, as follows:

- Create a job template in the automation controller for building container images and push them to the container registry. (Refer to the Ansible artifacts from *Chapter 10, Managing Containers Using Ansible.*)

- Add a build step in the `Jenkinsfile` and call the automation controller job to build the container image for the application.

- Create another job template in the automation controller for deploying applications in a Kubernetes cluster by using the container image that was created in the previous stage. (Refer to the Ansible artifacts from *Chapter 11, Managing Kubernetes Using Ansible.*)

- Add a deployment step to the `Jenkinsfile` and call the automation controller job to deploy the updated application with the latest container image.

In the next section, you will learn how to enable notifications in the automation controller and integrate it with the Slack messaging service.

Integrating an automation controller with Slack and notification services

In *Chapter 3, Automating Your Daily Jobs*, you learned how to use the `mail` module to send custom emails using Ansible. In the Ansible automation controller, it is possible to configure **Notifications** to send emails and messages based on job start, success, or fail status. The following notification types are supported in the automation controller:

- Email
- IRC
- Webhook
- Grafana
- Slack
- Mattermost
- PagerDuty

- Rocket.Chat

- Twilio

Multiple notifications can be created and required notifications can be enabled for the job template.

Creating email notifications in the automation controller

To create an email notification, open the **Notifications** tab from the dashboard and click **Add**. Select the type as **Email** and fill in the details, as shown in the following screenshot:

- If the email server is open (no authentication required), then leave the **Username** and **Password** fields empty.

- Enable TLS or SSL if the email server supports it.

- Enter an approved sender email (email whitelisting).

Save the form and create the notification:

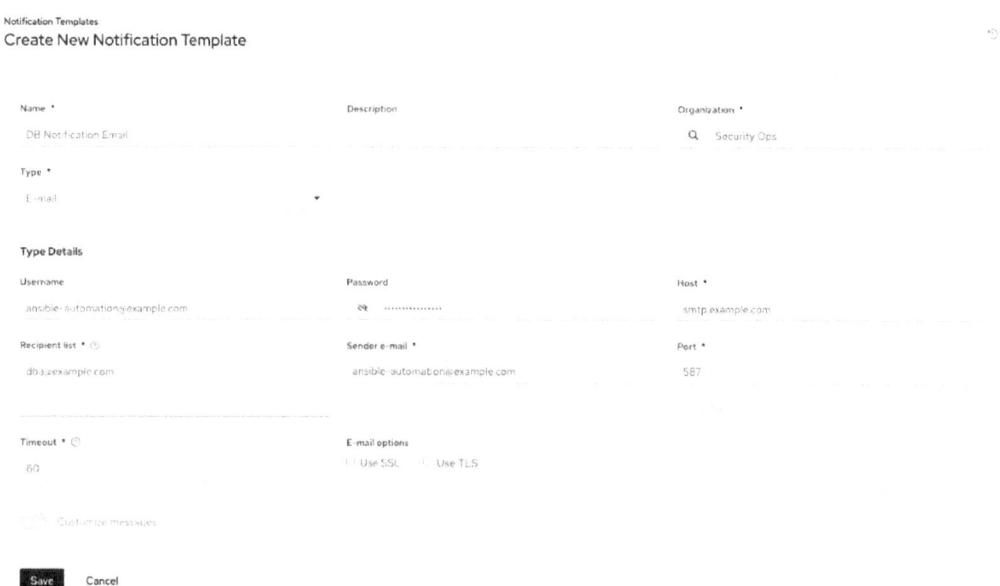

Figure 12.48 – Creating an email notification in the automation controller

Once you have created the notification, open the job template that you want to enable this notification for and select the **Notifications** tab. Toggle the switches for **Start**, **Success**, and **Failure** as needed:

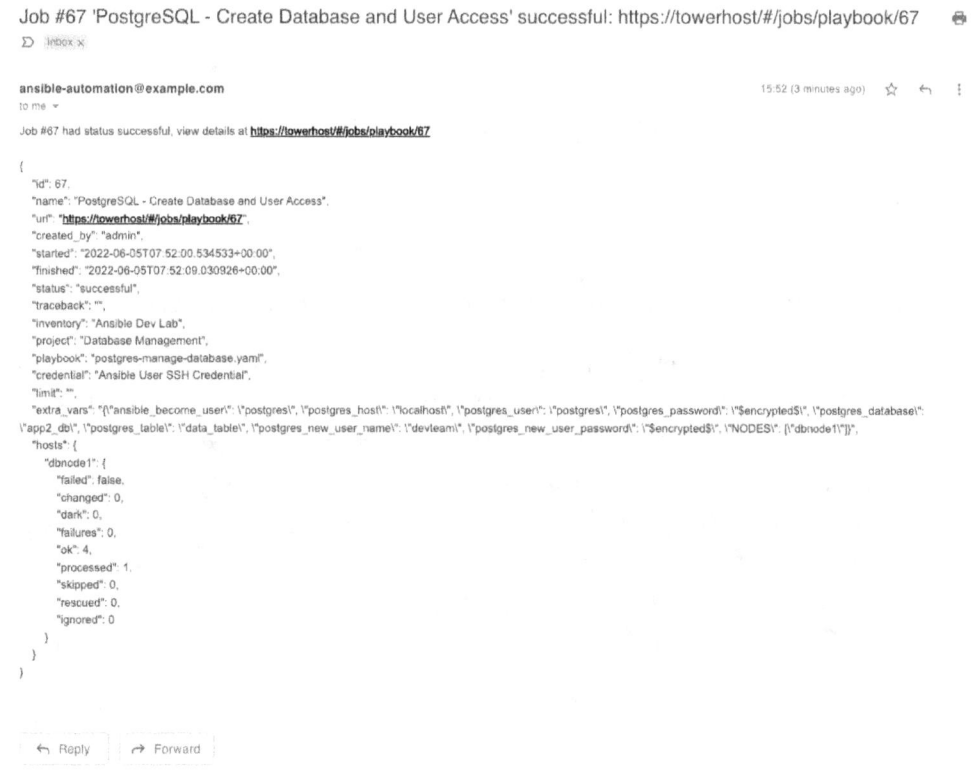

Figure 12.49 – Enabling notifications in the job template

In the preceding example, the notification will only be triggered if the job is successful.

Execute the job template and verify that you can receive the email on job success. A sample email is shown in the following screenshot:

Figure 12.50 – Sample email notification from the automation controller

Customize the notification message and its content by editing the notification entry, as shown here:

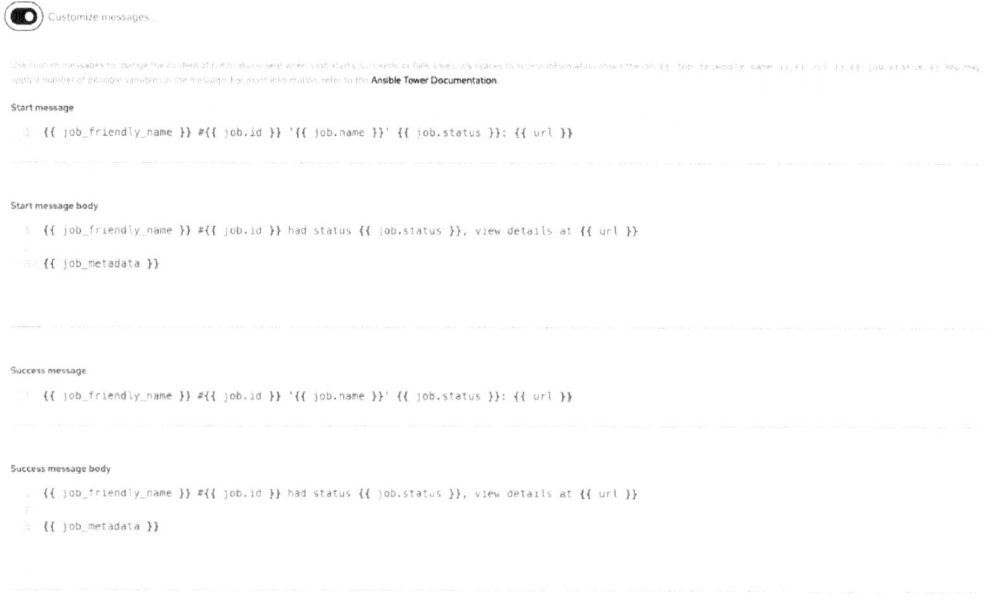

Figure 12.51 – Customizing the email notification in the automation controller

In the next section, you will learn how to integrate automation control with Slack to enable notifications in the Slack channel.

Sending Slack notifications from the automation controller

Slack is a workplace messaging platform (https://app.slack.com) in which users can send instant messages and create calls, video conferences, and more. Slack conversations are organized as Slack channels and multiple channels can be created in a workplace.

Slack is well known for its flexibility in integrating with your IT and application infrastructure. This is because it supports integration with about 2,400 applications (https://slack.com/apps):

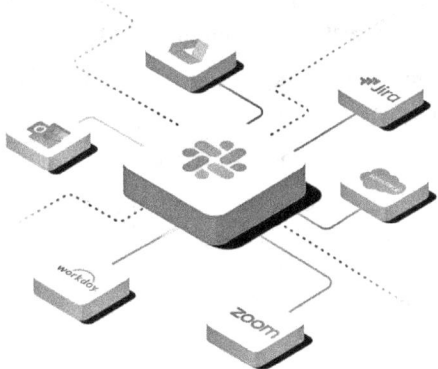

Figure 12.52 – Slack integration (source: https://slack.com/integrations)

In the following demonstration, you will learn how to create a Slack authentication token to send notifications from the automation controller. Follow these steps:

1. Sign up/log into Slack and access your existing workplace. (If you do not have a workplace to use, then create a new workplace by following the documentation at `https://slack.com/help/articles/206845317-Create-a-Slack-workspace`.)

2. Use an existing channel in your workplace (or create a new channel for testing purposes by following the documentation at `https://slack.com/help/articles/201402297-Create-a-channel`). For this demonstration, we have created a new channel called `#database-operations`.

3. Open `https://api.slack.com` in a web browser, click on the **Tutorials** menu (top right) and select **Publish interactive notifications**, as shown in the following screenshot:

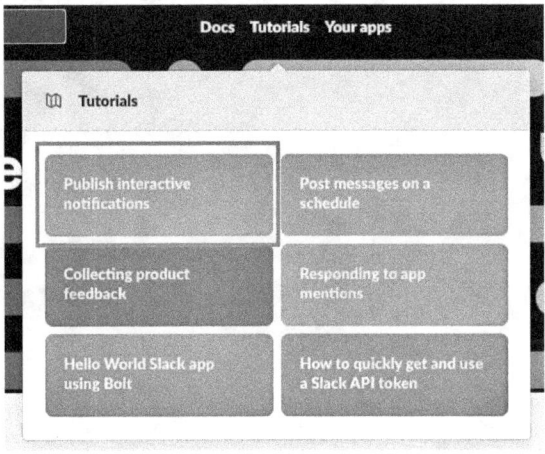

Figure 12.53 – Publish interactive notifications tutorial

4. The tutorial will open at `https://api.slack.com/tutorials/tracks/actionable-notifications`. Select **Create App** from the **Create a pre-configured app** section.

5. Slack will ask you to choose your workplace. Here, configure the app's details, such as its name, redirect URL, and other items.

6. Complete the instructions in the tutorial to create the app.

7. Once the app has been created, open the app configuration and copy the OAuth token, as shown in the following screenshot:

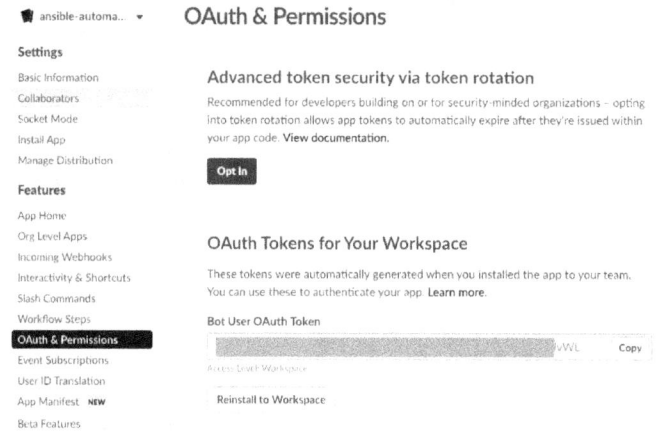

Figure 12.54 – Copying the Slack app OAuth token

8. Log into the Ansible automation controller and create a new notification with its **Type** set to **Slack** (*Figure 12.55*) and enter details about the token and the message channel (#database-operations). Customize the message as well if required:

Figure 12.55 – Creating a Slack notification in the automation controller

9. Open the notification settings for the job template and enable the newly created Slack notification, as shown in the following screenshot:

Figure 12.56 – Enabling the Slack notification for the job template

10. Execute the job again and verify the Slack channel to see the message:

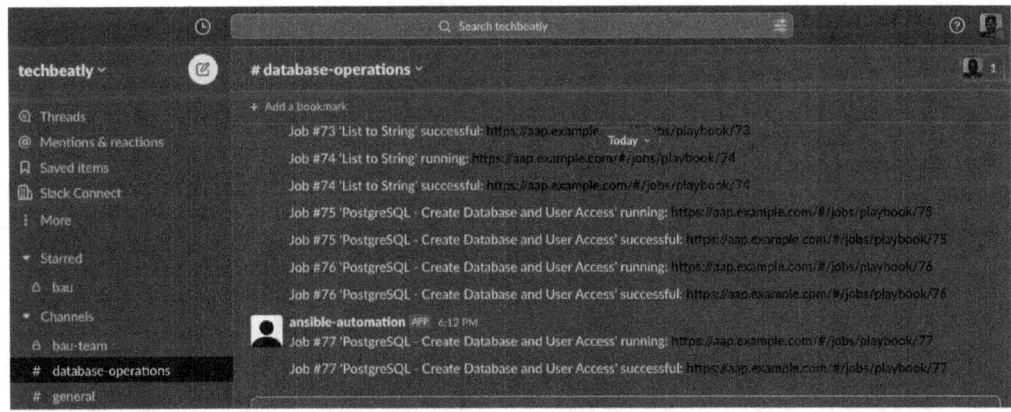

Figure 12.57 – Job notification in Slack from the automation controller

Explore the **Notifications** option with other applications to implement effective notifications for your team and workplace.

Slack Tutorials

Refer to `https://api.slack.com/tutorials` to learn more about Slack and its available tutorials.

With that, you have learned how to integrate Ansible with Slack to enable notifications. Implement the same for other notification and messaging services such as Mattermost, Rocket.Chat, and others.

Summary

In this chapter, you learned about the enterprise automation solution called Red Hat AAP. First, you learned about the benefits of using AAP and its features. You also learned about the different components of AAP, such as its execution environment, automation controller, automation mesh, and Automation Hub.

After that, you learned more about the automation controller by creating different resources such as organizations, projects, inventories, managed nodes and groups, credential job templates with survey forms and extra variables, and more. You also learned how to integrate the Jenkins CI/CD tool with Red Hat AAP to trigger the jobs automatically as part of the build and deployment pipeline.

Finally, you explored the notification options in the automation controller and tested them with different types of notifications such as email and Slack. All this knowledge will help you implement and manage automation using AAP and integrate AAP with different tools in your environment.

In the next chapter, you will learn how to use Ansible Vault to manage secrets and sensitive information for Ansible artifacts.

Further reading

To learn more about the topics that were covered in this chapter, take a look at the following resources:

- *Why choose Red Hat for automation?*: `https://www.redhat.com/en/topics/automation/why-choose-red-hat-for-automation`
- *Deploying Ansible Automation Platform 2.1*: `https://access.redhat.com/documentation/en-us/reference_architectures/2021/html-single/deploying_ansible_automation_platform_2.1/index`
- *Red Hat Ansible Automation Platform datasheet*: `https://www.redhat.com/en/resources/ansible-automation-platform-datasheet`
- *Control your content with private Automation Hub – Ansible blog*: `https://www.ansible.com/blog/control-your-content-with-private-automation-hub`
- *What is a Red Hat AAP automation execution environment?*: `https://www.redhat.com/en/technologies/management/ansible/automation-execution-environments`

13
Using Ansible for Secret Management

When we automate tasks, we need to implement them with little to no user interaction. However, we also know that there will be stages where Ansible needs inputs such as usernames, passwords, API keys, and secrets. Most of these details can be kept in a variable file and passed to playbooks without a user prompt or interaction but it is not a best practice to keep this kind of sensitive information in a plain text format as variables. There are external key vault services you can use but most of them require additional setup and configurations, which you need to integrate with Ansible.

Ansible Vault is an inbuilt feature of Ansible, using which we can safeguard the sensitive parts of our Ansible artifacts by encrypting our own vault passwords. Ansible Vault is installed together with Ansible and you can use it for Ansible ad hoc commands, playbooks, or within Red Hat Ansible Automation Platform.

In this chapter, you will learn about the following main topics:

- Handling sensitive data in Ansible
- Managing secrets using Ansible Vault
- Using secrets in an Ansible playbook
- Using Vault credentials in automation controller

We will start with basic Vault operations and learn how to use sensitive data in playbooks with the help of Vault.

Technical requirements

The following are the technical requirements to proceed with this chapter:

- One RHEL8/Fedora machine for an Ansible control node

- One or more Linux machines with Red Hat repositories configured. If you are using other Linux operating systems instead of **Red Hat Enterprise Linux** (**RHEL**) machines, then make sure you have the appropriate repositories configured to get packages and updates.

All the Ansible code, Ansible playbooks, commands, and snippets for this chapter can be found in the GitHub repository at `https://github.com/PacktPublishing/Ansible-for-Real-life-Automation/tree/main/Chapter-13`.

Handling sensitive data in Ansible

It is a known practice not to keep sensitive data in plain text format. The same rule applies to Ansible as well, as you will be dealing with different types of sensitive data in Ansible. The sensitive data could be anything, such as the following:

- System passwords

- API keys

- Port details of applications

- Database passwords

- SSL certificates or keys

- Cloud credentials

We have already learned that Ansible uses plain text format for playbooks, variables, and all other configurations. Hence, storing sensitive data in normal variable files is not desirable and we need to store such information using a more secure method.

Before we jump into the details of Ansible Vault, let us learn about some of the alternative secret management methods in the following sections.

Integrating with Vault services

One of the most common methods for storing sensitive information is using key vault software and services where we can access keys and secrets over GUIs, APIs, or CLIs. We need to add tasks in Ansible to contact the Vault store, authenticate, and retrieve secrets as needed. A sample integration is demonstrated in *Figure 13.1*:

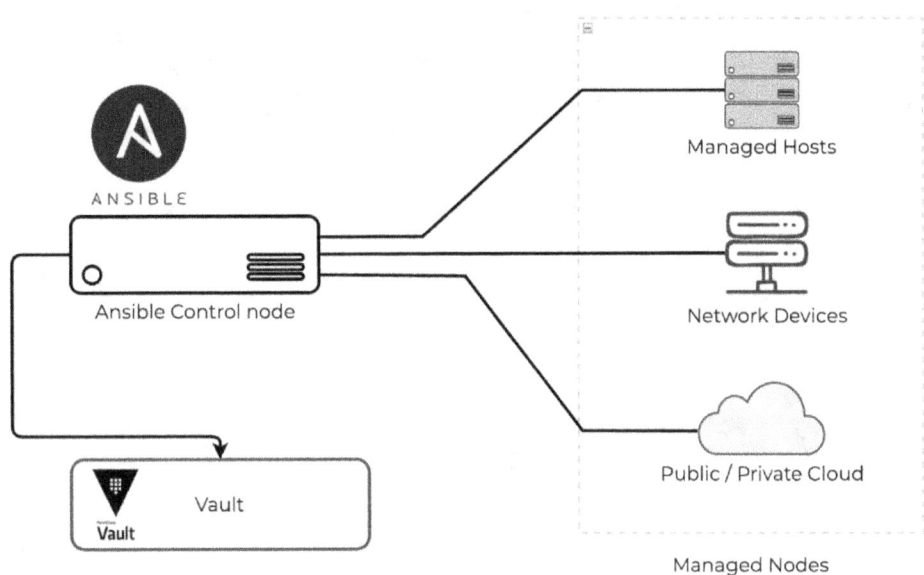

Figure 13.1 – Ansible integration with an external vault

There are managed and self-hosted external secret management solutions available to use, such as HashiCorp Vault (https://www.vaultproject.io), AWS Secrets Manager (https://aws.amazon.com/secrets-manager/), or Azure Key Vault (https://azure.microsoft.com/en-us/services/key-vault/). For retrieving keys or secrets, we can use the API calls, Ansible modules, or Ansible lookup plugins available, as shown in the following figure:

```
# Fetching database password from Hashicorp vault using hashi_vault lookup
- ansible.builtin.debug:
    msg: "{{ lookup('community.hashi_vault.hashi_vault', 'secret=secret/dbpass:value token=c975b780-d1be-8016-
866b-01d0f9b688a5 url=http://myvault:8200') }}"

# Fetching secret from AWS Secret manager using aws_secret lookup
- name: lookup secretsmanager secret in the current region
  debug: msg="{{ lookup('amazon.aws.aws_secret', '/path/to/secrets', bypath=true) }}"
```

Figure 13.2 – Using lookup plugins to retrieve vault keys

The external key vault services are good and useful but we also need to handle the overhead in terms of management and pricing. Refer to the *Further reading* section for more documentation and references on external key vault services.

Interactive input using prompts

One of the alternative methods of handling sensitive data in Ansible is collecting the data dynamically during playbook execution. Sensitive and non-sensitive inputs can be accepted using `vars_prompt`, as follows:

```yaml
---
- name: Accepting sensitive data using prompts
  hosts: node1
  gather_facts: no

  vars_prompt:
    - name: database_username
      prompt: Enter your username
      private: no

    - name: database_password
      prompt: Enter your password

  tasks:
    - name: Print a message
      ansible.builtin.debug:
        msg: 'Login to database as {{ database_username }}'
```

Figure 13.3 – An Ansible playbook accepting passwords using prompts in Ansible

When you execute the playbook, Ansible will ask for the input and, based on the `private` value, the input will either be visible or hidden on the prompt, as shown in *Figure 13.4*:

```
[ansible@ansible Chapter-13]$ ansible-playbook prompt.yaml
Enter your username: dbadmin
Enter your password:

PLAY [Accepting sensitive data using prompts] ************************************************************

TASK [Print a message] ************************************************************
ok: [node1] => {
    "msg": "Login to database as dbadmin"
}

PLAY RECAP ************************************************************
node1                      : ok=1    changed=0    unreachable=0    failed=0    skipped=0    rescued=0    ignored=0
```

Figure 13.4 – Accepting the user input using vars_prompt

However, as you can see in the preceding figure, this method is interactive and someone needs to input the details during the playbook execution. This means you will not be able to use these playbooks in a fully automated workflow where user interaction is not possible. Equally, `vars_prompt` will not work with the Ansible automation controller, as the controller does not interactively allow `vars_prompt` questions.

Encrypting data using Ansible Vault

As I mentioned at the beginning of this chapter, Ansible Vault is an inbuilt feature of Ansible, using which we can keep the sensitive parts of our Ansible artifacts secure by encrypting the data. We can use our own passwords as vault passwords for encrypting the content. It is possible to use Ansible Vault for Ansible ad hoc commands, playbooks, or within the Ansible Automation Platform.

It is a best practice to separate sensitive artifacts from non-sensitive artifacts and keep them in separate files. The first step of the process is to separate the sensitive data from regular variable files and store them in separate variable files, as shown in *Figure 13.5*.

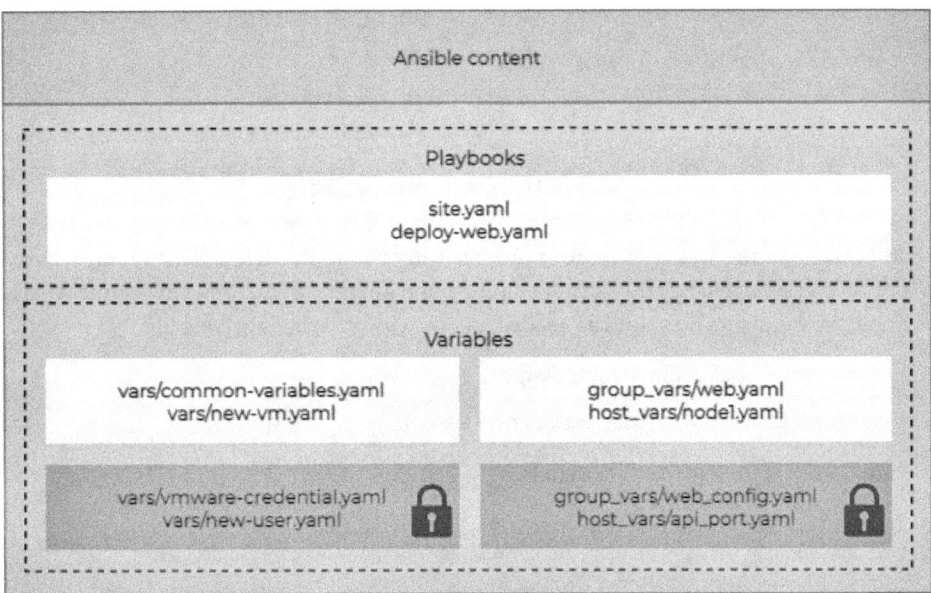

Figure 13.5 – Ansible artifacts with encrypted variables

With this practice, you will have the ability to store sensitive data in an encrypted format and the flexibility to modify normal variables at any time. When Ansible wants to read the content of these encrypted files, Ansible will use the vault password, which we can enter using the prompt (`--ask-vault-password`) or using special Vault password files. For the Ansible Automation Platform, the Vault password can be stored as a credential and assigned to the Job Template.

In the following sections, we will learn how to use Ansible Vault to encrypt variable files and access the encrypted data inside an Ansible playbook.

Managing secrets using Ansible Vault

Ansible Vault is very flexible, as we can encrypt, view, decrypt, or change the Vault password (as in, rekey it) at any time as needed. The Vault password must be stored safely, as you will not be able to retrieve the encrypted Vault content without the Vault password.

Creating Vault files

In the following exercise, we will learn how to create an encrypted file using Ansible Vault:

1. To create a Vault file from scratch, use the `ansible-vault create` command, as shown in *Figure 13.6*:

    ```
    [ansible@ansible Chapter-13]$ ansible-vault create vars/cloud-credential.yaml
    New Vault password:
    Confirm New Vault password:
    ```

    Figure 13.6 – Creating a Vault file

2. After we enter the Vault password, a new file will open in the default text editor, such as `vim` or `nano` (we can change the default editor by updating the `$EDITOR` environment variable). Enter the variables and values as needed, just as with a normal variable file:

    ```
    cloud_username: myusername
    cloud_password: mysecretpassword
    ```

    Refer to *Figure 13.7* for further details:

● ● ●

```
cloud_username: myusername
cloud_password: mysecretpassword
~
~
~
~
~
~
~
~
~
~
~
~
:wq
```

Figure 13.7 – Adding content to the Vault file

3. Save the file as per text editor actions (for example, :wq) and exit the editor.

    We can view the Vault file but the content will be encrypted (see *Figure 13.8*) and you will not be able to retrieve the data without the Vault password.

● ● ●

```
[ansible@ansible Chapter-13]$ cat vars/cloud-credential.yaml
$ANSIBLE_VAULT;1.1;AES256
6633663735323933373832343565623362386536346134323462333964653562653762633132
3833366432313965336566663864356662393030643238320a30663037326466316434623564313 7
6130383035386336303462363838323537646534613338363565343366646166613139373631643 7
6265646630653437300a306533643333313173562653439643736333763653734393666263353530
363663626263337306661346438633636162346239663839666237303630376137633334383666463
663563313035333163323938333633653235333036323632653036333323562353739363764623263 65
64326663335343966616431613633373838
```

Figure 13.8 – Encrypted content inside the Vault file

We have learned how to create an encrypted file using Ansible Vault in the preceding exercise. In the following section, we will learn how to encrypt existing files and content using Ansible Vault.

Encrypting existing files

If you have a variable file or file with sensitive content and you want to encrypt it using Ansible Vault, then you can do it using the encrypt command, as explained in the following exercise:

1. Verify the current content of the files as follows:

●●●

```
[ansible@ansible Chapter-13]$ cat vars/dbdetails.yaml
database_username: dbadmin
database_password: dbPassword
database_port: 5432
```

Figure 13.9 – Database details in a plain text format

2. Encrypt the content using the `ansible-vault encrypt` command:

●●●

```
[ansible@ansible Chapter-13]$ ansible-vault encrypt vars/dbdetails.yaml
New Vault password:
Confirm New Vault password:
Encryption successful
```

Figure 13.10 – Encrypting an existing file using Ansible Vault

3. Now, verify the encryption of the file as follows:

●●●

```
[ansible@ansible Chapter-13]$ cat vars/dbdetails.yaml
$ANSIBLE_VAULT;1.1;AES256
396231336433376466373731326538353039393337376533361623132326336643237633466356665
363164626435336337336562643238366630663763636302300a613035646533333631643835613463
653333364373637643261303136383333666326538353938363665633935636663133343739313366431
626636613233656164 0a656330376461323831363533336232373565335662323937373331313316563
336313536646363363386133623226639646530353766356664461535336334393663335666138623164 62
643431353035613736646330623038623565565666634303734623735623161626236393338373434
646366666613830376266663643863643566333963339303433353164336238663666346162343261
326338323232373633377363661333161326131346265363734303263333332383433663035386263 62
3330
```

Figure 13.11 – The plain text file after being encrypted

Now that we know how to encrypt the files, we will learn how to use Vault ID to handle multiple Vault passwords in the next section.

Adding Vault ID to an encryption

When we have many Vault files and multiple Vault passwords, we can use the **Vault ID** to identify the Vault content. A Vault ID is an identifier for one or more vaulted secrets and contents. Let's follow an example:

1. Create and encrypt the secret file with the `--vault-id` option as follows. The Vault ID will be visible in the prompt, as shown in *Figure 13.12*.

```
[ansible@ansible Chapter-13]$ ansible-vault create --vault-id mysecret@prompt vars/secret-with-id.yaml
New vault password (mysecret):
Confirm new vault password (mysecret):
```

Figure 13.12 – Create a Vault file with a Vault ID

2. The same Vault ID can be checked from the content of the Vault file as follows:

```
[ansible@ansible Chapter-13]$ cat vars/secret-with-id.yaml
$ANSIBLE_VAULT;1.2;AES256;mysecret
34336230626266393462346439313564333232376132616362393534323339303135633239323133
33356463613134656435626561666562623237653734613880a326431646361383336633233383366
31653330316538393666463303136646366613239646265303033656439393636333330366263663933
61636263326533663340a6566343061616230353535396666665633653661326661353863330343939
3162313032646336663333463323630313662373761363135343862373637373663431
```

Figure 13.13 – A Vault file with a Vault ID

3. The password for the Vault ID can be taken from the prompt (as we did in the preceding example) or from a configured path in `ansible.cfg` as follows:

```
# ansible.cfg
[defaults]
.
.
.
vault_identity_list = inline@~/ansible/.vault_pass , files@~/ansible/.secret_pass
```

Figure 13.14 – The Vault ID configured in ansible.cfg

Using a Vault ID will help you to manage multiple Vault files in a large environment and identify the Vaulted files and secrets.

Viewing the content of a Vault file

Once the content is encrypted, it is possible to display the Vault content using the `ansible-vault view` command, as shown in *Figure 13.15*. Ansible will prompt the existing Vault password and you will see the content in plain text:

```
[ansible@ansible Chapter-13]$ ansible-vault view vars/dbdetails.yaml
Vault password:
database username: dbadmin
database password: dbPassWord
database port: 5432
```

Figure 13.15 – Displaying the Vault content

Please note that the Vault file content is still in an encrypted state and you will therefore not be able to access it without the Vault password.

Editing a Vault file

To edit the encrypted Vault file, use the `ansible-vault edit` command as follows:

```
[ansible@ansible Chapter-13]$ ansible-vault edit vars/dbdetails.yaml
Vault password:
```

Figure 13.16 – Editing an encrypted file using Ansible Vault

The Vault file will open in the text editor in plain text format. Once editing is completed, save the file (for example, `:wq`) and exit the editor, as shown in *Figure 13.17*:

```
database_username: dbadmin
database_password: dbPassWord
database_port: 5432
database_ha: true
~
~
~
~
~
~
~
~
~
~
~
```

Figure 13.17 – Editing a Vault file in the text editor

Once editing is complete, the Vault files will be saved in an encrypted format without any additional actions required.

Decrypting a Vault file

There are situations where we need to decrypt the file back to plain text temporarily or permanently. In such cases, `ansible-vault decrypt` will help to decrypt the Vault file back to plain text format as follows:

```
[ansible@ansible Chapter-13]$ ansible-vault decrypt vars/dbdetails.yaml
Vault password:
Decryption successful
```

Figure 13.18 – Decrypting a Vault file

Verify the file content after decryption, as shown in *Figure 13.19*:

```
[ansible@ansible Chapter-13]$ cat vars/dbdetails.yaml
database_username: dbadmin
database_password: dbPassWord
database_port: 5432
database_ha: true
```

Figure 13.19 – The Vault file after decryption

When you encrypt the file again, you can use the same or a different Vault password; it does not matter.

Vault password rotation by rekeying

It is a common practice to rotate passwords, keys, and SSL certificates to ensure that the credentials are not compromised and also to follow the organization's password policies. The `ansible-vault rekey` command will help to change or rotate the Vault password for secret content. Ansible Vault will ask for the existing Vault password and, if successful, a prompt for a new Vault password will be displayed, as shown in *Figure 13.20*:

```
[ansible@ansible Chapter-13]$ ansible-vault rekey vars/cloud-credential.yaml
Vault password:
New Vault password:
Confirm New Vault password:
Rekey successful
```

Figure 13.20 – Rotating the Vault password

Remember to update the new Vault password wherever applicable, such as in local Vault password files or Ansible automation controller vault credentials, for example.

Encrypting specific variables

If you do not wish to encrypt an entire variable file, then encrypt a specific variable using the `ansible-vault encrypt_string` command as follows:

```
[ansible@ansible Chapter-13]$ ansible-vault encrypt_string mysecretpassword --name password
New Vault password:
Confirm New Vault password:
password: !vault |
          $ANSIBLE_VAULT;1.1;AES256
          6665643137396266343934366165396263356633666339616639376537623965353938636464643037
          3963343861383831623132343262636364633636363136610a3933613038353166363931396666637
          3931666234383362333232353373861616230363533353363066346662346635633337656163656396431
          37346464653762326630a393231303935623333731383336465393938373962653630326313063636535
          64353636303533666373293953383430333333326466613334333626232326136383232396636
Encryption successful
```

Figure 13.21 – Encrypting a string using Ansible Vault

The input can be taken from inline, as shown in the preceding example (*Figure 13.21*), or from standard input, as shown in *Figure 13.22*:

```
[ansible@ansible Chapter-13]$ ansible-vault encrypt_string --name password
New Vault password:
Confirm New Vault password:
Reading plaintext input from stdin. (ctrl-d to end input, twice if your content does not already have a newline)
this is a secret strng typed frm input.
!vault |
          $ANSIBLE_VAULT;1.1;AES256
          3664613339613762386137303363333303137346664336363637306630656563033343665313032238
          30643633626336633736333434376538643439326462646610a3331363364613836635363965376164
          336265393836623764343937636463636363383133613439376646433666363663343139326164323236
          39343662643764666640a62636133335623235383166383336636356335396363733713430303762383035
          6662303866336463646366643636373264376139616563616463343732386263663766623930396366
          38656464396361633565383032323037393661333886434653138
Encryption successful
```

Figure 13.22 – Ansible Vault encrypting the string using the input value

Use this encrypted string as a variable in Ansible playbooks or variable files as follows:

```
●●●

---
## Chapter-13/encrypted-string-playbook.yaml
- name: Using encrypted variables
  hosts: node1
  vars:
    password: !vault |
          $ANSIBLE_VAULT;1.1;AES256
          62386361656532643262336363363630326266373866313461343938393832633533362373303463
          61383230383736656431643035313434313662323663666350a6433623232643735323930336323361
          31393332613566303064343436361363035323533165303436323635643237386335326666235353930
          34666630303866634300a6333343386439656530663431343237626534623137326465363665643034
          64663932363236363939373356164373966333937313935653937303264353665326233


  tasks:

    - name: Print a message
      ansible.builtin.debug:
        msg: 'Password is: {{ password }}'
```

Figure 13.23 – Encrypted string inside the playbook

> **Encrypting content with Ansible Vault**
>
> You have more options with Ansible Vault, such as editing, encrypting, decrypting, and rekeying, for example. Refer to the documentation for more details at `https://docs.ansible.com/ansible/latest/user_guide/vault.html`.

In the following section, we will learn how to use encrypted Vault files in Ansible playbooks and retrieve the secret information.

Using secrets in Ansible playbooks

You have learned the basic usage of secrets in an Ansible playbook in *Chapter 3*'s *Automating notifications* section. In this section, we will learn more about their usage and different methods of passing the Vault password.

In the following exercise, we will develop Ansible content to create users in Linux, with their passwords retrieved from an Ansible Vault file:

1. Create a `Chapter-13/vars/users.yaml` Ansible Vault file as follows and enter the Vault password:

    ```
    [ansible@ansible Chapter-13]$ ansible-vault create vars/
    users.yaml
    ```

    Remember the password, as we need this information when executing the playbook.

2. Add content to the variable files as follows:

```
userlist:
  john:
    username: john
    password: StrongPassword
  leena:
    username: leena
    password: AnotherPassWord
```

Figure 13.24 – User details inside an Ansible Vault file

Save the file and exit the editor. The `userlist` variable contains details of multiple users and their passwords.

3. Verify the file content, as shown in *Figure 13.25*:

● ● ●

```
[ansible@ansible Chapter-13]$ cat vars/users.yaml
```
$ANSIBLE_VAULT;1.1;AES256
33666132363764303461393063623223065316261393637306166343264353636435383766383561
34323534316636663234383837313966366623036373233300a35373434313733366666133373632
32373865336266616235376461643130626234313731376234343032353334373839333934363263
36396634616637643100a64636231303131363365316633336163363663316634393935393364938
34343237333530646666363564363533363139363732396162303063306365313462313034366230
37313438393861616333633264633063636363623134313637386333333337346131653235656663161
32353936364303330326635366623661333034323935633231396303033333323031134626163366364
6461643232323961393439373165306364333263613765313561366653635363263363230303330
35343735386538666337306662323039333838656232333363534363732613466343062622326537131
6434383764343362323463337383566363164396339323461333765313339383938653433643303434
393163396339616363666613435386463666635
```

Figure 13.25 – A user's encrypted details

4. Create a `Chapter-13/manage-user.yaml` playbook with the following content:

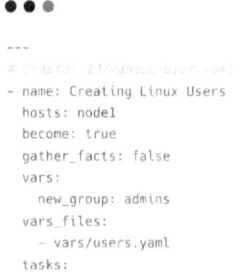

● ● ●

```
---
# Chapter-13/manage-user.yaml
- name: Creating Linux Users
  hosts: node1
  become: true
  gather_facts: false
  vars:
    new_group: admins
  vars_files:
    - vars/users.yaml
  tasks:
```

Figure 13.26 – A playbook to add users into
```

See the `vars_files` section because we have already included the encrypted variable file in the playbook.

```
# Chapter-13/manage-user.yaml....

  - name: Create new group
    ansible.builtin.group:
      name: "{{ new_group }}"
      state: present
  - name: Add the user
    ansible.builtin.user:
      name: "{{ item.value.username }}"
      password: "{{ item.value.password | password_hash('sha256') }}"
      shell: /bin/bash
      groups: admins
      append: yes
    loop: "{{ lookup('dict', userlist, wantlist=True) }}"
```

Figure 13.27 – Creating groups and users

> **Generating encrypted passwords in Ansible**
>
> The `password_hash('sha256')` filter has been used to encrypt the password and avoid sending a plain text password. Refer to `https://docs.ansible.com/ansible/latest/reference_appendices/faq.html#how-do-i-generate-encrypted-passwords-for-the-user-module` to learn more about password encryption in Ansible.

5. Now, execute the playbook using the `ansible-playbook` command as follows:

```
[ansible@ansible Chapter-13]$ ansible-playbook manage-user.yaml
ERROR! Attempting to decrypt but no vault secrets found
```

Figure 13.28 – An Ansible error due to there being no Vault secret

Since we have included Ansible Vault files inside the playbook, Ansible expects the Vault secret to be available and will fail if no appropriate secret is.

6. Execute the playbook by adding an `--ask-vault-password` argument:

```
[ansible@ansible Chapter-13]$ ansible-playbook manage-user.yaml --ask-vault-password
Vault password:

PLAY [Creating Linux Users] ******************************************************************

TASK [Create new group] **********************************************************************
ok: [node1]

TASK [Add the user] **************************************************************************
changed: [node1] => (item={'key': 'john', 'value': {'username': 'john', 'password': 'StrongPassword'}})
changed: [node1] => (item={'key': 'leena', 'value': {'username': 'leena', 'password': 'AnotherPassWord'}})

PLAY RECAP ***********************************************************************************
node1                      : ok=2    changed=1    unreachable=0    failed=0    skipped=0    rescued=0    ignored=0
```

Figure 13.29 – Executing the playbook with the Vault secret prompt

7. Verify the user creation success on node1 using an ad hoc command as follows:

    Figure 13.30 shows a sample output for the command:

    ```
    [ansible@ansible Chapter-13]$ ansible node1 -m shell -a "cat /etc/passwd |tail -2"
    node1 | CHANGED | rc=0 >>
    john:x:1003:1004::/home/john:/bin/bash
    leena:x:1004:1005::/home/leena:/bin/bash
    ```

    Figure 13.30 – An Ansible ad hoc command to check that a user has been created successfully

    In the preceding example, you have encrypted a variable file using Ansible Vault and retrieved the information inside the Ansible playbook by prompting the Vault secret. However, for automated operations, we need to skip this prompt. The Vault secret should be able to pass in the command line itself. In such scenarios, use the --vault-password-file argument and pass the Vault secret inside a file.

8. Create a file for storing your Vault secret. The Vault secret should be a plain text file but saved in a safe location, for example, a hidden file in your home directory, as shown in *Figure 13.31*:

    ```
    [ansible@ansible Chapter-13]$ echo "MyVaultSecret" > ~/.vault-secret

    [ansible@ansible Chapter-13]$ cat ~/.vault-secret
    MyVaultSecret
    ```

    Figure 13.31 – A Vault secret in a hidden file in your home directory

9. Execute the same playbook but pass the Vault secret file using the `--vault-password-file` argument. This time Ansible will not ask for a password, as shown in *Figure 13.32*:

```
[ansible@ansible Chapter-13]$ ansible-playbook manage-user.yaml --vault-password-file ~/.vault-secret
PLAY [Creating Linux Users] *******************************************************************

TASK [Create new group] ***********************************************************************
ok: [node1]

TASK [Add the user] ***************************************************************************
changed: [node1] => (item={'key': 'john', 'value': {'username': 'john', 'password': 'StrongPassword'}})
changed: [node1] => (item={'key': 'leena', 'value': {'username': 'leena', 'password': 'AnotherPassWord'}})

PLAY RECAP ************************************************************************************
node1                      : ok=2    changed=1    unreachable=0    failed=0    skipped=0    rescued=0    ignored=0
```

Figure 13.32 – The Ansible Vault secret from the password file

The Vault file can contain any type of data, whether a single variable, a string, a complex dictionary variable, or any other text content.

Ansible Vault will help us to encrypt the sensitive data in Ansible artifacts, but we need to safeguard such data from being captured within logs and we will learn about how to do so in the following section.

Hiding secrets from logs using no_log

You have learned how to keep sensitive content using Ansible Vault but we have a problem here, as Ansible will include the sensitive data content in plain text format when it carries out logging. Sometimes, it will not be visible in the default verbose mode but will be displayed in a higher-level verbose mode, such as `-vvv` or `-vvvv`. Refer to *Figure 13.33* for the details:

```
[ansible@ansible Chapter-13]$ ansible-playbook manage-user.yaml --ask-vault-password
Vault password:

PLAY [Creating Linux Users] *******************************************************************

TASK [Create new group] ***********************************************************************
ok: [node1]

TASK [Add the user] ***************************************************************************
changed: [node1] => (item={'key': 'john', 'value': {'username': 'john', 'password': 'StrongPassword'}})
changed: [node1] => (item={'key': 'leena', 'value': {'username': 'leena', 'password': 'AnotherPassWord'}})

PLAY RECAP ************************************************************************************
node1                      : ok=2    changed=1    unreachable=0    failed=0    skipped=0    rescued=0    ignored=0
```

Figure 13.33 – An Ansible playbook output displaying sensitive data

In such cases, use `no_log: True`, as shown in *Figure 13.34*, and any output in that task will be censored for safety reasons:

```
# Chapter-13/manage-user.yaml
  - name: Add the user
    ansible.builtin.user:
      name: "{{ item.value.username }}"
      password: "{{ item.value.password | password_hash('sha256') }}"
      shell: /bin/bash
      groups: admins
      append: yes
    loop: "{{ lookup('dict', userlist, wantlist=True) }}"
    no_log: True
```

Figure 13.34 – Disabling the logging of tasks using no_log

If you execute the playbook now, you will notice the change as follows:

```
[ansible@ansible Chapter-13]$ ansible-playbook manage-user.yaml --vault-password-file ~/.vault-secret
PLAY [Creating Linux Users] ********************************************************************

TASK [Create new group] ***********************************************************************
ok: [node1]

TASK [Add the user] ***************************************************************************
changed: [node1] => (item=None)
changed: [node1] => (item=None)
changed: [node1]

PLAY RECAP ************************************************************************************
node1                      : ok=2    changed=1    unreachable=0    failed=0    skipped=0    rescued=0    ignored=0
```

Figure 13.35 – The Ansible output with no_log applied to sensitive data

Even if you enable the high verbose mode, `-vvv`, Ansible will hide the information, as shown in *Figure 13.36*:

● ● ●

```
ased,publickey -o PasswordAuthentication=no -o 'User="devops"' -o ConnectTimeout=10 -o
ControlPath=/home/ansible/.ansible/cp/0726bd8bd1 192.168.56.25 '/bin/sh -c '"'"'rm -f -r
/home/devops/.ansible/tmp/ansible-tmp-1658050078.9681451-9038-58587566300946/ > /dev/null 2>&1 && sleep 0'"'"''
<192.168.56.25> rc=0, stdout and stderr censored due to no log
changed: [node1] => (item=None) => {
    "censored": "the output has been hidden due to the fact that 'no_log: true' was specified for this result",
    "changed": true
}
changed: [node1] => {
    "censored": "the output has been hidden due to the fact that 'no_log: true' was specified for this result",
    "changed": true
}
Read vars_file 'vars/users.yaml'
META: ran handlers
Read vars_file 'vars/users.yaml'
META: ran handlers

PLAY RECAP ************************************************************************************
node1                      : ok=2    changed=1    unreachable=0    failed=0    skipped=0    rescued=0    ignored=0
```

Figure 13.36 – A high verbose Ansible log with no_log applied

This is one of the best practices to safeguard the sensitive data being captured in system logs and job histories.

In the following section, we will learn how to keep the sensitive data inside other variable locations, such as group_vars and host_vars.

Ansible Vault for group_vars and host_vars

As we discussed earlier in the chapter, it is a best practice to separate variables and sensitive information wherever appropriate.

For the following exercise, we will reuse the PostgreSQL Ansible artifacts that we developed in *Chapter 8, Helping Database Team with Automation*. PostgreSQL is installed and configured on node2. We will create an additional database user account for accessing the db_sales database (refer back to the content of *Chapter 8* for more details.) Let's do the following:

1. Update the inventory (Chapter-13/hosts) by adding node2 as part of the postgres host group:

    ```
    [postgres]
    node2 ansible_host=192.168.56.24
    ```

2. Create group_vars and another subdirectory as follows:

```
[ansible@ansible Chapter-13]$ mkdir -p group_vars/postgres/vault

[ansible@ansible Chapter-13]$ ansible-vault create group_vars/postgres/vault/dbuser.yaml
New Vault password:
Confirm New Vault password:
```

Figure 13.37 – Creating the group_vars directory and Vault file

3. Add a database username and password for the new user:

```
postgres_app_user_name: appteam
postgres_app_user_password: 'AppPassword'
```

4. Save and verify the Vault file, as shown in *Figure 13.38*:

```
[ansible@ansible Chapter-13]$ cat group_vars/postgres/vault/dbuser.yaml
$ANSIBLE_VAULT;1.1;AES256
3939313361393033373465306165323732663930666432363162343166326531616263633139646 1
33343838633031333065363232663964393393365313164610a3330303336613162306438623132 37
3362326231643263333663234306536396662626566303263386337313532316439613362363731 36
6163306561646362360a653230333266393266653836343962383135633631646535613862306334
6565363131636666661343734323065313935666338336464363437393137363638343838383837373139
32383363323164363834663133334666639313965646439386136373565626361623838643130643 6
6566373532343533633535383932623437643437643232663030386663436373831383235353730356 2
3565666334643661303
```

Figure 13.38 – The database user information in the Vault file

5. Create another Vault file for storing the PostgreSQL admin password:

```
[ansible@ansible Chapter-13]$ ansible-vault create group_vars/postgres/vault/dbadmin.yaml
```

6. Add postgres_password inside and save the file:

```
postgres_password: 'PassWord'
```

7. Create a Chapter-13/postgres-create-dbuser.yaml playbook as follows:

```
# Chapter-13/postgres-create-dbuser.yaml
---
- name: Add new PostgreSQL Database user
  hosts: "{{ NODES }}"
  vars:
    ansible_become_user: postgres
    postgres_user: postgres
    #postgres_password: moved to Vault file
    postgres_host: localhost
    postgres_database: db_sales
    postgres_table: demo_table
  tasks:
```

Figure 13.39 – An Ansible playbook for managing PostgreSQL user information

8.   Add a task to create the PostgreSQL database user as follows:

```
# Chapter-13/postgres-create-dbuser.yaml...

    - name: Create user and grant access to database
      community.postgresql.postgresql_user:
        login_user: "{{ postgres_user }}"
        login_password: "{{ postgres_password }}"
        login_host: "{{ postgres_host }}"
        db: "{{ postgres_database }}"
        name: "{{ postgres_app_user_name }}"
        password: "{{ postgres_app_user_password }}"
        encrypted: yes
        priv: "CONNECT/{{ postgres_table }}:ALL"
        expires: "Dec 31 2022"
        comment: "Application user access"
        state: present
```

Figure 13.40 – The task to create a PostgreSQL user

All the sensitive variables are inside the group_vars Vault files now, for example, postgres_password, postgres_password, and postgres_app_user_password.

9.   Execute the playbook on the postgres host group (which we created in the first step), and the playbook will read the variables and create the new user safely, as shown in *Figure 13.41*:

```
[ansible@ansible Chapter-13]$ ansible-playbook postgres-create-dbuser.yaml --vault-password-file ~/.vault-secret -e
  @db-user-password.

PLAY [Add new PostgreSQL Database user] ******************************************************************

TASK [Gathering Facts] **********************************************************************************
ok: [node2]

TASK [Create user and grant access to database] *******************************************************
ok: [node2]

PLAY RECAP **********************************************************************************************
node2                      : ok=2    changed=0    unreachable=0    failed=0    skipped=0    rescued=0    ignored=0
```

Figure 13.41 – The creation of the PostgreSQL database user

As a best practice, keep all the sensitive details inside Vault files at appropriate locations and they will remain safe, even if you keep your Ansible content in Git repositories.

In the following section, we will learn how to use the Vault files and credentials in the Ansible Automation Platform.

Using Vault credentials in the Ansible Automation Platform

When you run your playbooks from the Web UI of the automation controller, then you have similar options to provide the Vault secret from the WebUI. We can either keep the Vault secret inside a Vault credential or we can select the **Prompt on launch** option. The latter involves interactive input (such as `--ask-vault-password` in the Ansible command-line execution) and will prompt for the Vault secret when you execute the Job Template from the automation controller's WebUI.

> **The Ansible automation controller**
>
> The Ansible automation controller is the control plane for the **Ansible Automation Platform** (**AAP**). When you migrate to AAP 2, the automation controller will be upgraded to include Ansible Tower. Refer to *Chapter 12, Integrating Ansible with Your Tools*, for more details.

In the following section, we will learn how to create Vault credentials in the Ansible automation controller GUI and attach them to the Job Template to retrieve encrypted content.

Creating Vault credentials

To store the Vault secret, create a new credential by following these steps:

1. Open the **Create New Credential** blade and set **Credential Type** to **Vault**, as shown in *Figure 13.42*. Enter the Vault secret (password) and add a Vault ID if required:

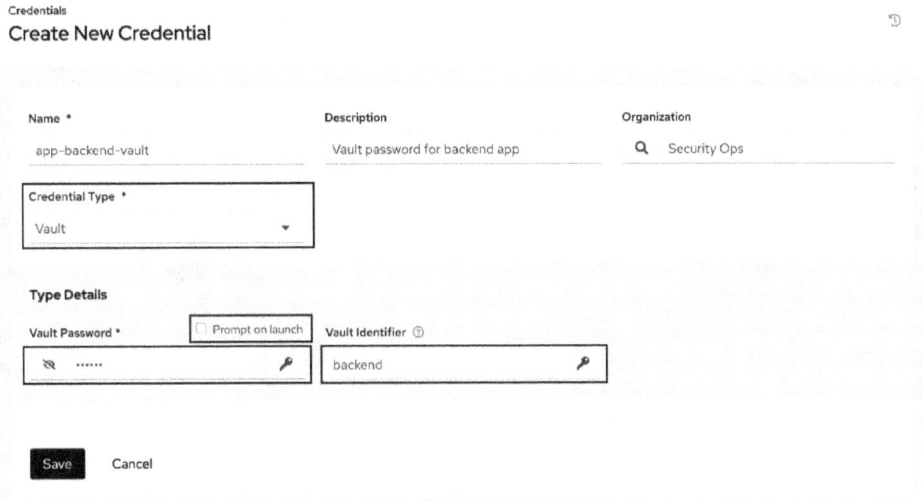

Figure 13.42 – Creating a new Vault credential in the automation controller

2. Once created, update your Job Template and add the new Vault credentials inside. Navigate **Job Template | Edit** and click on the *Search* button near **Credentials**.

3. Within the pop-up screen, set **Selected Category** to **Vault** and you will see the Vault credentials, as shown in *Figure 13.43*. Select the required Vault credentials and click the **Select** button.

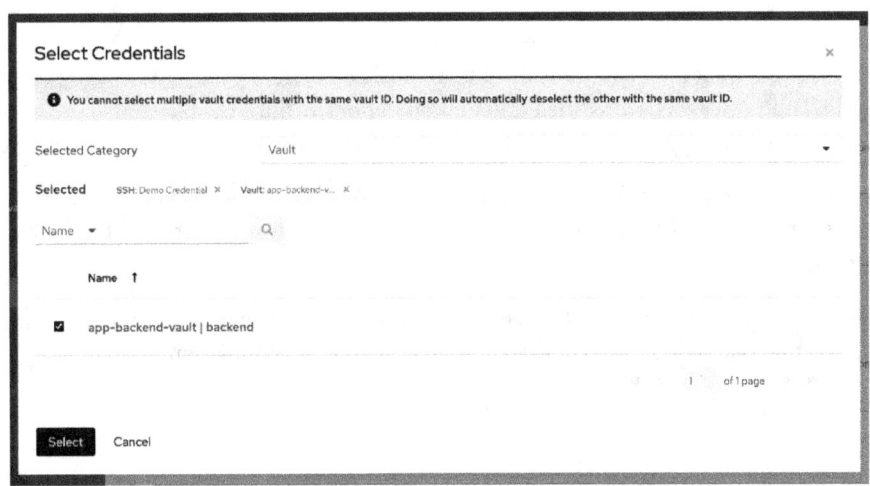

Figure 13.43 – Selecting the Vault credential for the Job Template

4.　Verify the credentials, as shown in *Figure 13.44*:

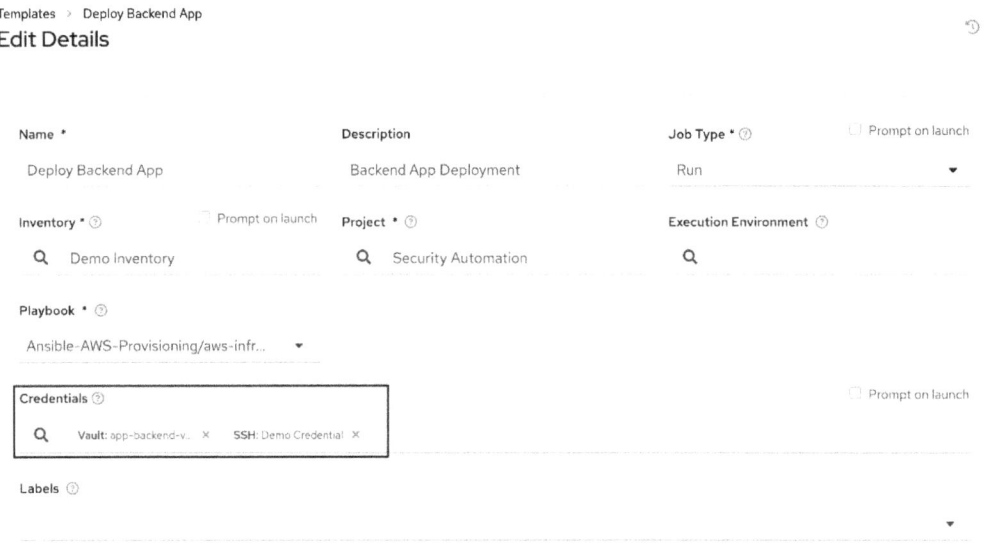

Figure 13.44 – The credentials added in the Job Template

If you have multiple Vault files with different Vault secrets, then add multiple Vault credentials as needed and add Vault IDs.

Summary

In this chapter, we learned the importance of keeping sensitive data secure within Ansible automation artifacts and the different methods available to do so, such as external Vault services, `vars_prompt`, and Ansible Vault. After this, we learned different operations within Ansible Vault, such as creating, modifying, viewing, decrypting, and rekeying Vault files and variables.

We also developed Ansible artifacts using Vault files for storing user information and database user credentials. We also discussed the Vault credentials in the automation controller GUI and how to use them with Job Templates.

In the next chapter, we will learn about different methodologies and approaches for developing Ansible automation artifacts and factors to consider throughout Ansible automation.

Further reading

To learn more about the topics covered in this chapter, please visit the following links:

- *Tutorial: Use Azure Key Vault to store VM secrets with Ansible* – `https://docs.microsoft.com/en-us/azure/developer/ansible/key-vault-configure-secrets?tabs=ansible`

- *How to send emails using Ansible and Gmail* – `https://www.techbeatly.com/ansible-gmail`

- *Logging Ansible output* – `https://docs.ansible.com/ansible/latest/reference_appendices/logging.html`

- *Keep vaulted variables safely visible* – `https://docs.ansible.com/ansible/latest/user_guide/playbooks_best_practices.html#keep-vaulted-variables-safely-visible`

- *Encrypting passwords in Ansible* – `https://docs.ansible.com/ansible/latest/reference_appendices/faq.html#how-do-i-generate-encrypted-passwords-for-the-user-module`

- *Vault IDs in Red Hat Ansible and Red Hat Ansible Tower* – `https://developers.redhat.com/blog/2020/01/30/vault-ids-in-red-hat-ansible-and-red-hat-ansible-tower`

Part 3

Managing Your Automation Development Flow with Best Practices

This part will describe the importance of continual assessment, monitoring, and security operations. You will learn about different monitoring technologies you can use to detect and protect your environment, as well as how to gain insights from them.

This part of the book comprises the following chapters:

14

Keeping Automation Simple and Efficient

Ansible is a simple and powerful automation tool. We can automate any kind of workflow using Ansible but if we increase complexity in automation, we decrease efficiency, which kills productivity. When you design an automated solution or use case, you must consider multiple factors, such as the capability of the tool and flexibility in adjusting the automation's flow or scalability.

For example, it is possible to write simple playbooks to monitor the service status in a system or to check the health of an application. But this is not efficient as you need other arrangements such as job schedulers to execute the job at regular intervals and monitor the execution. Instead of using Ansible natively for complex automation tasks, we can utilize the integration capabilities of the Ansible automation controller and other systems. We can use the existing tools for monitoring, logging, and security control, and use Ansible for remediation actions such as starting services and blocking ports and IP addresses.

In this chapter, we will cover the following topics:

- Utilizing surveys and automated inputs
- Integrating Ansible with monitoring tools
- Ansible for security automation
- Ansible workflow templates

We will start by looking at various survey features in the automation controller and continue with Ansible integration topics.

Technical requirements

The following are the technical requirements for this chapter:

- Basic knowledge of monitoring and logging platforms

- General knowledge about security platforms

- Basic knowledge about **IT Service Management** (**ITSM**) tools (Jira and ServiceNow)

- Access to the **Ansible Automation Platform** (**AAP**) environment

All the Ansible artifacts, commands, and snippets for this chapter can be found in this book's GitHub repository at `https://github.com/PacktPublishing/Ansible-for-Real-life-Automation/tree/main/Chapter-14`.

Utilizing surveys and automated inputs

In *Chapter 8*, *Helping the Database Team with Automation*, you learned the advantages of integrating Jira service management with Ansible to automate database operations. Instead of copying the input details from the Jira ticket to the Ansible automation controller, you learned how to integrate with Jira and pass the variables automatically. This enables zero-touch integration without needing to input the details for the automation job, such as the database name, server name, database username, or database tables.

You learned about similar samples in *Chapter 12*, *Integrating Jenkins with Ansible Automation Platform*, where Jenkins calls the Ansible automation controller API to execute the job template. From the Ansible automation controller, you used survey fields and elegant forms to pass such information:

Figure 14.1 – Survey form for the PostgreSQL – Create Database and User Access job template

When we integrate this job template with external tools such as Jenkins, we need to provide the input (database user, database, or table) in a non-interactive way. This is where we used Jenkins (or Jira in the previous example) to call the Job template with all the required information, including the database user, database name, table, and user details as extra variables.

Also, by integrating with the existing ITSM and request management tools, it is possible to enhance the automation system and offload the input form creation (survey form) overhead from the Ansible automation controller.

The following diagram shows the high-level communication that happens between the tools, AAP, and managed nodes or platforms. The external tools can pass all the required information to the AAP API as extra variables and implement the non-interactive workflow:

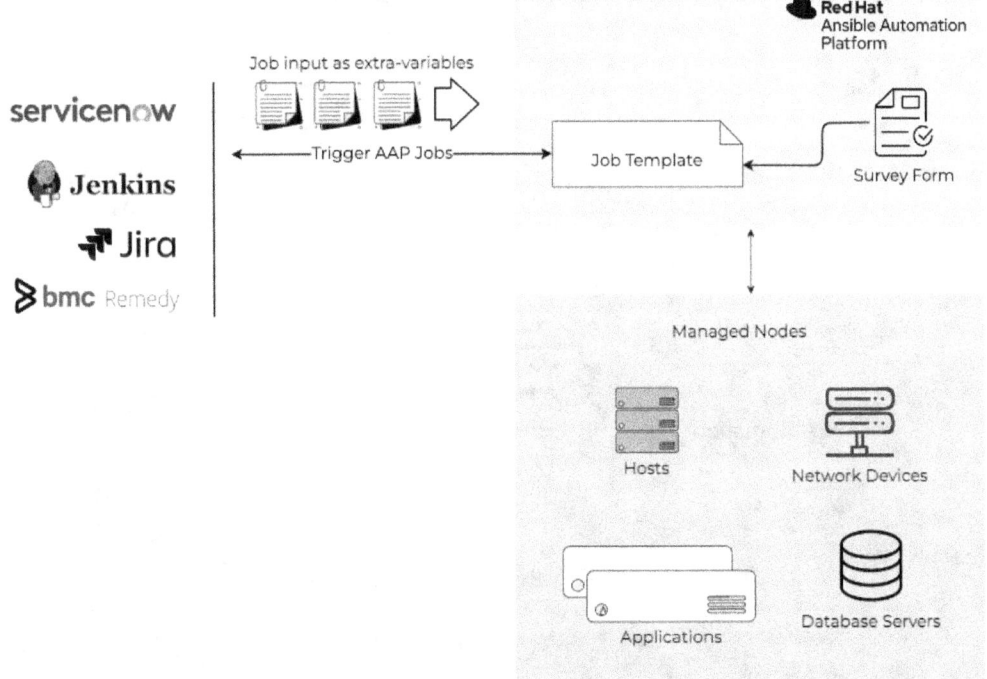

Figure 14.2 – Triggering AAP job templates with extra variables from other platforms

With that, you have learned about one of the best practices in Ansible, called *softcoding* (the opposite of hardcoding). This means accepting the parameters while executing the automation job.

In the following sections, you will learn about the advantages of integrating Ansible with other security and monitoring tools rather than using Ansible natively for such purposes.

Integrating Ansible with monitoring tools

Because Ansible is flexible and can automate most of your day-to-day jobs, it is a common practice to automate every possible use case, even if it is not efficient. One of the so-called non-standard use cases we have learned from the community is using Ansible for monitoring purposes, as follows:

- Monitoring the service or application status in a system

- Running health checks on endpoints (applications, web services, or clusters)

- Monitoring network and security device rules or status

The following diagram shows a typical scenario where Ansible automation jobs are scheduled to run health checks on managed nodes or applications. These jobs can be either running as **cron** jobs from an Ansible control node or as a scheduled job in an Ansible automation controller:

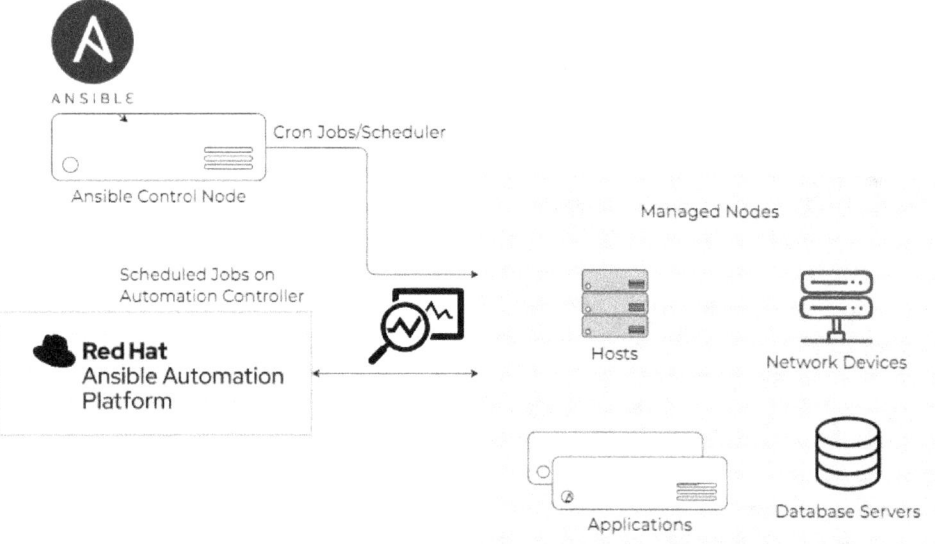

Figure 14.3 – Using scheduled automation jobs for monitoring

This method is possible and easy to implement but is not efficient. It has many disadvantages, as follows:

- You have less control over the frequency of monitoring.

- There's computing overhead on the Ansible control node or the automation controller. The number of monitoring targets will increase this overhead.

- You have less control over the execution time as some jobs may take more time and the next jobs in the queue may be delayed as well.

- You need to create a lot of Ansible playbooks, depending on the target items to be monitored.

Instead of using Ansible playbooks to execute the frequent monitoring job, we can use the standard monitoring tools that are already in place and do their jobs efficiently. The following diagram shows the high-level integration between monitoring and logging tools, ITSM software, and Ansible:

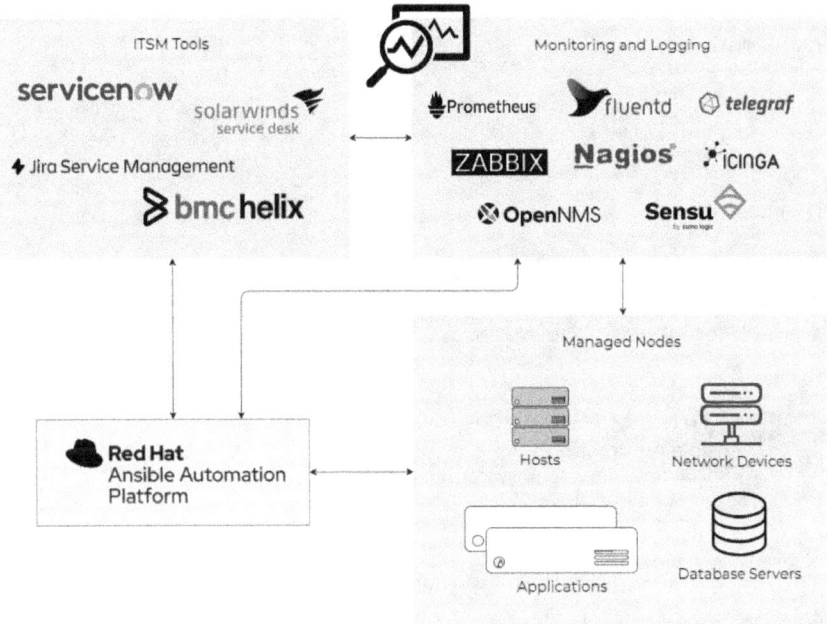

Figure 14.4 – Ansible integration with logging and monitoring tools

The monitoring and logging agents can effectively do their job and report the alerts, incidents, and issues to a central database system. Based on the organization's requirements and infrastructure landscape, the monitoring stack may contain different components and multiple moving parts. The monitoring and logging tools can integrate with your ITSM tools to log the incidents and alerts in the ITSM incident management system. Also, the ITSM tools can be integrated with Red Hat AAP for automated remediations. You will learn more about this in the next section.

The role of Ansible in monitoring

There are several use cases in the monitoring and alerting area for which we can use Ansible to automate the operations:

- Implement automatic job triggers on the automation controller via an API call.
- Automated remediation and deployment based on the API calls from the monitoring tools or ITSM tools.

- Deploy monitoring agents and configurations on the managed nodes.

- Implement configuration hardening based on alerts and API calls.

The following diagram shows the typical integration and workflow based on AAP and automated job executions:

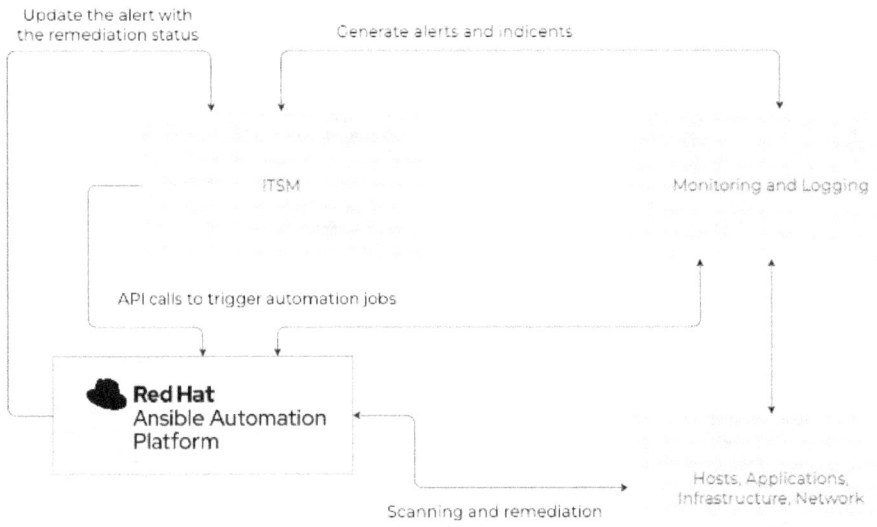

Figure 14.5 – Monitoring and alert workflow in Ansible

In the preceding diagram, the following actions occur:

- By monitoring the agents on the hosts or services, it is possible detect incidents and alerts such as system issues, security incidents, or application health issues.

- Monitoring and logging tools create incidents in ITSM and security monitoring tools.

- Monitoring and logging tools can also trigger alerts (or API calls) directly on the automation controller API to trigger the job.

- ITSM tools trigger API calls based on rules, workflows, and scripts.

- Based on the API call, the automation controller executes the configured job template and remediates the incident.

- The automation controller also updates the incident ticket automatically with the remediation job status and ticket status (for example, close the ticket or update the review status).

Ansible and ServiceNow Integration

To learn more about the integration between Ansible and ServiceNow, read *Ansible + ServiceNow Part 3: Making outbound RESTful API calls to Red Hat Ansible Tower* (`https://www.ansible.com/blog/ansible-servicenow-howto-part-3-making-outbound-restful-api-calls-to-ansible-tower`). Also check `https://www.ansible.com/integrations/it-service-management/servicenow` for additional references on this topic.

You will explore this integration with a use case in the next section.

ServiceNow, Ansible, and zero-touch incident fixes

In this section, you will learn about the important integration and connection points for a zero-touch incident fix with **ServiceNow** (also known as **SNOW**) and Ansible. A typical workflow can be seen in the following diagram:

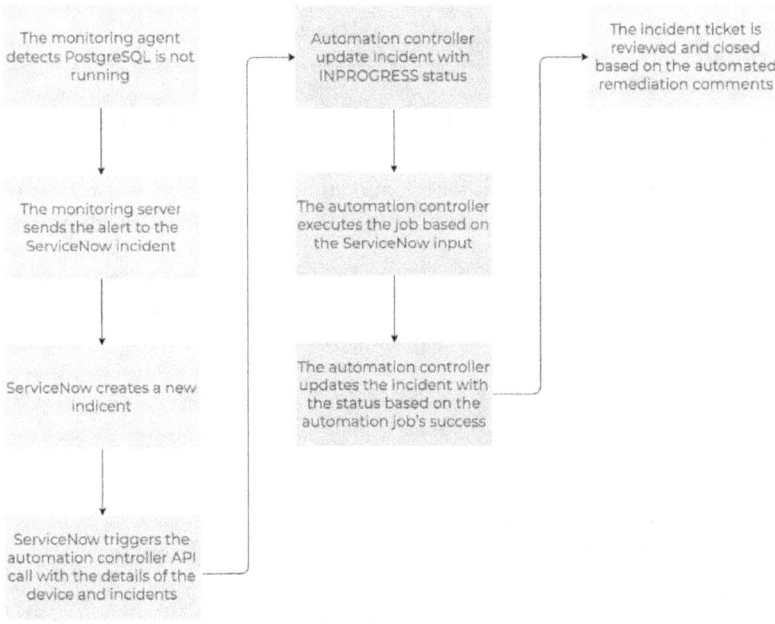

Figure 14.6 – Typical incident automation workflow

Configuring the monitoring agent and ServiceNow instances is outside the scope of this book, so we will only focus on the Ansible portion. It is possible to replace ServiceNow with other ITSM tools such as Jira service management or BMC Remedy, so long as the tool can trigger API calls to AAP for automation jobs.

ServiceNow Workflow

Refer to the ServiceNow documentation at `https://www.servicenow.com/products/workflow.html` to learn more about its workflow automation.

Once an incident has been created in the service management system, the Ansible automation controller will receive the API call based on the workflow's configuration. For example, we have configured a job template called **PostgreSQL - Service Start** in the automation controller, as shown here:

Figure 14.7 – The PostgreSQL – Service Start job template

The Ansible playbook contains a few tasks for updating the incident ticket and also for restarting the PostgreSQL service. The first step is to update the incident ticket with the In progress status, as shown here:

```
# postgresql-service-start.yaml
---
- name: Restarting PostgreSQL Database service
  hosts: "{{ NODES }}"
  vars:
  tasks:

    - name: Update ServiceNow incident as In progress
      servicenow.servicenow.snow_record:
        username: "{{ snow_username }}"
        password: "{{ snow_password }}"
        instance: "{{ snow_instance_name}}"
        state: present
        number: "{{ snow_incident_number }}"
        data:
          work_notes : "Updating PostgreSQL service"
          state: -3
```

Figure 14.8 – Task to update the ServiceNow ticket

Now, add a task that will start the PostgreSQL service:

```
- name: Start service postgresql, if not started
  ansible.builtin.service:
    name: postgresql
    state: started
  register: psql_service_status
```

Figure 14.9 – Task for starting the PostgreSQL service

Now, add a task that will update the ticket to show that it's been resolved:

```
- name: Update ServiceNOW incident
  servicenow.servicenow.snow_record:
    username: "{{ snow_username }}"
    password: "{{ snow_password }}"
    instance: "{{ snow_instance_name}}"
    state: present
    number: "{{ snow_incident_number }}"
    data:
      work_notes : "PostgreSQL Service has been started"
      state: 0
  when:
    - psql_service_status.state == 'started'
```

Figure 14.10 – Updating the ticket's status to resolved

Expand the playbook with more validations and ticket status updates to meet the ITSM processes in your organization and requirements. If required, you can even add the task of updating the ticket as closed.

> **The Life Cycle of an Incident in ServiceNow**
>
> Refer to the ServiceNow documentation (https://docs.servicenow.com/en-US/bundle/sandiego-it-service-management/page/product/incident-management/concept/c_IncidentManagementStateModel.html) to learn more about the incident life cycle. Also, refer to the Ansible blog post at https://www.ansible.com/blog/ansible-servicenow-opening-and-closing-tickets to learn more about ServiceNow integration with Ansible.

There will be cases where someone has stopped the PostgreSQL service for some activities; Ansible should be aware of this. It is possible to create a maintenance tracking system using your ITSM tool or third-party solutions, where system maintenance details can be updated with time slots to efficiently handle the incident ticket, as follows:

1. The ITSM tool detects the system maintenance window and updates the ticket as a false alarm.

2. Add tasks to the Ansible playbook to check against this maintenance window and update the alert as a false alarm if the system is in that maintenance window.

With that, you've explored some of the possibilities of integrating Ansible with various monitoring and logging systems. In the next section, you will learn how to use Ansible with security automation solutions.

Ansible for security automation

Security hardening is the practice of securing the hosts, networking devices, and applications by reducing the attack surface. There are multiple ways to implement security hardening, such as configuring the system appropriately, installing the latest version of the software (or firmware), or disabling unwanted configurations. Organizations use different security benchmarking methods and standards based on the requirements. **Center for Internet Security** (**CIS**) is one of the well-known organizations that provides the necessary enterprise standard benchmarks and CIS controls (`https://www.cisecurity.org/about-us`).

Log in and download the benchmark documents for operating systems or platforms for free, as shown in the following screenshot:

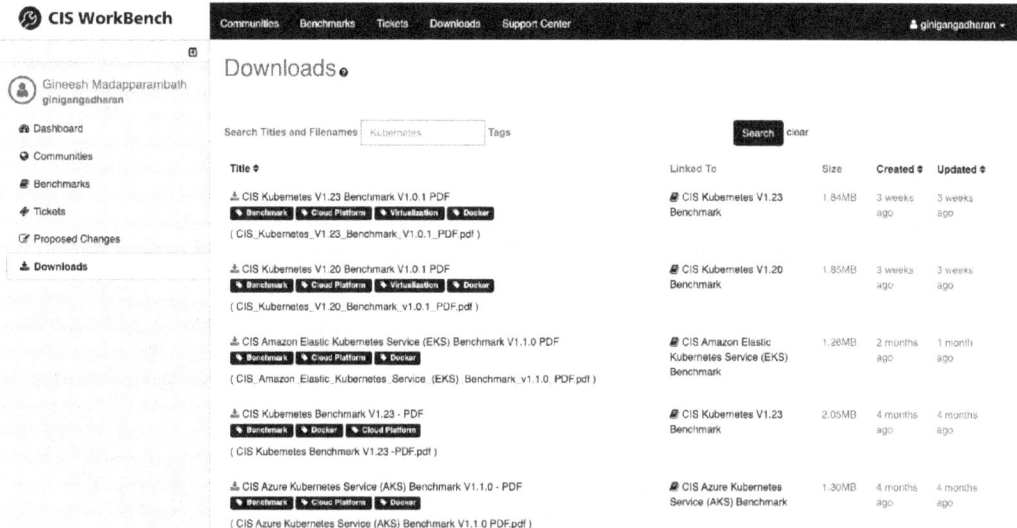

Figure 14.11 – CIS Benchmark download page for Kubernetes

It is not easy to configure the systems based on these benchmarks as hundreds of rules and configuration items must be executed to secure the system. Organizations use different methods to achieve these configurations, such as shell scripts, PowerShell scripts, and Ansible automation. Due to their flexibility and modularity, Ansible playbooks are much more efficient and maintainable.

It is possible to execute the CIS hardening playbooks frequently using Ansible or an automation controller (scheduled jobs) to ensure the system has a good security posture. However, this method will not be effective for detecting and remediating the threats and incidents on time.

As you learned in the previous section, it is also possible to utilize the power of security tools and solutions and integrate them with Ansible to automate various security operations, as follows:

- **Security information and event management (SIEM)**
- **Intrusion detection & prevention system (IDPS)** tasks
- **Privileged access management (PAM)** operations
- **Endpoint protection platform (EPP)** tasks
- Firewall operations

The following diagram shows a typical threat detection and automation workflow:

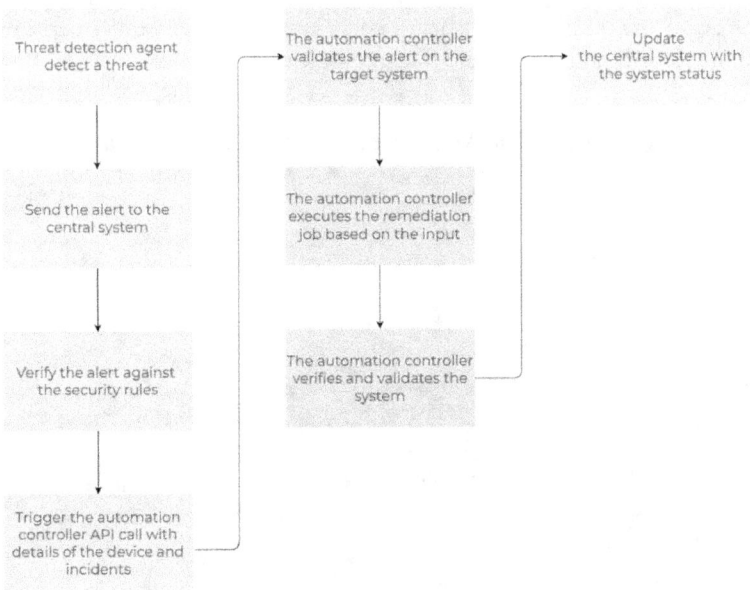

Figure 14.12 – Threat detection and automated remediation

By using the existing security solutions, it is possible to offload the detection and response overhead from Ansible but still use the power of Ansible to remediate the security issue.

Ansible has a large collection of community modules and collections available to support the automation of security devices and solutions, as follows:

- Cisco ASA

- FortiGate

- Palo Alto

- F5

- CheckPoint

- CyberArk

- Trend Micro

> **Ansible Collection for Security Automation**
>
> Refer to the Ansible collection index (`https://docs.ansible.com/ansible/latest/collections/index.html`) to find the security-related collections. Also, check out *Security Automation with Red Hat Ansible Automation Platform* (`https://access.redhat.com/articles/4001711`) to learn more about the platform's compatibility matrix. Finally, please read the blog post *Getting started with Ansible security automation: investigation enrichment* (`https://www.ansible.com/blog/getting-started-with-ansible-security-automation-investigation-enrichment`) to understand the security automation and integration with AAP.

The process of automatic IP address blocking can be seen in the following diagram:

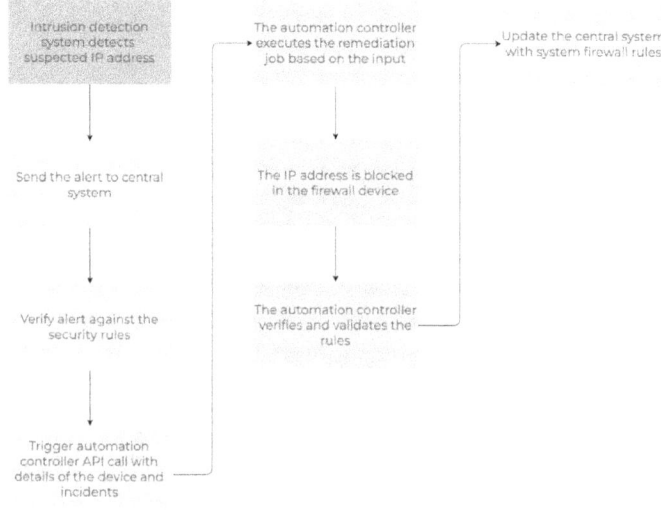

Figure 14.13 – Automated IP address blocking

With the help of Ansible modules, most of your security-related operations can be automated efficiently.

For example, adding an **access control list** (ACL) rule to the Cisco ASA device can be automated using the `cisco.asa.asa_acls` module, as shown in the following screenshot:

```
- name: Add new ACL Entry and Merge configuration with device configuration
  cisco.asa.asa_acls:
    config:
      acls:
        - name: "{{ acl_identifier }}"
          acl_type: "{{ acl_type }}"
          aces:
          - grant: "{{ acl_action }}"
            #line: 1
            protocol_options:
              tcp: true
            source:
              address: "{{ acl_entry_source_ip }}"
              netmask: "{{ acl_entry_source_mask }}"
            destination:
              object_group: "{{ asa_object_group_name }}"
            #log: default
    state: merged
  register: acl_status
```

Figure 14.14 – Adding an access control rule to the Cisco ASA device

The following screenshot shows a simple Ansible task using the `fortinet.fortios.fortios_firewall_address` module to block the IP address in the FortiGate device:

```
- name: Create {{ firewall_policy_address_entry_to_add }} Entry
  delegate_to: localhost
  fortinet.fortios.fortios_firewall_address:
    host:  "{{ fortigate_host_ip }}"
    username: "{{ fortigate_username }}"
    password: "{{ fortigate_password }}"
    vdom:  "{{ fortigate_vdom }}"
    https: "{{ fortigate_ssl_use }}"
    ssl_verify: "{{ fortigate_ssl_verify }}"
    state: "present"
    firewall_address:
      allow_routing: "disable"
      #color: "6"
      comment: "{{ firewall_policy_address_comment }}"
      name: "{{ firewall_policy_address_name }}"
      policy_group: "{{ firewall_policy_address_group }}"
      subnet: "{{ firewall_policy_address_entry_to_add }}/32"
      type: "ipmask"
      visibility: "enable"
```

Figure 14.15 – Blocking the IP address in the FortiGate device

Similarly, the following screenshot shows adding a security rule to the Palo Alto device using the `paloaltonetworks.panos.panos_security_rule` module:

```
- name: Create Security Rule
  paloaltonetworks.panos.panos_security_rule:
    provider: "{{ panos_provider }}"
    rule_name: "{{ panos_rule_name }}"
    source_ip: "{{ panos_source_ip_address.splitlines() | default('any')  }}"
    source_user: "{{ panos_source_user.splitlines() | default('any') }}"
    destination_ip: "{{ panos_destination_ip_address.splitlines() }}"
    category: "{{ panos_url_category.splitlines() | default('any') }}"
    application: "{{ panos_application_category.splitlines() |  default('any') }}"
    service: "{{ panos_service.splitlines() }}"
    group_profile: "{{ panos_group_profile | default('None') }}"
    action: "{{ panos_rule_action }}"
    rule_type: "{{ panos_rule_type }}"
    log_start: "{{ panos_log_start | bool }}"
    log_end: "{{ panos_log_end | bool }}"
```

Figure 14.16 – Adding a security rule to the Palo Alto device

Please refer to the *Further reading* section for more resources about security automation modules in Ansible.

In the next section, you will learn about the workflow templates in the automation controller.

Ansible workflow templates

It is possible to create any number of tasks in a single playbook and make it a long workflow. For example, a Linux operating system job template can include the following tasks:

1. Create a VM snapshot before you start patching.

2. Save the configuration file backups.

3. Stop the services inside the system.

4. Perform various Linux operating system patching tasks.

5. Reboot the system.

6. Wait for the system to boot up and start the necessary services.

7. Handle the VM snapshot restore operation in the same job if the VM reboot is not successful.

Note that most of the tasks can be reused as individual jobs for creating snapshots, stopping services, or configuration backup.

Instead of developing long, complex job templates, utilize the workflow templates in the automation controller to create modular job workflows and handle tasks based on success/failure status. Workflow templates are created by stitching multiple job templates together to achieve a bigger or more complex workflow.

The following diagram shows the high-level differences between a job template and a workflow template in an Ansible automation controller:

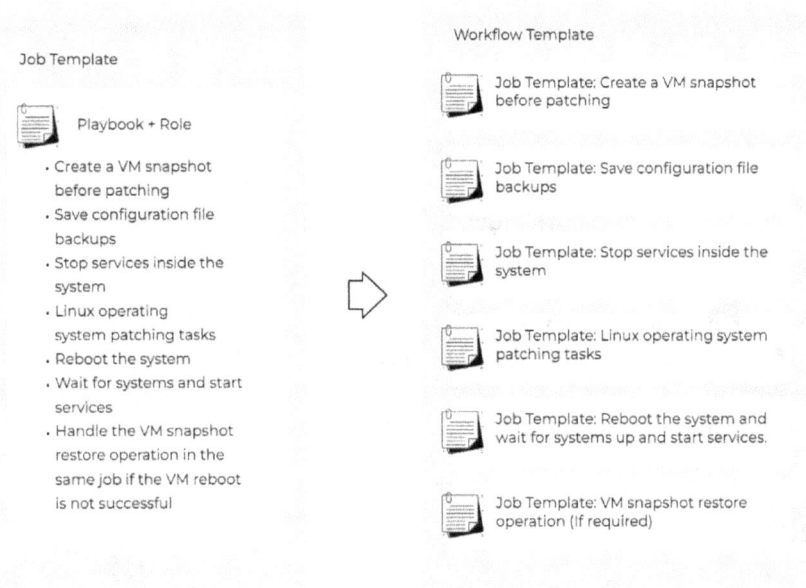

Figure 14.17 – Job templates versus workflow templates

The following screenshot shows a simple workflow template for Linux operating system patching. Here, we have multiple tasks, as follows:

1. Create a snapshot job before patching.

2. Create a job template to stop the important and dependency services inside the system.

3. Perform operating system patching tasks.

4. Reboot.

5. If the reboot succeeded, all the required services will be started.

6. If the reboot failed, the job template for restoring the snapshot will be executed:

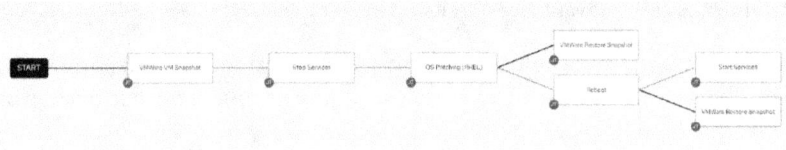

Figure 14.18 – Workflow template for operating system patching

This modular approach helps you implement complex automation workflows by reusing the job templates. The following screenshot shows a similar workflow where the operating system-specific job templates have been replaced and other job templates are being reused:

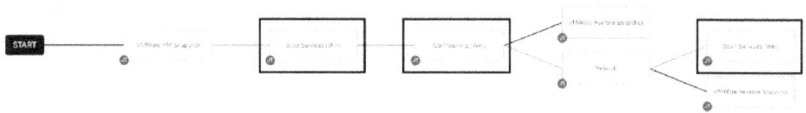

Figure 14.19 – Windows patching workflow

Ansible Workflow Job Templates

Please refer to `https://docs.ansible.com/automation-controller/4.1.2/html/userguide/workflow_templates.html` to learn more about the features of workflow templates, such as webhooks and integration options.

Summary

In this chapter, you learned about the importance of simplifying Ansible automation. You explored the advantages of survey features in the automation controller and soft coding opportunities to accept parameters dynamically. This method helps you reuse the same playbook and job templates for different devices and scenarios.

Then, you learned about the integration options between Ansible and other monitoring and alerting tools to implement zero-touch incident fixes. It is possible to utilize the existing ITSM tools and workflows to trigger Ansible automation controller jobs based on the rules and conditions. By utilizing the power of monitoring tools and the automation capabilities of Ansible, an efficient monitoring and remediation system can be implemented.

After that, you explored similar integration opportunities within the security domain for automating threat detection and automated remediation using an automation controller. The security modules and collections for Ansible automation help you implement complex and enhanced security automation solutions.

Finally, you learned about the workflow templates in an automation controller, the modular way of creating workflows, and the advantages of reusing job templates.

In the next chapter, you will learn how to automate non-standard and non-supported platforms using Ansible. You will explore some of the useful modules and Ansible custom module creation basics as well.

Further reading

To learn more about the topics that were covered in this chapter, take a look at the following resources:

- *Ansible for Security Automation*: `https://www.ansible.com/use-cases/security-automation`

- *Ansible blog on security automation*: `https://www.ansible.com/blog/topic/security-automation`

- *Security & Compliance Using Ansible*: `https://www.ansible.com/use-cases/security-and-compliance`

- *Ansible Collections*: `https://docs.ansible.com/ansible/latest/collections/index.html#list-of-collections`

- *Applications in Ansible automation controller*: `https://docs.ansible.com/automation-controller/latest/html/userguide/applications_auth.html`

- *Ansible for Network Automation*: `https://docs.ansible.com/ansible/latest/network/index.html`

15

Automating Non-Standard Platforms and Operations

So far, you have learned about the different systems and platforms (managed nodes) such as Linux systems, Microsoft Windows systems, network devices, and cloud platforms, all of which can be automated using Ansible. You have used different methods and protocols to connect to the target system or platform to execute the automated operations. You do not need to understand how complex operations are running inside these systems because Ansible will understand and take appropriate actions on the target systems to reach the desired state.

What if there are no standard Ansible modules or connection methods available for a specific operation on the managed node? Or what if the module is missing some feature you are looking for? For example, let's say you want to automate a machine that doesn't have the supported Python version installed, and you want to use Ansible to automate this Python installation or run raw commands on a firewall device that does not have an appropriate module available.

Since Ansible is flexible and modular, you have multiple ways to overcome these challenges. In this chapter, you will learn how to automate such non-standard operations using Ansible.

In this chapter, we will cover the following topics:

- Executing low-down raw commands
- Using raw commands for network operations
- Using API calls for automation
- Creating custom modules for Ansible

We will start with the Ansible `raw` module and the lifesaving situations where the `raw` module helps automate non-standard use cases. Then, you will learn how to use API calls to automate operations when Ansible modules or SSH-based connections are not available. You will also explore custom Ansible modules and how to develop simple custom modules using Python.

Technical requirements

You will need the following technical requirements for this chapter:

- A Linux machine for the Ansible control node

- Basic knowledge of commands for network devices (for example, FortiOS and Cisco ASA)

- Basic knowledge of REST API endpoints and their usages

- Basic knowledge of bash scripting and the Python programming language

All the Ansible artifacts, commands, and snippets for this chapter can be found in this book's GitHub repository at `https://github.com/PacktPublishing/Ansible-for-Real-life-Automation/tree/main/Chapter-15`.

Executing low-down raw commands

So far, you have learned about different Ansible modules and collections for developing automation content. Most of these modules expect a supported Python version on the target node (or PowerShell for Microsoft Windows) to execute the automation scripts in the backend (refer to Ansible's managed node requirements documentation for more details: `https://docs.ansible.com/ansible/latest/installation_guide/intro_installation.html#managed-node-requirements`). When there is no required Python version or packages available, then you need to install it manually (or using some scripts) on every node before Ansible can automate the tasks on that nodes. When several managed nodes need to onboard to Ansible, then it will be a tedious task to log into each node and install these dependencies manually. In such situations, it is possible to execute the raw commands (such as the ones you use inside a bash script or PowerShell script) to install and configure the systems, as follows:

```
sudo yum install python36
```

Before using the low-down raw commands, please note the following:

- Low-down raw commands are the same commands that you use inside the system to handle the operations manually.

- These low-down raw commands will not go through the Ansible module system; instead, they will be executed through the configured remote shell.

- You will have to assume the output, success, and error conditions of the commands.

It is possible to automate such operations via Ansible by executing low-down raw commands using the `ansible.builtin.raw` module.

> **The Difference in Ansible Network Automation**
>
> We cannot install Python on network devices. Because of that, network automation is different in the backend. Refer to *Chapter 6, Automating Microsoft Windows and Network Devices*, to learn more.

The `ansible.builtin.raw` module helps execute raw commands on the target nodes over an SSH connection. It is not a best practice to use raw commands on the target nodes as the command's result will not be predictable; therefore, Ansible will not be able to handle the validations and errors like other standard modules. However, the `ansible.builtin.raw` module is useful for some special cases, as you'll learn in the following sections.

Installing Python using the raw module

As I mentioned in the introduction of this chapter, what if your managed node doesn't have Python installed and you want to use Ansible to automate the Python installation? Let's learn how to create a playbook for installing Python using Ansible.

We assume that the required package repositories have been configured on the target node as per standard system configuration.

The following screenshot shows the sample playbook for installing and verifying the Python version:

```
# install-python.yaml
---
- name: Installing Python on target machine
  hosts: "{{ NODES }}"
  gather_facts: false
  tasks:
    - name: Install Latest Python package
      ansible.builtin.raw: sudo yum -y install python36

    - name: Verify Python version
      ansible.builtin.raw: python3 -V
      register: python_version

    - name: Display installed Python version.
      ansible.builtin.debug:
        msg: "Installed Python version: {{ python_version.stdout_lines }}"
```

Figure 15.1 – Playbook for installing and verifying Python using the raw module

Note the `gather_facts: false` line in the preceding screenshot; this is a fact-gathering task that relies on Python. The playbook will not work otherwise.

The playbook's installation output can be seen in the following screenshot:

```
[ansible@ansible Chapter-15]$ ansible-playbook install-python.yaml -e "NODES=node1"

PLAY [Installing Python on target machine] ***************************************************

TASK [Install Latest Python package] *********************************************************
changed: [node1]

TASK [Verify Python version] *****************************************************************
changed: [node1]

TASK [Display installed Python version.] *****************************************************
ok: [node1] => {
    "msg": "Installed Python version: ['Python 3.6.8']"
}

PLAY RECAP ***********************************************************************************
node1                      : ok=3    changed=2    unreachable=0    failed=0    skipped=0    rescued=0    ignored=0
```

Figure 15.2 – Output of the Python installation playbook

Once you have Python installed on the target nodes, use any other Ansible modules as usual. The `ansible.builtin.raw` module also supports Windows, network, and firewall devices for executing low-down raw commands.

Ansible Raw Module

The raw module is part of `ansible-core` and is included in your installation by default. Refer to the documentation at `https://docs.ansible.com/ansible/latest/collections/ansible/builtin/raw_module.html` to learn more. As mentioned earlier, the `raw` modules are only used in special situations; otherwise, use the `ansible.builtin.shell` or `ansible.builtin.command` module to execute shell commands (check out the alternative `ansible.windows.win_command` and `ansible.windows.win_shell` modules for Windows machines).

In the next section, you will learn how to use the `ansible.builtin.raw` module for network automation when there are no appropriate modules available to use.

Using raw commands for network operations

Ansible has a large collection of modules and plugins to support most of the well-known network and firewall devices. The following figure shows some of the supported network devices and brands for network automation:

Figure 15.3 – Supported brands for Ansible network automation
(source: https://www.ansible.com/integrations/networks)

However, there will be situations where you must overcome the following challenges:

- No supported Ansible module is available to execute a specific network or firewall operation.

- There is a module available for operation but the specific feature or parameter you are looking for doesn't exist in the current module version.

- The supported module for the operation is using a slow method compared to the CLI command on the device.

> **Network Collections in the Ansible Collection Index**
>
> Refer to the documentation page at `https://docs.ansible.com/ansible/latest/collections/index.html` to see the network collections in the Ansible collection index.

To overcome such situations, use the same `ansible.builtin.raw` module and execute the raw commands on the target network devices over an SSH connection.

In the next section, you will learn how to use the `ansible.builtin.raw` module to execute the necessary operations in FortiOS and Cisco devices, such as taking a configuration backup, upgrading software, and more.

Backup FortiOS configurations

The `fortinet.fortios` Ansible collection contains more than 600 modules for managing and operating devices or appliances with FortiOS. Let's look at a simple operation, such as taking a configuration backup from a FortiOS device.

The following screenshot shows the inventory variables for the FortiOS device connection:

```
[fortios]
fg01 ansible_host=192.168.57.125

[fortios:vars]
ansible_user=admin
ansible_ssh_pass='Admin#123'
ansible_host_key_checking=false
ansible_network_os=fortinet.fortios.fortios
ansible_connection=ansible.netcommon.httpapi
ansible_httpapi_use_ssl=True
ansible_httpapi_validate_certs=False
```

Figure 15.4 – Inventory variables for the FortiOS connection

We can use `fortinet.fortios.fortios_monitor_fact` to collect and save the backup to a system, as shown here:

```
- name: Backup global settings on FortiOS device
  fortinet.fortios.fortios_monitor_fact:
    selector: 'system_config_backup'
    vdom: 'root'
    params:
      scope: 'global'
```

Figure 15.5 – Backing up FortiOS using the fortios_monitor_fact module

> **Ansible fortinet.fortios Collection**
>
> Find the modules and plugin containers in the `fortinet.fortios` Ansible collection at `https://galaxy.ansible.com/fortinet/fortios`.

However, the configuration backup needs to be transferred to the file server (for example, a TFTP server) securely using another task in Ansible. Some organizations may want to follow the standard and legacy way of backing up, which involves backing up from the device itself without copying to the Ansible control node (or automation controller). In such cases, use the `raw` module to execute the raw commands, as shown here:

```
•••
- name: FortiGate Configuration Backup
  raw: |
        execute cfg save
        execute backup config tftp {{ backup_filename }} {{ tftp_server }}
  register: tftp_copy_status
```

Figure 15.6 – FortiOS backup using raw commands

In the preceding example, we used raw CLI commands to copy the configurations directly to the TFTP server.

Remember to modify the inventory variables to use the default SSH connection instead of ansible. netcommon.httpapi:

```
•••
[fortios]
fg01 ansible_host=192.168.57.125

[fortios:vars]
ansible_user=admin
ansible_ssh_pass='Admin#123'
ansible_host_key_checking=false
#ansible_network_os=fortinet.fortios.fortios
#ansible_connection=ansible.netcommon.httpapi
#ansible_httpapi_use_ssl=True
#ansible_httpapi_validate_certs=False
```

Figure 15.7 – Inventory variables for the FortiOS connection without httpapi

Ansible will use the default SSH connection and execute the raw commands on the target FortiOS device.

FortiOS software upgrade

Let's take a look at another example of a FortiOS software upgrade or patching. We have a module called fortios_system_federated_upgrade to do this task in the standard Ansible way. However, this module needs the new software image to be uploaded from the localhost (which is the Ansible control node or the Ansible execution environment), so first, you need to copy this image from some location to the localhost. Also, if there is a restriction in the network stopping you from copying large images such as firmware files, then the only choice is to execute the legacy raw commands inside the FortiOS device, as shown in the following screenshot:

```
● ● ●
- name: FortiGate Update Software
  raw: |
        execute restore image tftp {{ fortios_image_filename }} {{ tftp_server }}
        Y
  register: image_update_status
```

Figure 15.8 – Running the FortiOS software upgrade using the raw module

The `execute restore image` command will ask you to confirm the image upgrade; Y on the second line is the input to that confirmation question.

The `fortios_system_federated_upgrade` module will take care of all such confirmations and programmatic upgrades. Keep in mind that the raw command method should only be followed in special cases.

Raw commands on Cisco devices

Similarly, it is possible to use raw commands on any device that supports an SSH connection. Let's take a look at another example for Cisco ASA device backup.

The `cisco.asa.asa_config` module is part of the `cisco.asa` Ansible collection (`https://galaxy.ansible.com/cisco/asa`). Use this module to configure the Cisco ASA device and take a configuration backup in the standard Ansible way. However, the backup will be downloaded to your `localmachine` (Ansible control node or the Ansible execution environment), so it will need to be transferred to the destination file server (for example, a TFTP server). The backup may contain sensitive information, so the organization may prefer to copy the backup directly to the TFTP server (or other supported target file servers).

In such situations, utilize the same `ansible.builtin.raw` module to execute the Cisco ASA commands on the target Cisco devices. But before you use the `ansible.builtin.raw` module, you need to check the respective Ansible collection (for example, the `cisco.asa` collection here) for any modules to execute the raw network commands in the Ansible way. In this case, the `cisco.asa.asa_command` module can be used to execute the Cisco ASA raw commands, as shown here:

```
● ● ●
- name: Take Cisco ASA Backup
  cisco.asa.asa_command:
    commands:
    - write memory
    - copy /noconfirm running-config tftp://{{ tftp_server }}/{{ backup_filename }}
```

Figure 15.9 – Cisco ASA backup using raw commands

Similarly, you can utilize the `cisco.ios.ios_command` and `cisco.nxos.nxos_command` modules on Cisco IOS and Cisco Nexus OS devices, respectively. If there is no device-specific module available to execute raw commands, then utilize `ansible.builtin.raw`, as you learned earlier.

The raw module can be used on any device that supports SSH and command-line execution. But for devices or platforms that don't support SSH connections, or no such commands are available, then it is possible to use API calls. In the next section, you will learn how to use API calls to automate operations from Ansible.

Using API calls for automation

In *Chapter 6*, *Automating Microsoft Windows and Network Devices*, you learned about the different ways Ansible can talk to managed nodes, platforms, or applications. So as long as there is a supported method to access the target system, it is possible use Ansible to automate these tasks. In the previous chapters, you learned about the integration between Ansible and other tools such as Jira, ServiceNow, Kubernetes, public or private cloud platforms, and so on. For such platforms, most of those modules use HTTP/HTTPS API calls to execute operations. This means that if there are no modules available to automate your operations but there is an API method, you can use the same raw API calls from your Ansible playbook.

> **Python SDK and API Calls**
>
> Please remember that not all modules use direct or native API calls to execute the operations; some modules use Python libraries and **software development kits** (**SDKs**) to implement these tasks. For example, the FortiOS modules used to use the `fortiosapi` Python library, but now, `httpapi` is the preferred way to execute the playbooks. Read more about **the httpapi** plugins at `https://docs.ansible.com/ansible/latest/plugins/httpapi.html`.

For example, the `amazon.aws.ec2_instance_info` module (part of the Ansible `amazon.aws` collection) helps gather information about EC2 instances in AWS. Its usage is straightforward, as shown in the following screenshot, and you do not need to worry about the complex API calls to the AWS EC2 endpoints:

```
- name: Gather EC2 insance details
  amazon.aws.ec2_instance_info:

- name: Gather information about instances in Singapore
  amazon.aws.ec2_instance_info:
    filters:
      availability-zone: ap-southeast-1
```

Figure 15.10 – Gathering AWS EC2 information using amazon.aws.ec2_instance_info module

Let's assume you have a requirement to collect some additional requirement that is not available as part of the module's output. In this case, it is also possible to use the AWS EC2 API endpoints (`https://docs.aws.amazon.com/AWSEC2/latest/APIReference/Welcome.html`) to utilize the full features of API calls.

In the following sections, you will learn how to use API calls from Ansible to automate the platform or applications.

Automating a ToDo app using Ansible

In this section, you will automate a simple ToDo application using the API provided. This demonstration will help you understand how to make API calls using Ansible and how to handle the API call output appropriately. This section will help you understand how to handle API calls for any other service (for example, the cloud, network devices, software applications, and so on) by following the product's API documentation.

Introducing the Ansible uri module

The `ansible.builtin.uri` module is used to interact with the HTTP and HTTPS endpoints and implement web-related operations using Ansible. The `ansible.builtin.uri` module supports multiple authentication mechanisms, such as Digest, Basic, and WSSE HTTP (`https://docs.ansible.com/ansible/latest/collections/ansible/builtin/uri_module.html`). The `ansible.builtin.uri` module can be used for several use cases, as follows:

- Calling an AWS API service to fetch EC2 instance details
- Calling a Jira API to update a ticket
- Fetching details from a web server
- Verifying a web service health check

Now, let's learn how to use the `ansible.builtin.uri` module to interact with a ToDo application API.

Assume you have a ToDo application running, and the API is available at `http://todo-app.example.com:8081/api` without any authentication. Test the API using the default `curl` command, as shown here:

```
$ curl http://todo-app.example.com:8081/api/todos
[{"id":1,"title":"Send weekly report to team","description":"Weekly health check report","completed":false},
{"id":2,"title":"Arrange team dinner","description":"Check for places","completed":false},{"id":3,"title":"Schedule
meeting with John for security audit","description":"Pending long time","completed":false}]
```

Figure 15.11 – Testing the ToDo API's access

Let's use Ansible and the `ansible.builtin.uri` module (`https://docs.ansible.com/ansible/latest/collections/ansible/builtin/uri_module.html`) to handle the ToDo application.

Here, you have `Chapter-15/todo-app.yaml` (refer to the GitHub repository) whose content is as follows:

```
# todo-app.yaml
---
- name: Managing todo application using API
  hosts: localhost
  gather_facts: false
  become: false
  vars:
    todo_app_ur: 'http://todo-app.example.com:8081'
    todo_app_healthcheck: 'health'
```

Figure 15.12 – ToDo app playbook details

In the preceding screenshot, we can see the following:

- `hosts: localhost`: This is used because we are running this API call from a localhost machine.

- `gather_facts: false`: This is used because we do not require any facts from `localhost` (enable this if there is a requirement to use Ansible facts).

- `become: false`: This is used because the API calls don't require privileged access.

The `method` parameter uses GET as the default value in the `uri` module. Since we are fetching the data from a URL, the default GET HTTP method will be used in the following example. The first task is to fetch the health status of the API and then print the output with a debug module, as shown in the following screenshot:

```
tasks:
  - name: Check that you can connect (GET) to a page and it returns a status 200
    uri:
      url: "{{ todo_app_ur }}/{{ health_check }}"
      return_content: yes
      status_code: 200
    register: health_status

  - name: Display health check status
    debug:
      msg: "{{ health_status.content }}"
```

Figure 15.13 – API health check task

The `status_code: 200` parameter helps validate the API call and task success since you are expecting `OK success status`; you do not need to add additional validation tasks. If the status code is anything other than `200`, the task will fail.

The following screenshot shows the sample output for when you execute the playbook:

```
...<omitted>...
TASK [Display health check status]
*********************************************************************************
ok: [localhost] => {
    "msg": {
        "changed": false,
        "connection": "close",
        "content": "{\"uptime\":2438.676111528,\"message\":\"OK\",\"timestamp\":1655004678873}",
        "content_length": "66",
        "cookies": {},
        "cookies_string": "",
        "date": "Sun, 12 Jun 2022 03:31:18 GMT",
        "elapsed": 0,
        "failed": false,
        "msg": "OK (66 bytes)",
        "redirected": false,
        "status": 200,
        "url": "http://todo-app.example.com:8081/health"
    }
}
...<omitted>...
```

Figure 15.14 – API health check sample output

If you only need the returned content (result) of the API call, then modify the `msg` parameter by setting it to `msg: "{{ health_status.content }}"`. By doing this, you will get a more accurate result, as shown here:

```
TASK [Display health check status]
*********************************************************************************
ok: [localhost] => {
    "msg": {
        "message": "OK",
        "timestamp": 1655004693105,
        "uptime": 2452.908586769
    }
}
```

Figure 15.15 – API call returned content

The `uri` call will return detailed JSON output. Filter out the content as needed.

> **HTTP Request Methods**
>
> Depending on the API or the web endpoint, different HTTP methods will be available, such as GET, POST, PUT or DELETE. Refer to `https://www.w3schools.com/tags/ref_httpmethods.asp` to understand more about HTTP methods.

The next task is to fetch the items in the ToDo list, as shown here:

```
- name: Get ToDo Items
  uri:
    url: "{{ todo_app_ur }}/api/todos"
    return_content: yes
    status_code: 200
  register: todo_items

- name: Display items
  debug:
    msg: "{{ todo_items.content }}"
```

Figure 15.16 – Fetching the items in the ToDo app using the API

When you execute the playbook, you will get the list of ToDo items, as follows:

```
<omitted>...
TASK [Display items]
**************************************************************************************************************
****
ok: [localhost] => {
    "msg": [
        {
            "completed": false,
            "description": "Weekly health check report",
            "id": 1,
            "title": "Send weekly report to team"
        },
        {
            "completed": false,
            "description": "Check for places",
            "id": 2,
            "title": "Arrange team dinner"
        },
        {
            "completed": false,
            "description": "Pending long time",
            "id": 3,
            "title": "Schedule meeting with John for security audit"
        }
    ]
}
<omitted>...
```

Figure 15.17 – ToDo items fetched using the API call

The individual items can be handled by filtering the JSON output. This will allow you to retrieve the exact content as needed.

For creating or updating new items, use the POST HTTP method, along with content to post and other details as needed. Let's add a new item to the ToDo list by using the POST method and the body content. The first step is to prepare the data to post. In this case, you have a dictionary variable called new_item that contains item details, as shown in the following screenshot:

```
vars:
  todo_app_ur: 'http://todo-app.example.com:8081'
  health_check: 'health'

  new_item:
    title: Learn API call using Ansible
    description: A new task added by Ansible
    completed: false
```

Figure 15.18 – Variable for the new ToDo item

The task to POST the content to the ToDo list requires a few more parameters, as shown here:

```
- name: Add a new item in ToDo list
  uri:
    url: "{{ todo_app_ur }}/api/todos"
    method: POST
    return_content: yes
    status_code: 201
    body_format: json
    body: "{{ new_item }}"
  register: item_add_status

- name: Display items
  debug:
    msg: "{{ item_add_status }}"
```

Figure 15.19 – Adding a new task to the ToDo app via an API call

In the preceding screenshot, we can see the following:

- status_code: 201–201 is the status code for created. If you do not mention status_code, Ansible will use a value of 200 (default) and the task will show as failed (for example, "msg": "Status code was 201 and not [200]: OK (unknown bytes)").

- body_format: json: The default value is raw. This is where you are passing the JSON formatted input.

The following screenshot shows the sample output for the tasks when you execute the playbook:

```
● ● ●
<omitted>...
TASK [Add a new item in ToDo list]
*******************************************************************************************************
ok: [localhost]

TASK [Display items]
*******************************************************************************************************
****
ok: [localhost] => {
    "msg": {
        "changed": false,
        "connection": "close",
        "content": "{\"id\":12,\"title\":\"Learn API call using Ansible\",\"description\":\"A new task added by
Ansible\",\"completed\":false}",
        "content_type": "application/json",
        "cookies": {},
        "cookies_string": "",
        "date": "Sun, 12 Jun 2022 04:21:50 GMT",
        "elapsed": 0,
        "failed": false,
        "json": {
            "completed": false,
            "description": "A new task added by Ansible",
            "id": 12,
            "title": "Learn API call using Ansible"
        },
        "msg": "OK (unknown bytes)",
        "redirected": false,
        "status": 201,
        "transfer_encoding": "chunked",
        "url": "http://todo-app.example.com:8081/api/todos"
    }
}
<omitted>...
```

Figure 15.20 – Output of the tasks

This simple demonstration explains the different ways to interact with an API endpoint using Ansible. Depending on the API endpoint you want to manage, explore more automation use cases and workflows.

Some of the API calls may require you to prepare complex body content, and in such cases, use Ansible Jinja2 templates; this will be covered in the next section.

> **The Ansible uri Module and REST APIs**
>
> The uri module is a multipurpose module with parameters available for most API operations, including credentials, certificates, agent configuration, and more. Check out the module documentation at https://docs.ansible.com/ansible/latest/collections/ansible/builtin/uri_module.html to learn more.

Interacting with the Akamai API

Let's assume there's a situation where you want to automate a DNS management device such as Akamai and no supported Ansible modules are available. Fortunately, the Akamai device provides a simple API endpoint for managing the DNS entries. Use the same in Ansible to automate the necessary operations.

The API needs to be input in a complex body format. For that, a Jinja2 template called `akamai-url-block-format.j2` can be used, as shown here:

```
{"add":[
{% for dns in dns_list %}
{"name": "{{ dns }}" },
{% endfor %}
]}
```

Figure 15.21 – The Jinja2 template for preparing the Akamai API call body

The `akamai-dns-block.yaml` playbook contains a few variables, as shown in the following screenshot:

```
# akamai-dns-block.yaml
---
- name: Block DNS in Akamai Device
  hosts: localhost
  gather_facts: false
  become: false
  vars:
    akamai_api_endpoint: 'http://10.1.10.100:8080'
    akamai_list_path: '/list/blacklist/nodes'
    akamai_api_username: 'admin'
    akamai_api_password: 'secretpassword'

    dns_list:
      - blockthisurl.com
      - antherwebsite.com
      - notagoodwebsite.com
```

Figure 15.22 – Variables for Akamai DNS blocking

Remember to keep the sensitive items (such as credentials) in encrypted format using Ansible Vault (or **Credentials** in the Ansible automation controller).

The list of DNS entries (`dns_list`) can be converted into the desired format using the `template` lookup plugin and passed to the `uri` module (body: "`{{ dns_list_templated }}`"), as shown here:

```
tasks:
  - name: Template the DNS List to block
    ansible.builtin.set_fact:
      dns_list_templated: "{{ lookup('template', 'akamai-url-block-format.j2') }}"

  - name: "{{ akamai_list_name }} - Create substitute records Akamai"
    uri:
      url: "{{ akamai_api_endpoint }}{{ akamai_list_path }}"
      method: POST
      return_content: yes
      user: "{{ akamai_api_username }}"
      password: "{{ akamai_api_password }}"
      status_code:
        - 201
        - 200
      headers:
        Accept: application/json
        Content-Type: application/json
      body_format: json
      body: "{{ dns_list_templated }}"
    register: akamai_add_out
```

Figure 15.23 – The Akamai API call using the uri module

You need to add all validation and verification tasks as required before executing the API POST calls. Refer to the Akamai documentation (`https://techdocs.akamai.com/home/page/products-tools-a-z`) for the products to learn more.

In this section, you learned how to automate non-supported operations using raw commands and API calls. In the next section, you will explore Ansible modules and the basic steps to create custom modules for Ansible.

Creating custom modules for Ansible

In the previous sections, you learned how to automate operations if the standard modules are not available for specific tasks. But modules are the standard way of implementing automation and help you develop Ansible playbooks without worrying about the complex operations in the backend. If you know the backend operations and how to execute the tasks in the backend, then create a module for Ansible to execute a specific operation. Finally, contribute it back to the community via Ansible collections. That is the way the open source community grows.

Facts to check before creating a custom Ansible module

You can use any programming language (which can be called by the Ansible API, the `ansible` command, or the `ansible-playbook` command), libraries, and methods for your new Ansible module. Most of the Ansible modules that you are using now are written in the Python programming language. Before developing a new module, check yourself on the following facts:

- If the specific task can be completed using an Ansible role, then develop a role instead of an Ansible module.

- If the actual execution is happening on the control node (instead of a managed node) then create an action plugin instead of a module (read *Developing plugins* at `https://docs.ansible.com/ansible/latest/dev_guide/developing_plugins.html` for more details).

- If there are similar modules available, then modify that module, add features, and contribute it back to the community.

- If the module you are planning to develop requires a lot of dependencies, then try to create it as an Ansible collection with all the dependencies, custom libraries, and plugins as required.

> **Should You Develop an Ansible Module?**
>
> Refer to the Ansible documentation at `https://docs.ansible.com/ansible/latest/dev_guide/developing_modules.html` to learn more.

If you are using your custom module without an Ansible collection (or as a standalone module), then make sure the module script has been copied to the right location. The following are the common locations and methods we can use to store custom modules:

- Use `DEFAULT_MODULE_UTILS_PATH` in the Ansible configuration and mention all the directories to look for modules.

- Add the modules to the user's directory: `~/.ansible/plugins/modules/`.

- Add the modules to system directory: `/usr/share/ansible/plugins/modules/`.

- The `ANSIBLE_LIBRARY` environment variable can be configured to set the custom module directory.

  The following screenshot shows the configured module paths for Ansible:

  ```
  [ansible@ansible Chapter-15]$ ansible-config dump |grep DEFAULT_MODULE_PATH
  DEFAULT_MODULE_PATH(default) = ['/home/ansible/.ansible/plugins/modules', '/usr/share/ansible/plugins/modules']
  ```

  Figure 15.24 – Ansible module path

- If the module is used for a specific Ansible role, then store the module inside the `library` directory of the role. Ansible will automatically detect the module.

- If the module is being used by playbooks and different roles, then store it in the `library` directory of the project directory. The following screenshot shows an example of `ansible.cfg` with `library` configured:

```
• • •
[defaults]

library = ./library
```

Figure 15.25 – Library path in ansible.cfg

Ansible Module Directory Configuration

Read the documentation at `https://docs.ansible.com/ansible/latest/ dev_guide/developing_locally.html` to learn more about adding modules and plugins locally.

In the next section, you will learn how to use simple bash scripts for Ansible modules and explore more with Python-based Ansible modules.

Developing Ansible modules using bash scripts

Modules are simply reusable scripts for executing a specific task. Before moving on, you will need to create a simple bash script and use it as an Ansible module.

`library/customhello.sh` is a simple Bash script that displays the operating system, hostname, and a custom message, as shown in the following screenshot:

```
• • •
[ansible@ansible Chapter-15]$ cat library/customhello.sh
#!/bin/bash
#
# This script accepts two inputs
# 1. application_name
# 2. application_version

changed="false"
display="This is a simple bash module"
OS="$(uname)"
HOSTNAME="$(uname -n)"

source $1
display="Application Name: $application_name (version: $application_version)"
if [ "$application_name" == "bash" ]; then
  changed="true"
  display="$display - This is a bash App"
fi

printf '{"changed": %s, "msg": "%s", "operating system": "%s", "hostname": "%s"}' "$changed" "$display" "$OS"
 "$HOSTNAME"
exit 0
```

Figure 15.26 – Bash script for an Ansible module

Notice the echo line, where the output is formatted as JSON. The module should provide a defined interface (that also accepts arguments) and should return the result or information to Ansible by printing a JSON string to stdout before exiting.

We have a playbook with a task for calling this custom module, as shown in the following screenshot:

```
● ● ●
---
- name: Testing Custom Module
  hosts: node1
  gather_facts: false
  vars:
    app_name: "bash"
    app_version: "1.0"
  tasks:
    - name: Application Name and Version
      customhello:
        application_name: "{{ app_name }}"
        application_version: "{{ app_version }}"
      register: custom_value

    - debug:
        msg: "{{ custom_value }}"
```

Figure 15.27 – Ansible playbook with the custom hello module task

When you execute this playbook, the bash script will be executed in the backend, and Ansible will get information from the script, as shown here:

```
● ● ●
<omitted>...
TASK [debug]
*********************************************************************************************************
ok: [node1] => {
    "msg": {
        "changed": true,
        "failed": false,
        "hostname": "node-1",
        "msg": "Application Name: bash (version: 1.0) – This is a bash App",
        "operating system": "Linux"
    }
}
<omitted>...
```

Figure 15.28 – Ansible playbook output for the custom module

This is a very basic concept of an Ansible module and its parameters and variables. Expand it as required.

> **Developing Ansible Modules**
>
> Following the best practices will help you create quality Ansible modules that can be enhanced when required. Refer to the documentation (`https://docs.ansible.com/ansible/latest/dev_guide/developing_modules_general.html`) and Ansible module architecture (`https://docs.ansible.com/ansible/latest/dev_guide/developing_program_flow_modules.html`) to learn more about developing Ansible modules.

In the next section, you will learn more about custom modules by following the Ansible module development guidelines.

Developing Ansible modules using Python

When you create a module, follow the best practices as much as possible and increase the reusability of the module by adding appropriate documentation, examples or expected results.

Check the `library/hello_message.py` file in the `Chapter-15` directory of this book's GitHub repository and explore the standard module components inside the Python script:

- The script starts with a **Python shebang** that allows `ansible_python_interpreter` to work.

- Add the copyright and other contact information as needed after that.

- The DOCUMENTATION block is a very important part of the module since it helps users understand the usages of this module, such as its available parameters, options to use, field types, and so on. The following screenshot shows the documentation portion of our custom `hello_message.py` module:

```
DOCUMENTATION = '''
---
module: hello_message
short_description: A Hello Message Module
version_added: "2.10"
description:
    - "A Hello Message Module"
options:
    message:
        description:
            - The message to be printed.
        required: true
        type: string
...<omitted>...

author:
    - Gineesh Madapparambath (@ginigangadharan)
'''
```

Figure 15.29 – Module documentation

- The EXAMPLES block contains the sample usage of the module:

```
EXAMPLES = '''
# Simple Custom Hello App
- name: Calling hello_message module
  hello_message:
    message: "Hello"
    name: "John"
'''
```

Figure 15.30 – EXAMPLE part of the module

- The RETURN block should contain the sample output or result of the module being returned after being executed successfully:

```
RETURN = '''
greeting:
    description: Hello Response
    returned: success
    type: str
    sample: Hello World
os_version:
    description: Operating System Information
    returned: success
    type: str
    sample: Linux 4.18.0-305.el8.x86_64 #1 SMP Thu Apr 29 08:54:30 EDT 2021
'''
```

Figure 15.31 – The module's RETURN block

- Finally, there's the actual script. This contains all the necessary libraries and dependencies, just like a normal Python script:

```
from ansible.module_utils.basic import AnsibleModule, platform

def main():
    module_args = dict(
        message=dict(type='str', required=True),
        name=dict(type='str', required=False),
    )
    result = dict(
        changed=False,
        greeting='Sample Message',
        os_version='',
    )
```

Figure 15.32 – The script portion of an Ansible module

> **Module Format and Documentation**
>
> Following the best practices will help those who are using the module to understand how it's used. Refer to https://docs.ansible.com/ansible/latest/dev_guide/ developing_modules_documenting.html#developing-modules- documenting to learn more about the module format and documentation.

Verify the module using the `ansible-doc` command, as shown in the following screenshot:

```
● ● ●

[ansible@ansible Chapter-15]$ ansible-doc hello_message
> HELLO_MESSAGE    (/home/ansible/ansible-book-packt/Chapter-15/library/hello_message.py)

        A Hello Message Module

OPTIONS (= is mandatory):

= message
        The message to be printed.

        type: string

- name
        The name of the person.
        [Default: (null)]
        type: string


AUTHOR: Gineesh Madapparambath (@ginigangadharan)
```

Figure 15.33 – Custom module details after using the ansible-doc command

When scrolling down using the keyboard, you will see the EXAMPLES and RETURN VALUES sections of the module's documentation, as shown here:

```
<omitted>...
EXAMPLES:

# Simple custom hello msg
- name: Calling hello_message module
  hello_message:
    message: "Hello"
    name: "John"


RETURN VALUES:
- greeting
      Hello Response

      returned: success
      sample: Hello World
      type: str

- os_version
      Operating System Information
<omitted>...
```

Figure 15.34 – Ansible custom module documentation details after using the ansible-doc command

Let's use the module inside the `hello-python.yaml` playbook and pass the `message` and `name` module parameters:

```
---
- name: Testing Custom Module
  hosts: localhost
  gather_facts: false
  vars:
    custom_message: "Hello"
    custome_name: "John"
  tasks:
    - name: Calling custom module
      hello_message:
        message: "{{ custom_message }}"
        name: "{{ custome_name }}"
      register: custom_value

    - debug:
        msg: "{{ custom_value }}"
```

Figure 15.35 – Using the hello_message module in the playbook

Execute the playbook and verify its output, as shown here:

```
● ● ●
<omitted>...
TASK [debug]
**********************************************************************************************************************
ok: [localhost] => {
    "msg": {
        "changed": false,
        "failed": false,
        "greeting": "Hello John",
        "os_version": "Linux 4.18.0-305.el8.x86_64 #1 SMP Thu Apr 29 08:54:30 EDT 2021"
    }
}
<omitted>...
```

Figure 15.36 – Verifying the playbook's execution and the hello_message module

Check the Chapter-15 directory in this book's GitHub repository to learn more about Ansible custom modules. Also, refer to the *Further reading* section at the end of this chapter for more resources on Ansible custom module development.

Using Ansible collections and contributing back

As mentioned in the previous section, store and distribute modules, roles, and libraries as an Ansible collection. This collection can be distributed to the public via Ansible Galaxy (https://galaxy.ansible.com/) or internally using Red Hat Ansible **Private Automation Hub (PAH)**.

In this section, you will learn how to export a collection to Ansible Galaxy.

Preparing the collection directory

The following screenshot shows a typical Ansible collection's directory structure:

```
collection/
├── docs/
├── galaxy.yml
├── meta/
│   └── runtime.yml
├── plugins/
│   ├── modules/
│   │   └── module1.py
│   ├── inventory/
│   └── .../
├── README.md
├── roles/
│   ├── role1/
│   ├── role2/
│   └── .../
├── playbooks/
│   ├── files/
│   ├── vars/
│   ├── templates/
│   └── tasks/
└── tests/
...<omitted>...
```

Figure 15.37 – Ansible collection directory structure

In our scenario, we will make various adjustments without moving the original playbooks and module directories. (This is only for demonstration purposes; it is possible to keep the collection directories and files in the root of your GitHub repository.)

Similar to the `ansible-galaxy role init` command, use the `ansible-galaxy collection init` command to initialize the collection with the base directories and structure. In this example, we will be manually creating the collection and subdirectories to demonstrate the required directory structure and files for the collection. This will also allow us to reuse the existing playbooks and modules:

1. Create the `collection` directory.
2. Create the `collection/playbook` directory and copy the playbooks into it.
3. Create the `collection/plugins/modules` directory and copy the custom modules into it.
4. Create `collection/meta/runtime.yml`.
5. Create `collection/galaxy.yml`, as shown in the following screenshot:

```
● ● ●

---
namespace: ginigangadharan
name: custom_modules_demo
version: 1.0.4
readme: README.md
authors:
  - Gineesh Madapparambath <gini@iamgini.com>
description: Ansible Custom Module Demo for Ansible Book
license:
  - GPL-2.0-or-later
license_file: ''
tags:
  - demos
  - ansible
  - devops
dependencies: {}
repository: https://github.com/PacktPublishing/Ansible-for-Real-life-Automation/
documentation: https://github.com/PacktPublishing/Ansible-for-Real-life-Automation/tree/main/Chapter-15/collection
homepage: https://github.com/PacktPublishing/Ansible-for-Real-life-Automation/tree/main/Chapter-15/collection
issues: https://github.com/PacktPublishing/Ansible-for-Real-life-Automation/issues
```

Figure 15.38 – galaxy.yml for the Ansible collection

6. Create `collection/README.md` with the necessary documentation and details. Add/ update the `.gitignore` file (in the root of the GitHub repository) and add the following content (this is for ignoring unwanted files when you sync the content to the GitHub server):

    ```
    *.tar.gz
    test/results
    ```

Once you have prepared the collection's content, you must build and publish it to Ansible Galaxy. This will be covered in the next section.

Publishing an Ansible collection to Ansible Galaxy

Publish the collection to Ansible Galaxy via a GUI or using the CLI. Follow these steps to publish the collection using the CLI:

1. Log in to Ansible Galaxy and get the Ansible Galaxy API key (token) from `https:// galaxy.ansible.com/me/preferences`.

2. Export the token to an environment variable:

    ```
    $ export ANSIBLE_GALAXY_TOKEN='YOUR_ANSIBLE_GALAXY_API_ TOKEN'
    ```

3. Build the collection archive using the `ansible-galaxy collection build` command:

```
[ansible@ansible Chapter-15]$ cd collection
[ansible@ansible collection]$ ansible-galaxy collection build
Created collection for ginigangadharan.custom_modules_demo at /home/ansible/ansible-book-packt/Chapter-
15/collection/ginigangadharan-custom_modules_demo-1.0.0.tar.gz
[ansible@ansible Chapter-15]$
```

Figure 15.39 – Building the Ansible collection archive

Use `--force` to overwrite the collection archive if there's an existing archive with the same version.

4. Now, publish the collection to Ansible Galaxy using the `ansible-galaxy collection publish` command:

```
[ansible@ansible collection]$ ansible-galaxy collection publish \
> --token $ANSIBLE_GALAXY_TOKEN \
> ./ginigangadharan-custom_modules_demo-1.0.0.tar.gz
Publishing collection artifact '/home/ansible/ansible-book-packt/Chapter-15/collection/ginigangadharan-
custom_modules_demo-1.0.0.tar.gz' to default https://galaxy.ansible.com/api/
Collection has been published to the Galaxy server default https://galaxy.ansible.com/api/
Waiting until Galaxy import task https://galaxy.ansible.com/api/v2/collection-imports/20104/ has completed
Collection has been successfully published and imported to the Galaxy server default https://galaxy.ansible.com/api/
```

Figure 15.40 – Publishing the collection to Ansible Galaxy

5. Verify the published collection in the Ansible Galaxy portal (`https://galaxy.ansible.com/my-content/namespaces`), as shown in the following screenshot:

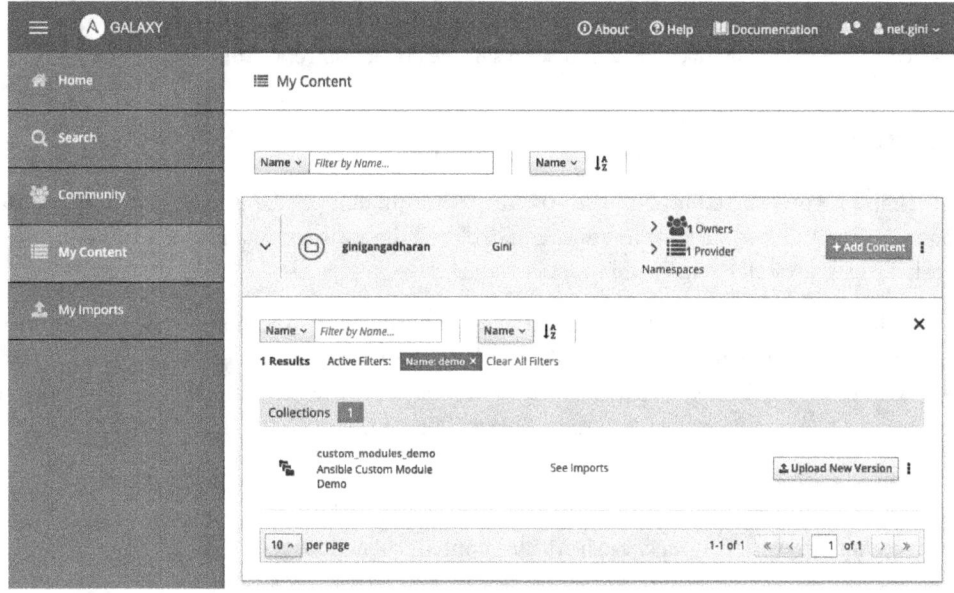

Figure 15.41 – New collection published in Ansible Galaxy

6.   Open the collection and view its content:

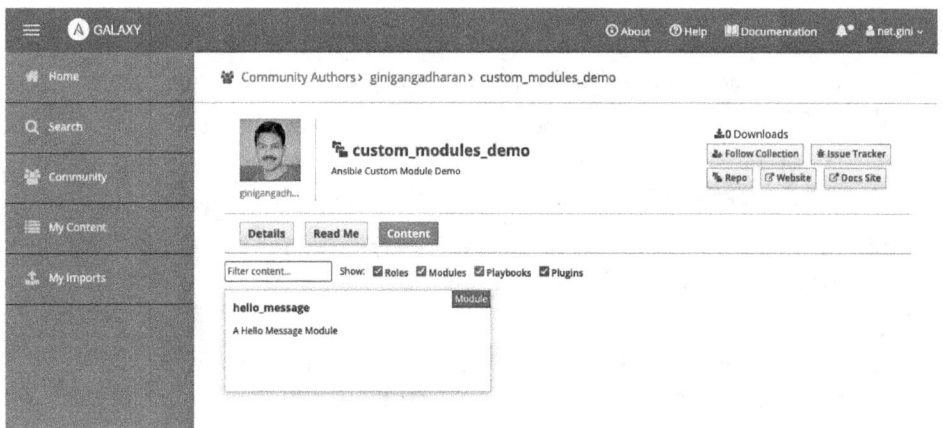

Figure 15.42 – Ansible collection content

In this section, you created a subdirectory (`collection`) where you can keep the collection's content and build the archive. As I mentioned earlier, keep the collection's content in the root directory of the GitHub repository. Refer to `https://github.com/ginigangadharan/ansible-collection-custom-modules` to see a sample collection repository.

Summary

In this chapter, you learned how to use Ansible to automate non-supported and non-standard operations using the Ansible `raw` module. You explored the `raw` command's execution on servers, network devices, firewall devices, and more.

Then, you learned how to interact with the devices that provide API-based operations. The Ansible `uri` module was used to interact with a ToDo application; you explored the options for fetching and adding items to the application via APIs. You also learned about the API-based operations for devices and explored some sample usage using the Akamai DNS API.

In addition to the `raw` command and API-based operations, you learned about Ansible custom modules and how to create custom modules using bash and Python. In the end, you distributed the custom modules to Ansible Galaxy as an Ansible content collection.

With that, you have explored a common production use case where you can use Ansible as a perfect automation tool and Red Hat Ansible Automation Platform as an enterprise automation solution.

In the next chapter, you will learn about the best practices in Ansible, such as storing playbooks and their content, organizing inventories, the YAML style guide, and more.

Further reading

To learn more about the topics that were covered in this chapter, take a look at the following resources:

- *Network Automation with Ansible*: `https://www.ansible.com/integrations/networks`

- *Developing Ansible modules*: `https://docs.ansible.com/ansible/latest/dev_guide/developing_modules_general.html`

- *Ansible module best practices – conventions, tips, and pitfalls*: `https://docs.ansible.com/ansible/latest/dev_guide/developing_modules_best_practices.html`

- *Control your content with a private Automation Hub (Ansible blog)*: `https://www.ansible.com/blog/control-your-content-with-private-automation-hub`

Ansible Automation Best Practices for Production

Ansible can be used to automate IT infrastructure and DevOps tasks. Because of its flexible and modular architecture, we can implement large, complex automation use cases using Ansible. But at the same time, we need to keep the simplicity and reusability of the automation artifacts and methods.

In this chapter, you will learn about the important and well-known best practices for implementing efficient automation solutions.

First, you will learn how to organize the playbooks, roles, collections, and inventories in an Ansible project. After that, we will discuss the best practices for storing managed node information in the inventory and different methods for storing and maintaining multiple inventories. You can store the remote nodes separately based on their function, criticality, or location; these details will be explained in the upcoming sections.

You will also learn about the most efficient ways to store the variables in dynamic methods and how to store host variables and group variables to maintain them appropriately.

Another critical component in Ansible automation is handling credentials such as usernames and passwords, API keys, and secrets. Therefore, you will explore the best practices for Ansible credential management, such as how to store sensitive data for an Ansible playbook.

Finally, you will learn about the best practices, methods, and optimization techniques for developing and executing Ansible playbooks.

In this chapter, we will cover the following topics:

- Organizing Ansible automation content
- Storing remote host information – inventory best practices
- Ansible host variables and group variables
- Ansible credentials best practices
- Ansible playbook best practices

This chapter will start by covering various Ansible content organization methods and different inventory organization methods.

Technical requirements

You will need the following technical requirements to complete this chapter:

- A Linux machine for the Ansible control node.
- One or more Linux machines as managed nodes with Red Hat repositories configured (if you are using non-RHEL machines, then make sure you have the appropriate repositories configured to get packages and updates).

All the Ansible artifacts, Ansible playbooks, commands, and snippets for this chapter can be found in this book's GitHub repository at `https://github.com/PacktPublishing/Ansible-for-Real-life-Automation/tree/main/Chapter-16`.

Organizing Ansible automation content

In *Chapter 4, Exploring Collaboration in Automation Development*, you learned about **version control systems** (**VCSs**) and **source control management** (**SCM**) and how to use GitHub services to store Ansible artifacts.

It is the best practice to create project-specific directories (that is, repositories) to keep all related items at a single location, such as project-specific `ansible.cfg` files, playbooks, roles, collections, or libraries. If there are external roles or collections dependencies, then mention the details inside the `requirements.yaml` (or `requirements.yml`) file.

Use the `tree` command in Linux to list the directories and files recursively and understand the structure of the directory's content. A sample project directory can be organized like so:

```
[ansible@ansible Chapter-16]$ tree ./
./
├── ansible.cfg              # ansible configuration
├── deploy-web.yml           # a playbook
├── group_vars               # directory for group level variables
│   ├── dbnodes.yaml         # variables for inventoy group dbnodes
│   └── web.yaml             # variables for inventoy group web
├── hosts                    # another inventory file
├── host_vars                # directory for host level variables
│   ├── node1.yaml           # variables for node1
│   └── node2.yaml           # variables for node2
├── nodes_development        # inventory for development nodes
├── nodes_production         # inventory for production nodes
├── nodes_staging            # inventory for staging nodes
├── README.md
```

Figure 16.1 – Typical Ansible project directory

Your roles will be under the `roles` directory, as shown in the following screenshot:

```
├── roles                         # roles directory
│   ├── deploy-web-server         # web deployment role
│   │   ├── defaults
│   │   │   └── main.yml
│   │   ├── tasks
│   │   │   └── main.yml
│   │   ├── templates
│   │   ├── tests
│   │   │   ├── inventory
│   │   │   └── test.yml
│   │   └── vars
│   │       └── main.yml
│   ├── security-baseline-rhel8   # security hardening role
│       ... output omitted...
├── site.yml
├── system-info.yml
├── system-reboot.yml

38 directories, 56 files
```

Figure 16.2 – Ansible roles directory

Depending on the projects and use cases, you may have more or fewer directories and files. We will explore the best practices for storing an inventory in the next section.

Storing remote host information – inventory best practices

Managed nodes or remote host information is critical data in Ansible automation since, without the proper host details, Ansible will not be able to execute the automation tasks. You learned about the Ansible inventory and its basic details in *Chapter 1, Ansible Automation – Introduction*. In *Chapter 4, Exploring Collaboration in Automation Development*, you learned about the importance of storing an inventory in a GitHub repository for version control and better management. If your managed nodes are hosted in cloud platforms, then it is a best practice to use Ansible dynamic inventories, as you learned in *Chapter 5, Expanding Your Automation Landscape*.

Using meaningful hostnames

When you create your Ansible static inventory files, use meaningful and user-friendly names for your managed nodes instead of complex **Fully Qualified Domain Names** (**FQDNs**) or IP addresses. It will help you while executing the Ansible playbook and troubleshooting it if that's required.

For example, the following is a generic Ansible static inventory file:

```
10.1.10.100
192.168.1.25
10.1.10.25
10.2.100.40
dbserver-101.example.com
prod-app-101.example.com
```

Figure 16.3 – Sample static inventory file

The same static inventory can be rewritten with user-friendly names and `ansible_host` information, as shown in the following screenshot:

```
web01 ansible_host=10.1.10.100
app02 ansible_host=192.168.1.25
lb101 ansible_host=10.1.10.25
db201 ansible_host=10.2.100.40
web102 ansible_host=sglxwp-101.example.com
app301 ansible_host=sllxmkp-app-101.example.com
```

Figure 16.4 – Ansible inventory with user-friendly names

Using `ansible_host` means you don't have to rely on your DNS name (FQDN) by using the IP address to access the managed nodes.

This practice will not only help you troubleshoot output and logs but also help you manage your inventory with simple and meaningful names.

> **How to Build Your Inventory**
>
> Refer to the official documentation for more details: `https://docs.ansible.com/ansible/latest/user_guide/intro_inventory.html`.

In the next section, you will learn how to separate the inventory based on environments such as production, staging, and development.

Storing production, staging, and development hosts separately

You need to organize your inventory at the project level or overall inventory level. If you are using the same managed nodes for multiple Ansible projects (the same nodes but different automation and use cases), then keep your inventory somewhere in a central GitHub repository as a single source of truth. This will help you organize your managed node information in a better way so that it can be used for different automation playbooks. The following diagram shows a scenario where the inventory is stored in a dedicated GitHub repository:

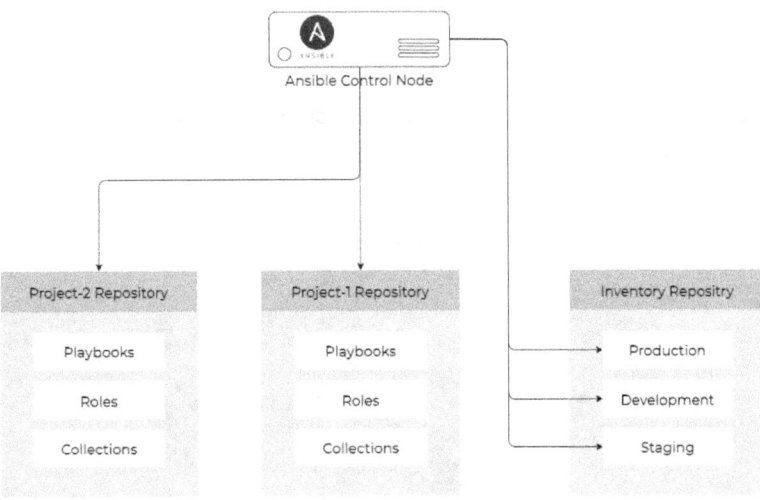

Figure 16.5 – Ansible inventory in a dedicated repository

In the following example, we have created a separate directory for the inventories and placed the production, development, and staging managed nodes in separate directories (refer to the `Chapter-16` directory in this book's GitHub repository):

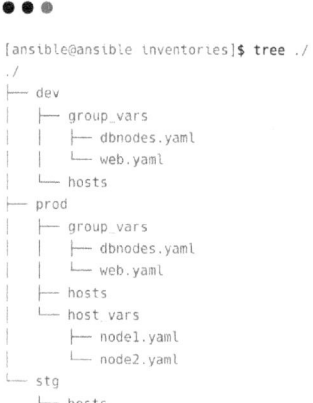

```
[ansible@ansible inventories]$ tree ./
./
├── dev
│   ├── group_vars
│   │   ├── dbnodes.yaml
│   │   └── web.yaml
│   └── hosts
├── prod
│   ├── group_vars
│   │   ├── dbnodes.yaml
│   │   └── web.yaml
│   ├── hosts
│   └── host_vars
│       ├── node1.yaml
│       └── node2.yaml
└── stg
    └── hosts
```

Figure 16.6 – Ansible inventory organized based on the environment

When you execute the playbook, you must mention which inventory file will be used, as follows:

```
[ansible@ansible Chapter-16]$ ansible-playbook site.yml -i
inventories/prod/host
```

The inventory can also be categorized based on location, criticality, server type, and more. It is possible to do the same categorization inside the inventory file using host groups, as shown in the following screenshot:

```
# file: dev/hosts
# singapore web servers
# group variables in dev/group_vars/web.yaml
[web]
web101.example.com
web102.example.com
web103.example.com

# singapore db servers
# group variable in dev/group_vars/dbnodes.yaml
[dbnodes]
db201.example.com
db202.example.com
db203.example.com

# backup nodes in Malaysia
[backupnodes]
bkp101.example.com
bkp102.example.com

# Singapore servers in a parent group
[sgnodes:children]
web
dbnodes
```

Figure 16.7 – Host groups and group variables for managed nodes

Verify the grouping of managed nodes using the `ansible-inventory` command, as shown in the following screenshot:

```
[ansible@ansible inventories]$ ansible-inventory -i dev/hosts --list
{
    "_meta": {
        "hostvars": {}
    },
    "all": {
        "children": [
            "backupnodes",
            "sgnodes",
            "ungrouped"
        ]
    },
    ... output omitted...
    "sgnodes": {
        "children": [
            "dbnodes",
            "web"
        ]
    },
    "web": {
        "hosts": [
            "web101.example.com",
            "web102.example.com",
            "web103.example.com"
        ]
    }
}
```

Figure 16.8 – Listing the hosts and host groups using the ansible-inventory command

In the preceding examples, we created multiple inventory files in the same or different folders and grouped-managed nodes based on function or location. In the next section, you will learn how to maintain host-specific and group-specific variables using the group_vars and host_vars variables, respectively.

Ansible host variables and group variables

As you learned previously, like many other automation tools, Ansible allows you to use variables for dynamically executing playbooks. It is possible to configure the same playbook so that it can be executed for different desired states using variables and values. We can keep the variables inside the playbooks, external variable files, inventory files, and many other places. You learned more about variables in *Chapter 6, Automating Microsoft Windows and Network Devices*.

The same variable can be specified in multiple places but depending on the location of your variable and variable precedence, Ansible will apply the appropriate value for the variable.

Ansible uses the appropriate variable values and executes the playbooks based on them; the following diagram shows the typical flow where Ansible combines the variable values with the playbook:

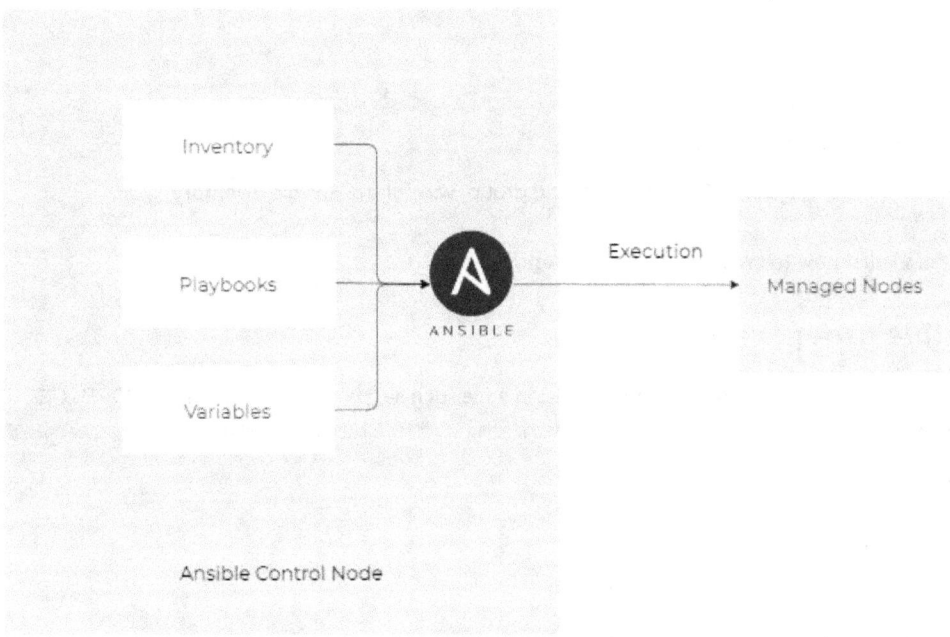

Figure 16.9 – Ansible combines playbooks and variables for the final execution

> **Understanding Variable Precedence**
>
> Refer to `https://docs.ansible.com/ansible/latest/user_guide/playbooks_variables.html#understanding-variable-precedence` to learn and understand more about variable precedence.

It is best practice to use host-specific variables and group-specific variables in the `host_vars` and `group_vars` directories, respectively, as shown in the following screenshot:

```
● ● ●
[ansible@ansible Chapter-16]$ tree inventories/stg/
inventories/stg/
├── group_vars
│   ├── dbnodes.yaml
│   └── web.yaml
├── hosts
└── host_vars
    ├── node1.yaml
    └── node2.yaml

2 directories, 5 files
```

Figure 16.10 – host_vars and group_vars for the Ansible inventory

Now, let's learn how to create host_vars and group_vars.

Ansible group_vars

The variables for a group should be mentioned in a file such as group_vars/INVENTORY_GROUP_NAME.yaml or a subdirectory such as group_vars/INVENTORY_GROUP_NAME/VAR_FILE.

For example:

- group_vars/web.yaml

- group_vars/web/vars.yaml

- group_vars/web/credential.yaml

Ansible host_vars

The variables for the host should be mentioned in a file such as host_vars/INVENTORY_HOSTNAME or a subdirectory such as host_vars/INVENTORY_HOSTNAME/VAR_FILE.

For example:

- host_vars/node1.yaml

- host_vars/node1/vars.yaml

- host_vars/node2.yaml

It is possible to create multiple variable files for the same managed node so that you can manage related variables separately.

Keeping your secret variables in a safe location

If you have credentials or secrets as part of your host variables, then keep such variables in a separate variable file and encrypt them using Ansible Vault. The following are some examples of this:

- `host_vars/node1/vars.yaml`
- `host_vars/node1/credentials.yaml`
- `group_vars/web/vault.yaml`

It is also possible to use other vault services such as HashiCorp Vault or CyberArk instead of Ansible Vault. Refer to *Chapter 13, Using Ansible for Secret Management*, to learn more about Ansible Vault.

Managing group_vars and host_vars in Ansible

In this exercise, you will use host variables and group variables to control the values of multiple web servers. Follow these steps:

1. Create a `hosts` inventory file inside the staging inventory directory (`Chapter-16/inventories/stg`) with the following content (do not worry about `node1`, `node2`, or `node3` as we are not going to connect to these machines):

```
# file: stg/hosts
[web]
node1 ansible_host=192.168.56.25
node2 ansible_host=192.168.56.24
node3 ansible_host=192.168.56.60

[all:vars]
ansible_ssh_private_key_file=/home/ansible/.ssh/id_rsa
```

Figure 16.11 – Ansible inventory with a web group

2. Create the `group_vars` and `host_vars` directories for storing group variables and host variables, respectively:

```
[ansible@ansible Chapter-16]$ mkdir inventories/stg/group_vars
[ansible@ansible Chapter-16]$ mkdir inventories/stg/host_vars
```

Figure 16.12 – Creating directories for group variables and host variables

3. Create a group variable file called `inventories/stg/group_vars/web.yaml` with the following content:

Figure 16.13 – Creating a group variable file

4. Configure a different value for `web_server_port` (8080) for node1. Create a host variable file called `inventories/stg/host_vars/node1.yaml` with the following content:

Figure 16.14 – Creating a host variable file

5. Configure a different value for `web_server_port` (8081) for node2. Then, create a host variable file for node2 in `inventories/stg/host_vars/node2.yaml` with the following content:

Figure 16.15 – Creating a host variable file for node2

6. Now that all the variable files have been created, you must verify them, as shown in the following screenshot:

```
[ansible@ansible Chapter-16]$ tree inventories/stg/
inventories/stg/
├── group_vars
│   ├── dbnodes.yaml
│   └── web.yaml
├── hosts
└── host_vars
    ├── node1.yaml
    └── node2.yaml

2 directories, 5 files
```

Figure 16.16 – Project directory structure with group variables and host variables

7. Now, verify the variable values for each host with the `ansible-inventory` command:

```
[ansible@ansible Chapter-16]$ ansible-inventory --list -i inventories/stg/
{
    "_meta": {
        "hostvars": {
            "node1": {
                "ansible_host": "192.168.56.25",
                "ansible_ssh_private_key_file": "/home/ansible/.ssh/id_rsa",
                "web_server_port": 8080
            },
            "node2": {
                "ansible_host": "192.168.56.24",
                "ansible_ssh_private_key_file": "/home/ansible/.ssh/id_rsa",
                "default_web_page_content": "Welcome to node2",
                "web_server_port": 8081
            },
            "node3": {
                "ansible_host": "192.168.56.60",
                "ansible_ssh_private_key_file": "/home/ansible/.ssh/id_rsa",
                "web_server_port": 80
            }
        }
    },
    ...output omitted...
}
```

Figure 16.17 – Verifying the inventory and variables using the ansible-inventory command

8.   Now, you must verify and understand the host variables that have been assigned, as follows:

   • node1 has been assigned with `web_server_port:  8080`, which is coming from `host_vars/node1.yaml`.

   • node2 has been assigned with `web_server_port:  8081`, which is coming from `host_vars/node2.yaml`.

   • node3 has been assigned with `web_server_port:  80`, which is coming from `group_vars/web.yaml`.

Based on variable precedence, the nodes will get different values for the same variable, as shown in the following diagram:

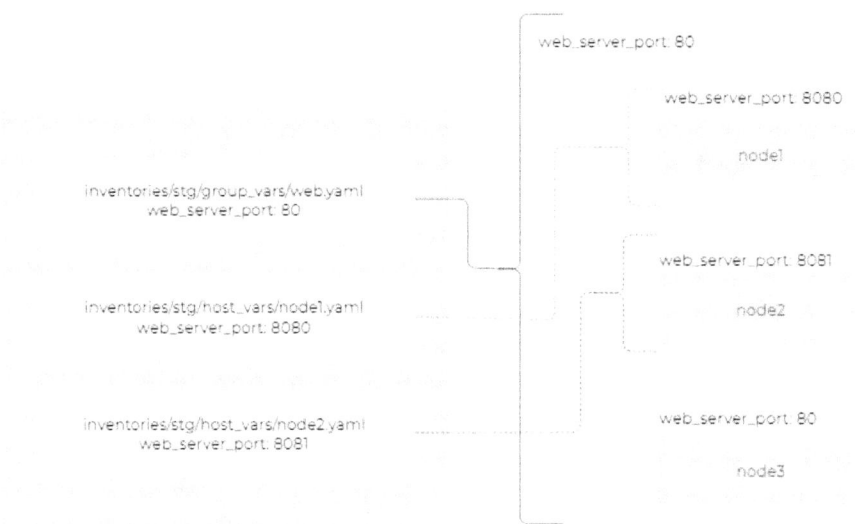

Figure 16.18 – Ansible group variables and host variables on target nodes

It is important to configure the host-specific variable in `host_vars` or `group_vars` so that you don't miss important values at the playbook level. If you have the same variable and values being shared by multiple hosts in the same group, then configure them under `group_vars`.

In the next section, you will learn about the best practices for storing credentials in Ansible.

Ansible credentials best practices

Ansible supports multiple credentials and authentication methods, such as username and password, SSH keys, API tokens, webhooks, and even the ability to create custom credentials. You should use a simple authentication mechanism as a starting point, but you need to consider the best practices to ensure security and safety are in place.

Avoid using default admin user accounts

It is common for engineers to configure the default administrator accounts as a `remote_user` such as `root` in Linux or as an **administrator** in Microsoft Windows. This is not a best practice; you should create dedicated accounts for Ansible and configure them for managed nodes.

Split the login credentials for environments and nodes

In the previous examples, you created user accounts in Linux and Microsoft Windows for Ansible to log in and execute tasks. It is possible to create the same user account for all of your nodes, but this is not required or recommended. It is possible to create different user accounts for different managed nodes since you have the option to specify `remote_user` or `ansible_user` for every managed node or host group, as shown here:

```
[ansible@ansible Chapter-06]$ ansible-inventory web --list
{
    "_meta": {
        "hostvars": {
            "node1": {
                "ansible_host": "192.168.56.25",
                "ansible_ssh_private_key_file": "/home/ansible/.ssh/id_rsa",
                "ansible_user": "ansibleadmin"
            },
            "node2": {
                "ansible_host": "192.168.56.24",
                "ansible_user": "user1"
            },
            "node3": {
                "ansible_host": "192.168.56.60",
                "ansible_user": "devops"
            },
            "win2019": {
...output omitted...
                "ansible_user": "ansible",
                "ansible_winrm_server_cert_validation": "ignore",
                "ansible_winrm_transport": "basic"
            }
        }
    },
...output omitted...
```

Figure 16.19 – A different user account for remote nodes

In the preceding inventory output, notice different `ansible_user` instances have been configured for different nodes, such as `ansibleadmin` for `node1`, `user1` for `node2`, `devops` for `node3`, and more.

Avoid passwords in plain text

If you are using password-based authentication, then the password should be encrypted and saved separately. Refer to *Chapter 3's, Encrypting Sensitive Data Using Ansible Vault* section, to learn more about Ansible Vault and secret management. Once encrypted using Ansible Vault, the password file will be safe and cannot be read by anyone else, as shown in the following screenshot:

```
[ansible@ansible Chapter-03]$ ansible-vault create vars/secrets
New Vault password:
Confirm New Vault password:
[ansible@ansible Chapter-03]$ cat vars/secrets
$ANSIBLE_VAULT;1.1;AES256
38393063373031356638353866353937306462663565366266323166363130356435326564343735
30616638313262373564303533616462353966616635383310a37333737363393835613537623 56265
39363830316465346166303666373064353061343536361373434333665363065653339373964323 8
313630613063337616610a646138326130333435373836303832343335373737303535353663656165430
32323537303765356366383930623631666561393661626535666313531636232613462306662 3234
313731386161373461326262306264643430343066673716636633539663530303338396163666131
383237626162626334376133663039366331
```

Figure 16.20 – Encrypting sensitive files using Ansible Vault

When you execute the playbook, it is also possible to instruct Ansible to prompt for a password using the `--ask-pass` switch:

```
[ansible@ansible Chapter-06]$ ansible-playbook password-promt.yaml --ask-pass
SSH password:
```

Figure 16.21 – Ansible Vault password prompt

Based on your organization's best practices and compliance requirements, add more restrictions and best practices for handling sensitive data in Ansible.

In the next section, you will learn about some of the best practices for Ansible playbooks.

Ansible playbook best practices

It is important to develop your Ansible playbooks with reader-friendliness and reusability in mind. Since the YAML format is human readable, it is easy to develop and follow some style guides for your Ansible playbooks.

In *Chapter 15, Using Raw Commands for Network Operations*, you learned when to use the `raw` module and commands. Always check the documentation and see if there are modules available for your task. The `command`, `shell`, `raw`, and `script` modules can be used if no suitable modules are available for the task. But always keep in mind that the `command`, `shell`, `raw`, and `script` modules are not idempotent and will always report as `changed` when executed.

Always give your tasks names

Even though the `name` parameter is an optional component, it is a best practice to provide an appropriate and meaningful name for the plays, tasks, blocks, and other components in your Ansible playbooks. Refer to *Figure 16.22*, where you can see the sample names that were used for the tasks.

Use the appropriate comments

Adding comments to your playbooks will help you troubleshoot when there is an issue. Comments are also useful when further developments or enhancements are required so that the original author and other developers can easily understand the task or steps that are required in the Ansible playbook.

The following screenshot shows that comments have been added before the tasks:

```
# Task to send a notification email before the reboot operation.
- name: Email notification before reboot
  include_role:
    name: send-email
  vars:
    email_report_body: "Alert: {{ inventory_hostname }} is rebooting as per schedule. Please do not use the
server. Notification will be sent after the reboot activity is completed."
    email_smtp_subject: "Weekly System Reboot - {{ inventory_hostname }} - Initiated"
  tags:
    - email
    - notification

# You may add your pre-reboot tasks here
# such as taking backups, configure maintainance mode,
# disable monitoring and so on.
- name: Running Pre-reboot tasks
  debug:
    msg: "Taking backup and snapshot"
  tags:
    - pretasks
    - backup
```

Figure 16.22 – Ansible playbook with comments, extra lines, and tags

Extra lines and whitespaces

Adding whitespaces and extra lines in a playbook will increase its readability. As shown in the preceding screenshot, adding an extra line after each task will help you identify the individual tasks easily.

Implement your own style guide and follow the best practices for YAML writing to achieve better readability and reusability of Ansible artifacts.

Add tags to the tasks

When you have large or complex playbooks, you may need to run some tasks specifically instead of executing every task in the playbooks and roles. It is possible to achieve this by using **Ansible tags** in the playbooks. Tags can be added to plays, tasks, blocks, or roles. *Figure 16.22* shows how to use tags in an Ansible playbook. Once added, selectively execute the tasks by calling the `---tags` argument, as follows:

```
$ ansible-playbook site.yml --tags=pretasks
```

These tasks can be skipped by using the `--skip-tags` argument, as follows:

```
$ ansible-playbook site.yml --skip-tags=email
```

Refer to the Ansible tags documentation (`https://docs.ansible.com/ansible/latest/user_guide/playbooks_tags.html`) to learn more about the usage and methods of tags in Ansible playbooks.

Use explicit declarations

The modules may have default parameter values and these values may apply automatically if we do not mention them in the playbook. But declaring such parameters explicitly in your playbooks will help you identify the desired result of the task. For example, in the `ansible.posix.firewalld` module, the default value for `immediate` is **no**, as shown in the documentation at `https://docs.ansible.com/ansible/latest/collections/ansible/posix/firewalld_module.html`:

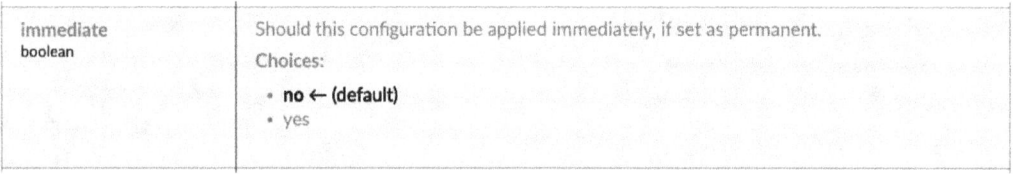

Figure 16.23 – firewalld module showing immediate parameter details

Leave it as-is or declare it explicitly as `immediate: yes` if you need to apply the firewall entry immediately. The following screenshot shows an example `firewalld` task:

```
- name: Enable and Run Firewalld
  ansible.builtin.service:
    name: firewalld
    enabled: true
    state: started

- name: Firewalld permit httpd service
  ansible.posix.firewalld:
    service: http
    permanent: true
    state: enabled
    immediate: yes
```

Figure 16.24 – firewalld task with explicit declarations

Always declare the desired result in the Ansible playbook so that it is possible to troubleshoot and remediate issues when they occur.

Use native YAML for playbooks

It is possible to write tasks in any acceptable YAML format, so long as Ansible can read and understand it. The following screenshot shows some example tasks in an Ansible playbook:

```
tasks:
  - name: Copy a file to managed hosts
    copy: name=demo.txt dest=/tmp/demo.txt owner=ansible group=ansible
  - name: Create a new directory if it does not exist
    file: path=/home/ansible/new-dir state=directory mode='0755'
```

Figure 16.25 – Ansible tasks in the non-native YAML format

The same playbook can be written in native YAML, which is tidier and more readable. This can be seen in the following screenshot:

```
● ● ●
tasks:
  - name: Copy a file to managed hosts
    copy:
      src: files/demo-text-file.txt
      dest: /home/ansible/demo-text-file.txt
      owner: ansible
      group: ansible

  - name: Create a new directory if it does not exist
    file:
      path: /home/ansible/new-dir
      state: directory
      mode: '0755'
```

Figure 16.26 – Ansible tasks written in native YAML format

Refer to the Ansible YAML syntax (`https://docs.ansible.com/ansible/latest/reference_appendices/YAMLSyntax.html`) and advanced syntax (`https://docs.ansible.com/ansible/latest/user_guide/playbooks_advanced_syntax.html`) documentation to learn more about YAML for Ansible.

Avoid hardcoding variables and details

The following screenshot shows a play where the target nodes and package details have been mentioned (hardcoded) inside the playbook:

```
● ● ●
- name: Installing Web Packages
  hosts: webservers
  tasks:
    - name: Installing Web
      yum:
        name: httpd
        state: present
```

Figure 16.27 – Ansible playbook with hardcoded values

If you need to execute the playbook for other target nodes, then you must modify the playbook file and update its values. The same playbook can be written like so:

```
●  ●  ●
- name: Installing Web Packages
  hosts: "{{ nodes }}"
  tasks:
    - name: Installing Web
      yum:
        name: "{{ web_package }}"
        state: present
```

Figure 16.28 – Ansible playbook with dynamic variables

Now, pass the variable while executing the playbook, as follows:

```
$ ansible-playbook site.yaml --extra-vars "nodes=webservers
web_package=httpd"
```

By avoiding hardcoding, it is possible to dynamically use the same playbook for different target nodes with different values.

Use blocks in Ansible playbooks

Blocks are a logical grouping of tasks in Ansible playbooks and help handle errors during execution. Instead of validating the success rate of tasks, use block in a playbook, as shown here:

```
●  ●  ●
tasks:
  - block:
      - name: Show Message
        debug:
          msg: "Trying httpd"
      - name: Install Package
        yum:
          name: httpd-wrong
          state: present

    rescue:
      - name: Show error
        debug:
          msg: "Unknown Package"
      - name: Install nginx
        yum:
          name: nginx
          state: latest

    always:
      - name: Message
        debug:
          msg: "Playbook Done"
```

Figure 16.29 – Using blocks in an Ansible playbook

If any of the tasks in `block` fail, Ansible will execute the tasks under the `rescue` block. The tasks under the `always` block will be executed regardless of the failure or success of the block and rescue tasks.

Refer to the blocks documentation (`https://docs.ansible.com/ansible/latest/user_guide/playbooks_blocks.html`) to learn more about how to use blocks in Ansible.

Use roles and subtasks

When you develop large and complex automation use cases, you should split the playbook into small subtask files and roles. This practice will improve the modularity and flexibility of Ansible artifacts and also help troubleshoot the playbook easily:

```
- name: "Patching Pre-tasks"
  include_role:
    name: linux-patching
    tasks_from: linux-patching-pre-tasks.yaml

- name: "Patching Tasks"
  include_role:
    name: linux-patching

- name: "Patching Post-tasks"
  include_role:
    name: linux-patching
    tasks_from: linux-patching-post-tasks.yaml
```

Figure 16.30 – An Ansible playbook calling roles and subtask files

The preceding screenshot shows an Ansible playbook calling the `linux-patching` role and some of the specific task files from the role.

Use meaningful names for variables

In the previous chapters, you learned about Ansible variables and their different usages. It is possible to use multiple variables in your playbooks and roles, so it is important to use meaningful names for the variables. The following screenshot shows both good and bad examples of how to name variables:

```
● ● ●
# variable names with shortnames
myvar: something
webport: 8080
dbpath: /opt/mysql
fwpackage: firewalld
fg_api: 10.1.10.10

# variables with meaningful names
user_location: /home/devops/
httpd_web_port: 8080
mysql_database_home: /opt/mysql
firewall_package: firewalld
fortigate_api_ip: 10.1.10.10
```

Figure 16.31 – Ansible variables with short and meaningful names

Naming your variables appropriately will help you avoid duplicating variable names and complexity in playbook development.

Learn playbook optimization

There are multiple ways to optimize Ansible playbooks and speed up their execution. Some of them are as follows:

- Use parallelism.

- Use the appropriate execution strategy as needed.

- Use the appropriate value for `forks`.

- Use `serial` to execute in batches.

- Use `order` to control execution based on inventory.

- Use `throttle` for high CPU-intensive tasks.

Read *8 ways to speed up your Ansible playbooks* (`https://www.redhat.com/sysadmin/faster-ansible-playbook-execution`) and *5 ways to make your Ansible modules work faster* (`https://www.redhat.com/sysadmin/faster-ansible-modules`) to learn more about Ansible optimization techniques. To expand your learning on Ansible best practices, refer to the official Red Hat course *Advanced Automation: Red Hat Ansible Best Practices* (`https://www.redhat.com/en/services/training/do447-advanced-automation-ansible-best-practices`).

Summary

In this chapter, you learned about some of the best practices that can be implemented in your Ansible development workflow. You explored the best practices for organizing Ansible artifacts, including playbooks, roles, variables, inventories, and other Ansible content. Then, you learned about the importance of storing the inventory separately based on the managed node environment, criticality, and other facts. You also learned how to use host variables and group variables to organize variables.

After that, you learned about some of the best practices for storing and managing credentials in Ansible, such as avoiding plain text passwords and separating secrets from regular variable files. Finally, you learned about the different best practices and optimization techniques for improving the efficiency of Ansible playbooks. Refer to the *Further reading* section to learn more about Ansible best practices.

Congratulations! With this chapter, you have reached the end of this book on Ansible automation for real-life use cases.

First, you were introduced to Ansible and learned how to install and deploy it. Based on that knowledge, you learned about Ansible commands, modules, and managed nodes. After that, you learned about Ansible playbooks and developed basic automation use cases, such as collecting system information, weekly system reboots, and system report generation. You also learned about the importance of version control systems and practiced how to use them to store Ansible artifacts.

After that, you expanded your learning by understanding how to find Ansible automation use cases. You learned how to automate Microsoft Windows and network devices such as VyOS and Cisco ASA using Ansible. You also learned how to use Ansible to manage virtualization platforms, cloud platforms (AWS, GCP, and VMware), and database operations.

Later, you learned how to use Ansible in DevOps practices and workflows and practiced container management using Ansible. You also learned how to use Ansible for Kubernetes management by deploying and scaling applications on Kubernetes. To expand your knowledge, you learned about Ansible Automation Platform and its various integration methods. After that, you learned how to manage sensitive information in Ansible using Ansible Vault.

In the last few chapters, you learned how to manage non-standard platforms and operations using raw commands, API calls, and modules. You also learned the best practices for developing and storing Ansible artifacts.

Before moving on, remember to join the Ansible community, real-time chat groups, and mailing lists. Refer to the Ansible community page (`https://www.ansible.com/community`) to find details about meetup events. Contact me on LinkedIn (`https://www.linkedin.com/in/gineesh`) if you have any questions or feedback on the content of this book.

If you are looking for official Ansible training, then check out the courses from Red Hat (`https://www.ansible.com/products/training-certification`).

Raise issues in the book repository (`https://github.com/PacktPublishing/Ansible-for-Real-life-Automation/issues`) if you have any issues while practicing the exercises in this book.

Thank you for your interest and your dedication to completing this book!

Further reading

To learn more about the topics that were covered in this chapter, take a look at the following resources:

- *Reusing Ansible artifacts (Include and Import)*: `https://docs.ansible.com/ansible/latest/user_guide/playbooks_reuse.html#playbooks-reuse`
- *Ansible tips and tricks*: `https://docs.ansible.com/ansible/latest/user_guide/playbooks_best_practices.html`
- *Ansible Best Practices (Presentation Deck archive)*: `https://aap2.demoredhat.com/decks/ansible_best_practices.pdf`
- *10 habits of great Ansible users*: `https://www.redhat.com/sysadmin/10-great-ansible-practices`

Index

Z

Packt.com

Subscribe to our online digital library for full access to over 7,000 books and videos, as well as industry leading tools to help you plan your personal development and advance your career. For more information, please visit our website.

Why subscribe?

- Spend less time learning and more time coding with practical eBooks and Videos from over 4,000 industry professionals

- Improve your learning with Skill Plans built especially for you

- Get a free eBook or video every month

- Fully searchable for easy access to vital information

- Copy and paste, print, and bookmark content

Did you know that Packt offers eBook versions of every book published, with PDF and ePub files available? You can upgrade to the eBook version at packt.com and as a print book customer, you are entitled to a discount on the eBook copy. Get in touch with us at customercare@packtpub.com for more details.

At www.packt.com, you can also read a collection of free technical articles, sign up for a range of free newsletters, and receive exclusive discounts and offers on Packt books and eBooks.

Other Books You May Enjoy

If you enjoyed this book, you may be interested in these other books by Packt:

Learning DevOps - Second Edition

Mikael Krief

ISBN: 9781801818964

- Understand the basics of infrastructure as code patterns and practices

- Get an overview of Git command and Git flow

- Install and write Packer, Terraform, and Ansible code for provisioning and configuring cloud infrastructure based on Azure examples

- Use Vagrant to create a local development environment

- Containerize applications with Docker and Kubernetes

- Apply DevSecOps for testing compliance and securing DevOps infrastructure

- Build DevOps CI/CD pipelines with Jenkins, Azure Pipelines, and GitLab CI Explore blue-green deployment and DevOps practices for open sources projects

Mastering Ansible - Fourth Edition

James Freeman, Jesse Keating

ISBN: 9781801818780

- Gain an in-depth understanding of how Ansible works under the hood
- Get to grips with Ansible collections and how they are changing and shaping the future of Ansible
- Fully automate the Ansible playbook executions with encrypted data
- Use blocks to construct failure recovery or cleanup
- Explore the playbook debugger and Ansible console
- Troubleshoot unexpected behavior effectively
- Work with cloud infrastructure providers and container systems

Packt is searching for authors like you

If you're interested in becoming an author for Packt, please visit `authors.packtpub.com` and apply today. We have worked with thousands of developers and tech professionals, just like you, to help them share their insight with the global tech community. You can make a general application, apply for a specific hot topic that we are recruiting an author for, or submit your own idea.

Share Your Thoughts

Now you've finished *Ansible for Real Life Automation*, we'd love to hear your thoughts! Scan the QR code below to go straight to the Amazon review page for this book and share your feedback or leave a review on the site that you purchased it from.

`https://packt.link/r/1803235411`

Your review is important to us and the tech community and will help us make sure we're delivering excellent quality content.